Separation Technology
ICoST 2025

5th International Conference on Separation Technology (ICoST2025), Universiti Teknologi Malaysia, Putrajaya Malaysia, 27-28 September 2025

Editors

Muhammad Abbas Ahmad Zaini[1],
Syed Anuar Faua'ad Syed Muhammad[1]

[1]Centre of Lipids Engineering and Applied Research (CLEAR), Universiti Teknologi Malaysia, 81310 Johor Bahru, Johor, Malaysia

Peer review statement

All papers published in this volume of "Materials Research Proceedings" have been peer reviewed. The process of peer review was initiated and overseen by the above proceedings editors. All reviews were conducted by expert referees in accordance to Materials Research Forum LLC high standards.

Published under License by **Materials Research Forum LLC**
Millersville, PA 17551, USA

Published as part of the proceedings series
Materials Research Proceedings
Volume 59 (2026)

ISSN 2474-3941 (Print)
ISSN 2474-395X (Online)

ISBN 978-1-64490-394-0 (Print)
ISBN 978-1-64490-395-7 (eBook)

This book contains information obtained from authentic and highly regarded sources. Reasonable efforts have been made to publish reliable data and information, but the author and publisher cannot assume responsibility for the validity of all materials or the consequences of their use. The authors and publishers have attempted to trace the copyright holders of all material reproduced in this publication and apologize to copyright holders if permission to publish in this form has not been obtained. If any copyright material has not been acknowledged please write and let us know so we may rectify in any future reprint.

Distributed worldwide by

Materials Research Forum LLC
105 Springdale Lane
Millersville, PA 17551
USA
https://mrforum.com

Manufactured in the United State of America
10 9 8 7 6 5 4 3 2 1

Table of Contents

Preface

Committees

Application of Response Surface Methodology in Optimizing Thiourea Leaching
for Metal Extraction from E-Waste
Norul Fatiha Mohamed NOAH, Norasikin OTHMAN, Izzat Naim Shamsul KAHAR,
Sazmin Sufi SULIMAN, Shuhada Atika IDRUS-SAIDI, Aishah ROSLI................ 1

Effect of Temperature and Time on the Conversion of Defatted Microalgae
Biomass into Bio-Oil through Hydrothermal Liquefaction
Muhammad Khairul Hilmi MOHD ZAKI, Shafirah SAMSURI, Usman BELLO,
Nor Adilla RASHIDI.................... 10

Eco-Friendly Activation of Palm Kernel Shell Carbon using Calcium Chloride from
Cockle Shells
Mohamad Razlan MD RADZI, Nor Adilla RASHIDI.............. 17

Pectin-Functionalized Iron Oxide Nanoparticles for Enhanced Removal of Cationic
Dyes: Influence of Pectin Methoxylation on Surface Charge and Adsorption
Efficiency
Yen Yan NG, Peck Loo KIEW, Lian See TAN, Man Kee LAM, Wei Ming YEOH,
Muhamad Ali MUHAMMAD YUZIR 25

One-Step Phosphoric Acid Activated Pomegranate Peel Powder for Adsorptive
Removal of Tetracycline in Aqueous Solution: Synthesis, Adsorption and
Mechanism
Mohammed Awwal SULEIMAN, Muhammad Abbas AHMAD ZAINI,
Nuhu Dalhat MU'AZU 33

Biogas Production Potential from Animal Farm Waste: A Preliminary Study at
Sirukam Dairy Farm, Indonesia
Erda Rahmilaila DESFITRI, Adillah Rahmi PUTRI, Bunga Karuni PUTRI, Ellyta SARI,
Reni DESMIARTI.................. 41

Electrocoagulation Process of Palm Oil Mill Effluent: Effect of Applied Voltage on
Removal of Organic Content
Nofri NALDI, Ariadi HAZMI, Reni DESMIARTI, Primas EMERALDI,
Maulana Yusup ROSADI, Nofrizon RAHMAN, Erda Rahmilaila DESFITRI........... 48

CBD-Grown MoS_2 Thin Films on Plastic Optical Fiber for NH_3 Sensing:
Fabrication and Performance
Nor Akmar MOHD YAHYA, Mohd Rashid YUSOF HAMID, Abdul Hadi ISMAIL,
Nurul Atiqah Izzati MD ISHAK, Mohd Hanif YAACOB, Saidur RAHMAN........... 56

Screening and Integration of Watermelon Rind Extract (WMRE) for Functional
Chitosan Thin Films
Rozaini ABDULLAH, Sharifah Zati Hanani SYED ZUBER,
Noor Amirah ABDUL HALIM, Muhammad Zafri Aziman MOHD SALLEH 64

Parametric Evaluation of Subcritical Water Extraction of Oleoresin from Syzygium
Aromaticum via Factorial Design Approach
Noor Amirah ABDUL HALIM, Sharifah Zati Hanani SYED ZUBER,
Rozaini ABDULLAH, Ain Nur Najwa ABDULLAH 72

Kinetic Modelling of Supercritical CO_2 Extraction from Swietenia macrophylla
Seeds Extracts Using the Single Sphere Model
Mohammad Lokman HILMI, Liza Md SALLEH, Hasmida Mohd NASIR,
Noor Azwani Mohd RASIDEK.. 80

Optimization of Amygdalin Extraction from Prunus Armeniaca Kernels for
Antioxidant and Anti-Inflammatory Potentials
Sayed Ibrahim WAFA, Lee Suan CHUA, Roshafima RASIT ALI 87

Zinc Chloride Recovery for Synthesis of Porous Adsorbent from Lipid Condensate
for Dye Removal
Nurul Aishah ABDUL RAHIM, Muhammad Abbas AHMAD ZAINI, Sariah ABANG,
Norlisa MILI.. 98

Simulation-Based Comparative Study of CO_2 Absorption Techniques for Biogas
Purification Using Aspen HYSYS and Aspen Plus
Aishah ROSLI, Aziatul Niza SADIKIN 106

Supercritical–CO_2 Dried Biodegradable Alginate/Zirconia Aerogels: Synthesis and
Characterization
Nur Hazwani Dalili MOHAMAD, Ana Najwa MUSTAPA, Suhaiza Hanim HANIPAH 113

Influence of Feed Temperature on Progressive Freeze Concentration of Magnesium
Sulphate Solutions
Hafizuddin SUTIMIN, Aishah ROSLI, Mazura JUSOH... 121

Phenolic Compounds Recovery from Palm Oil Sterilizer Condensate using
Synergistic Formulation
Norela JUSOH, Norasikin OTHMAN, Shuhada A. IDRUS-SAIDI,
Muhammad Abbas AHMAD ZAINI, Izzat Naim Shamsul KAHAR,
Norul Fatiha Mohamed NOAH.. 130

Influence of Temperature, Flow Rate and Extraction Time on Subcritical Water
Extraction of Eugenol and Hydroxychavicol from Piper Betel Leaves: A
Comparative Study with Soxhlet Extraction
Zuhaili IDHAM, Mohd Sharizan MD SARIP, Noor Azwani MOHD RASIDEK,
Siti Alyani MAT, Aishah ROSLI, Nik Zetti Amani NIK FAUDZI............................ 138

Subcritical Water Extraction of Protein from Trichanthera Gigantea: Optimization
using Response Surface Methodology
Tengku Zarith Hazlin TENGKU ZAINAL ABIDIN, Nur Hidayah ZAINAN,
Ahmad Syahmi ZAINI, Asiah Nusaibah MASRI, Yanti Maslina MOHD JUSOH,
Nur Farzana AHMAD SARNADI, Mohd Azan MOHAMMED SAPARDI 145

Optimization of Mimosa Pudica Linn Extraction at Varied Feed to Solvent Ratios
and Solvent Concentrations using Response Surface Methodology
Nurul Aishah ABDUL RAHIM, Sariah ABANG, Mohd Raziman ISMAIDI,
Muhammad Abbas AHMAD ZAINI, Rubiyah BAINI, Sherena SAR-EE 154

Prediction of Solvent Component and Composition for Absorption-Based Acid Gas
Removal Unit using Optuna-LightGBM and K-means
Rafi Jusar WISHNUWARDANA, Madiah OMAR, Haslinda ZABIRI, Kishore BINGI,
Rosdiazli IBRAHIM.. 162

Physicochemical Analysis of Neem Extract Loaded Polyvinylpyrrolidone/2-Hydroxypropyl-ß-Cyclodextrin Nanofibers for Wound Dressing Application
Crystal Hui Man TIONG, Sharol CHIA,
Nur Iman Batrisyia MOHD SHAHRUL NIZAM, Nurizzati MOHD DAUD 170

Removal of Cadmium from Simulated Wastewater through Synergistic Reactive Extraction
Fadzlin Qistina FAUZAN, Izzat Naim SHAMSUL KAHAR,
Norisya Balqis MOHD AMIN, Norasikin OTHMAN, Shuhada A. IDRUS-SAIDI 177

Microalgae-Mediated Biological Synthesis of Silver Nanoparticles: Optimization and Morphological Characterization
Hui Ying THE, Man Kee LAM, Voon-Loong WONG, Wai Hong LEONG, Inn Shi TAN,
Henry Chee Yew FOO 185

Green Valorization of Palm Oil Solid Condensate using Supercritical CO₂ Ethanol Extraction for Rich-Bioactive Compounds
Noor Azwani MOHD RASIDEK, Liza MD SALLEH, Nurizzati MOHD DAUD,
Noor Sabariah MAHAT, Zuhaili IDHAM, Mohammad Lokman HILMI,
Muhammad Abbas AHMAD ZAINI 193

Computer-Aided Design and Fabrication of Advanced Membranes for Nitrogen/Methane Separation
Jimoh K. ADEWOLE, Mohammed S. AL-AJMI, Amna S. AL-JABRI,
Faisal R. AL MARZUQI, Habeebllah B. OLADIPO, Faruq B. Owoyale 201

Keyword Index
About the Editors

Preface

Over the past editions, ICoST has grown into a dynamic platform that brings academics, scientists and industry practitioners to share and explore emerging frontiers within the broad and rapidly evolving fields in Separation Technology. This year, the 5th International Conference on Separation Technology (ICoST) is held in Putrajaya, Malaysia from 27–28 September 2025. The conference offers a unique convergence of inspiring plenary and keynote speeches, in-depth discussions on current research and professional networking, designed to foster collaboration and innovation. With the conference theme, 'Exploring Frontiers in Separation and Purification', the 2025 edition continues this mission by showcasing cutting-edge contributions across Advanced Separation Technology, Environmental Engineering and Advanced Material Engineering tracks. By bringing together multidisciplinary perspectives, the conference aims to catalyze new insights and accelerate the translation of research into impactful technologies.

This conference sought to address the global challenges by fostering interdisciplinary collaboration and promoting solutions that are innovative and sustainable. The compilation of papers in the proceedings showcase the recent breakthroughs, novel methodologies and transformative ideas that collectively push the boundaries of separation and purification technology. The papers reflect the diversity and depth of the discussions held throughout the event, highlighting new advances in separation and purification technologies.

We extend our sincere appreciation to all authors for their high-quality submissions, the reviewers for their rigorous evaluation and constructive feedback and the organizing committee for their unwavering commitment in ensuring a successful conference. We also express our gratitude to our sponsors and participants whose contributions enriched the intellectual discourse of this event. We hope that this volume will serve as a valuable reference to inspire knowledge explorations and collaborations across disciplines. May the knowledge shared at ICoST 2025 continue to drive scientific advancements and contribute meaningfully to a sustainable future. Thank you.

Editors,

Muhammad Abbas Ahmad Zaini

Syed Anuar Faua'ad Syed Muhammad

Committees

Honorary Advisor
Prof. Dr. Rosli bin Md. Illias (TNCPI, UTM)
Program Advisor
Assoc. Prof. Ir. Dr. Muhammad Abbas bin Ahmad Zaini (CLEAR Director)
Prof. Dr. Norasikin binti Othman
Prof. Ir. Dr. Rahmat bin Mohsin

Program Director
Dr. Aishah binti Rosli

Vice Program Director
Prof. Madya Dr. Chua Lee Suan

Treasurer
Miss Siti Zulfarina Fadzli (Chief)
Mrs. Wong Yah Jin (Deputy Chief)

Secretary
Dr. Nurizzati binti Mohd Daud (Chief)
Mrs. Munirah binti Onn

Program and Technical Secretariat
Dr. Shuhada Atika binti Idrus Saidi (Chief)
Dr. Mohammad Sukri bin Mohamad Yusof
Mr. Sazmin Sufi Suliman
Mr. Izzat Naim Shamsul Kahar
Mr. Abdul Khalil bin Abdollah
Mr. Mohd Syafiq bin Md Hidiah
Mr. Mohd Shazaril bin Mohd Zain

Publication Secretariat
Assoc. Prof. Ir. Dr. Muhammad Abbas Ahmad Zaini (Chief)
Dr. Syed Anuar Faua'ad Syed Mohd
Dr. Zuhaili binti Idham

Invited Speaker and Protocol Secretariat
Assoc. Prof. Dr. Mazura binti Jusoh (Chief)
Mrs. Nik Zetti binti Nik Faudzi
Mrs. Sharifah Iziuna binti Sayed Jamaludin
Dr. Nur Syazwani binti Mohd Ali
Dr. Nurul Adilah binti Manshor
Mr. How Chee Yang

Logistic Secretariat
Ts. Dr. Muhammad Syafiq Hazwan bin Ruslan
Pn. Noor Sabariah binti Mahat

Separation Technology - ICoST 2025
Materials Research Proceedings 59 (2026) 1-9

Materials Research Forum LLC
https://doi.org/10.21741/9781644903957-1

Application of Response Surface Methodology in Optimizing Thiourea Leaching for Metal Extraction from E-Waste

Norul Fatiha Mohamed NOAH[1,a], Norasikin OTHMAN[1,2,b*],
Izzat Naim Shamsul KAHAR[1,2,c], Sazmin Sufi SULIMAN[1,c],
Shuhada Atika IDRUS-SAIDI[1,2,d*], Aishah ROSLI[1,2,e]

[1]Faculty of Chemical and Energy Engineering, Universiti Teknologi Malaysia, 81310 Skudai, Johor, Malaysia

[2]Centre of Lipids Engineering and Applied Research (CLEAR), Ibnu Sina Institute for Scientific and Industrial Research, Universiti Teknologi Malaysia, 81310 Skudai, Johor, Malaysia

[a]norulfatiha.mn@gmail.com, [b]norasikin@cheme.utm.my, [c]izzatnaim.sk@utm.my,
[d]sazmin5821@gmail.com, [e]shuhada.atika@utm.my, [f]aishahrosli@utm.my

Keywords: Leaching, Gold Recovery, Thiourea Leachate, Waste Printed Circuit Boards, Electronic Waste, Box-Behnken Design

Abstract. Electronic waste (e-waste) is an escalating environmental concern that poses significant risks to human health and ecosystems. Despite its hazardous nature, e-waste is rich in valuable metals, particularly gold (Au), which has a high economic value and is experiencing a rising market price trend. Consequently, Au recovery from e-waste is crucial for long-term environmental sustainability as well as economic feasibility. This work describes a less harmful thiourea leachate method for removing gold from used printed circuit boards (PCBs). The study looks at a number of variables, such as the concentration of thiourea, the concentration of acid, and oxidizing agents, that affect how effective the leaching process is. To improve Au recovery, Box-Behnken design and Response Surface Methodology (RSM) are used to optimize these parameters. The maximum predicted Au extraction performance was found to be 2.81ppm, achieved at specific conditions of 57.971 mL H_2O_2, 0.502M thiourea, and 1.006M H_2SO_4. The observed extraction value closely matched this prediction at 2.87ppm, indicating a deviation of less than 5%. The results highlight thiourea as the most significant variable influencing Au recovery, with the interaction between thiourea and H_2SO_4 playing a critical role in the leaching process. Also, the separation factors (β) indicate that Au can be efficiently separated from other metals, particularly copper and aluminium, with high separation factors of 31.10 and 18.59, respectively.

Introduction

Gold is a precious metal with a bright yellow color that has been used for jewelry, currency, and electronic devices such as computers, laptops, cell phones, audio devices, etc. Unfortunately, the natural ores of Au are becoming depleted, and the extraction methods of Au from natural ores are inefficient and environmentally hazardous. Over 6800 tons of Au (16 % of the world reserves) are part of e-waste, 25–250 times more than the global average of the primary Au mine deposits (1–10 g/ton) [1]. Waste printed circuit boards (WPCBs) have a high concentration of Au and should be viewed as a valuable secondary source of Au [2]. Therefore, recovering Au from e-waste is crucial for its economic value and sustainability, as Au in e-waste can exceed that in primary Au deposits by several folds. Hydrometallurgy is extensively used in the downstream separation and refining of pyrometallurgical products due to its high selectivity, low environmental impact, and low energy consumption [3]. In the first step of hydrometallurgical processing, metals are extracted from their solid matrix by leaching using either acidic or alkaline solutions. Several compounds, including cyanides, halides, thiourea, and thiosulfates are available for this step. However, this process uses harmful reagents for instance cyanidation and aqua regia leaching. It is a dangerous process with a negative impact on the environment, the operation cost is high and requires careful handling and disposal [4]. In contrast, the use of mild acidic leaching such as acidic thiourea has been shown to be an efficient and environmentally friendly method for the extraction of Au from e-waste [5]. To oxidize native Au in e-waste, different oxidants, such as sodium peroxide, hydrogen peroxide, ferric ion, manganese dioxide, and dichromate have been added [6]. Among the oxidants studied, the addition of ferric ions to thiourea has been shown to be the most effective [7]. However, the resulting leaching solution often contains multiple metals, which makes the recovery of individual metals challenging. Therefore, there is a need for a novel technique to extract Au from e-waste leached solution.

In general, the concentration of Au in the leaching solution is much lower than that of other co-leached metals such as Cu, effective process optimization is crucial [8]. Thus, different approaches have been conducted for the recovery of Au from e-waste. Some of these studies have focused on chemical systems or process optimization for Au leaching leaching [5,9]. Others have developed separation/purification techniques, including adsorption, for the efficient separation of Au from various leaching solutions [10,11]. However, these studies are often fragmented or non-linked, leading to the lack of an overall picture of Au recovery from e-waste, which requires future focus. Response surface methodology (RSM) provides a statistical tool to study variable interactions and identify optimum operating conditions in multivariable systems [12]. In this study, a Box–Behnken design (BBD) with RSM was applied to optimize thiourea leaching of metals from WPCBs. The oxidizing agent, leachate, and acid concentration were selected as key variables because they represent the fundamental components of thiourea leaching chemistry: the oxidant promotes metal oxidation and dissolution, thiourea acts as the primary lixiviant, and acid concentration controls solution stability and reactivity. Although the leaching behavior of multiple metals was evaluated, the optimization was performed with respect to Au due to its higher economic value and critical importance in e-waste recycling.

This work also serves as a preliminary study towards developing a liquid–liquid extraction (LLE) process for selective metal separation, as optimizing the leaching step is essential for producing suitable feed solutions for downstream solvent extraction. After obtaining the optimum Au leaching conditions, the separation factors between Au and co-leached metals will also be compared to assess selectivity. The novelty of this work lies in applying RSM for systematic optimization of thiourea leaching, while at the same time establishing the groundwork for selective separation and sustainable recovery of valuable metals from e-waste.

Methodology

E-waste in the form of Au-based WPCBs (mobile phone) used in this work was procured from MEP Enviro Technology Sdn Bhd (MEPSB) in Penang, Malaysia. Nitric acid (HNO_3, 65% purity) hydrochloric acid (HCl, 37% purity), and Sulfuric acid (98 % pure) were purchased from Merck. Thiourea as a leachate and hydrogen peroxide (H_2O_2) as an oxidant was obtained from Sigma Aldrich. All the chemicals used is analytical reagent grade.

Prior to the leaching process, the WPCB samples were pre-processed by cutting them into pieces of approximately 1.0×1.0 cm^2 for initial size reduction. The cut pieces were mechanically crushed using a mill cutter (Retsch, Germany) for further comminution and then fractionated to a size less than 500 µm using a sieve machine. The WPCB samples were characterized by aqua regia digestion to determine the amount of metals in the sample and inductively coupled plasma-optical emission spectroscopy (ICP-OES) (PerkinElmer, USA) was used for chemical analysis. 5 g of comminute WPCBs was digested with hot aqua regia which is the combination of these acids (HNO_3:3HCl). The digestion was performed at 80 °C under magnetic stirring (300 rpm) with the solid/liquid (S/L) ratio of 1:10 (g/ml) and the digestion time was approximately 120 min. The

Separation Technology - ICoST 2025
Materials Research Proceedings 59 (2026) 1-9

Materials Research Forum LLC
https://doi.org/10.21741/9781644903957-1

digested fraction was filtered, and the leachate was diluted using the same method as suggested by Hubau et al. [14].

Thiourea was used to leach out the metals from the WPCBs sample. In this study, the batch leaching tests of comminuted WPCBs were carried out in a 500 ml jacketed glass reactor equipped with a thermometer and mechanical agitator (300 rpm) as in Fig. 1. The leaching condition is 1/10 solid/liquid ratio, 40°C temperature, 180 minutes leaching time and 373 rpm stirring speed. The required amount of WPCBs sample was added to the reactors after the leaching solution reached the desired temperature. Also, a circulating water bath was used to maintain the desired temperature throughout the experiments, and the reactors were covered with lids during the leaching process. The leachate was filtered (Whatman 41 filter paper) and was diluted with distilled water before Au and other metals concentration was analyzed by ICP-OES. The parameters studied by the previous work such as such as agitation speed, leaching time, acid concentration, and temperature were optimized using response surface methodology (RSM). Through RSM method, box behnken design was used. Metals extraction was calculated using Eq. 1, whereas the separation factor (β) of Au over other metals was calculated as in Eq. 2 considering the metal concentration in the leaching process and aqua regia digestion [17].

$$Au\ Leached\ (\%) = \frac{Au\ leached\ per\ gram\ of\ WPCBs}{Total\ Au\ per\ gram\ of\ WPCBs} \times 100 \tag{1}$$

$$\beta_{\frac{Au}{metal}} = \frac{D_{Au}}{D_{metal}} \tag{2}$$

Where $\beta_{\frac{Au}{metal}}$ indicates the separation factor between Au and each metal while D_{Au} and D_{metal} represent the distribution ratio of each metal respectively.

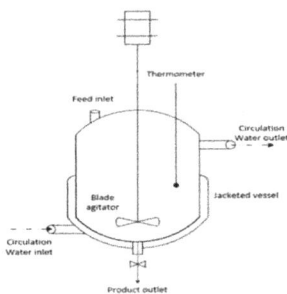

Figure 1 Au extraction via acidic thiourea leaching process.

The Box Behnken Design (BBD) in the software Design Expert 13 was used for analysis. Variables such as H_2O_2 (A), Thiourea (B) and H_2SO_4 (C) were employed in the design. The symbol Y basically represent the response of the experimental design which is the leaching value (%). In the study, a total of 15 experiments were carried out randomly throughout the process. Symbols of the variables and values range of their three coded levels are presented in Table 1.

Table 1 Symbols, coded levels and values of experimental independent variables.

Factors	Symbols	Level	
		Min (-1)	Max (+1)
Hydrogen peroxide (H_2O_2) (mL)	A	20	60
Thiourea (M)	B	0.5	1.5
H_2SO_4 (M)	C	1	3

Separation Technology - ICoST 2025   Materials Research Forum LLC
Materials Research Proceedings 59 (2026) 1-9   https://doi.org/10.21741/9781644903957-1

ANOVA was performed using Design-Expert. A significance level of $p < 0.05$ was applied to evaluate the effects of the variables on extraction efficiency. Statistical validation was estimated by comparing the F-calculated and the F-tabulated. Meanwhile, the significant effects of the variables on the Au extraction were studied by observing contour plot.

Results and Discussion

Optimization of Au leached using RSM method

The effect of independent variables which are H_2O_2 (A), Thiourea (B) and H_2SO_4 (C) on the response variable that is Au extraction was identified using the Box-Behnken design in RSM. The obtained results in Table 2 were then subjected to the response surface methodology (RSM) to evaluate the relationship between the parameters. Each effect was estimated independently due to the orthogonality of the design.

Table 2 *Box–Behnken design for the effects on Au leaching.*

	Factor 1	Factor 2	Factor 3	Response
Run	A:H_2O_2 (mL)	B:Thiourea (M)	C:H_2SO_4 (M)	Leaching Au (ppm)
1	60	1	3	1.45
2	60	1.5	2	0.87
3	40	1	2	0.55
4	40	1	2	1.23
5	60	0.5	2	1.44
6	40	0.5	3	0.35
7	20	0.5	2	1.84
8	20	1.5	2	0.70
9	40	1.5	1	0.49
10	60	1	1	2.44
11	40	0.5	1	2.14
12	40	1	2	1.12
13	20	1	3	1.97
14	20	1	1	1.58
15	40	1.5	3	1.37

The determination coefficient (R^2) of the model is value of 0.9166 indicates that about 91.66% of the variability in the response variable can be explained by the model. Table 3 presents the analysis of variance (ANOVA), which further confirms the model's significance.

Table 3 *ANOVA for Quadratic model.*

Source	df	Mean Square	F-value	p-value	Remark
Model	9	0.5677	6.11	0.0302	significant
A-H_2O_2	1	0.0015	0.0163	0.9035	
B-THIOUREA	1	0.6845	7.36	0.0421	
C-H_2SO_4	1	0.2850	3.07	0.1404	
AB	1	0.0812	0.8736	0.3929	
AC	1	0.4761	5.12	0.0731	
BC	1	1.78	19.17	0.0072	
A^2	1	0.9572	10.30	0.0238	
B^2	1	0.2560	2.75	0.1579	
C^2	1	0.5449	5.86	0.0601	
Residual	5	0.0930			
Lack of Fit	3	0.0661	0.4964	0.7212	not significant
Pure Error	2	0.1332			
Cor Total	14				

Separation Technology - ICoST 2025
Materials Research Proceedings 59 (2026) 1-9

Materials Research Forum LLC
https://doi.org/10.21741/9781644903957-1

According to the analysis, the computed F-value (6.11) surpasses the tabulated F-value ($F_{9,5}$=4.77) demonstrating the significance of the model and strong agreement between the model and experimental data. Besides, the p-value of this model is 0.0302 indicating there is only a 3.02% chance that an F-value this large could occur due to noise and its statistically significant. This is in agreement with Zhao et al. [15] who indicates that the models are considered significant if their p-values are less than 0.05 (p<0.05). In this case B, BC, A² are significant model terms. Values greater than 0.1000 indicate the model terms are not significant. If there are many insignificant model terms, model reduction may improve the model. The Lack of Fit F-value of 0.50 implies the Lack of Fit is not significant relative to the pure error. There is a 72.12% chance that a Lack of Fit F-value this large could occur due to noise. Non-significant lack of fit is good because it indicates that the model fits well. Based on the RSM analysis, the relationship between Au extraction and the three parameters in the actual form is represented in Eq. 3.

$$Y\,(\%) = 6.28167 + -0.0808958A + -1.71833B - 2.37042C + 0.01425AB - \\ 0.01725AC + 1.335BC + 0.00127292A^2 + -1.05333B^2 + 0.384167C^2 \qquad (3)$$

The quadratic regression equation in terms of actual factors can be used to make predictions about the response for given levels of each factor. Here, the levels should be specified in the original units for each factor.

Effect of variables interaction on response
The effects of variables interaction on the response were discussed in detail. The interactions between the variables on the response are shown in Fig. 2. The thiourea concentration provides a significant effect on amount of the leached of Au. The result shows that when the thiourea concentration increases from 0.5 to 1.5 M, the leached of Au decreases. This is because at higher thiourea concentrations, the thiourea can start reacting with the oxidant itself, reducing the amount available for Au leaching. This competition for oxidant reduces the effectiveness of the leaching process that in agreement with Li et al. [16]. Thus, 0.502 M of thiourea concentration was selected to ensure the high extraction of Au. Fig. 2(a) shows the interaction between H_2O_2 and thiourea. Although the circular contour plot shows no significant interaction between the variables, the amount of Au leached was still considered in the analysis. The presence of H_2O_2 as an oxidizing agent remains critical for the leaching process. It's important to maintain H_2O_2 within a specific concentration range to avoid thiourea degradation and ensure that enough oxidizing power is available for effective Au dissolution. The non-significant interaction between H_2O_2 and thiourea in the contour plot could suggest that H_2O_2 concentration alone does not strongly influence thiourea behaviour or the Au leaching efficiency under the conditions tested. While H_2O_2 and thiourea may not show a significant interaction in this plot, their combined effects could be more strongly influenced by the presence of other factors like H_2SO_4.

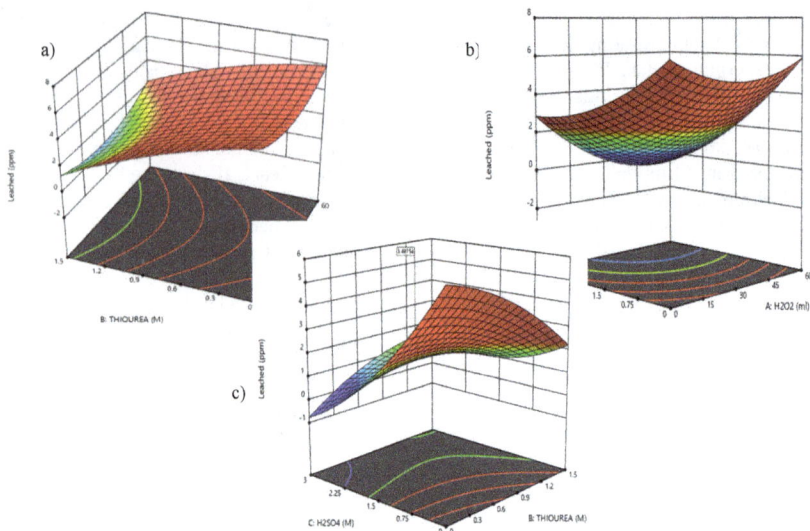

Figure 2 *Response surface plots for (a) Thiourea and H_2O_2, (b) H_2SO_4 and H_2O_2, (c) Thiourea and H_2SO_4.*

The interaction of H_2SO_4 and H_2O_2 is illustrated in Fig. 2(b). The circular contour plot indicates no significant interaction between the variables. Although sulfuric acid (H_2SO_4) and hydrogen peroxide (H_2O_2) play critical roles in the Au leaching process, the non-significant effect observed between them in the plot suggests that their direct interaction might not substantially impact the outcome under the tested conditions. One possible explanation is that both reagents act independently in the leaching mechanism, with H_2SO_4 primarily providing an acidic environment [17] and H_2O_2 serving as an oxidant [18]. This lack of significant interaction may indicate that their individual effects are sufficient for maintaining the necessary reaction conditions, but any combined or synergistic impact may be overshadowed by other dominant factors, such as the thiourea concentration or the leaching time . Additionally, the efficiency of H_2O_2 may already be maximized at the given H_2SO_4 concentrations, limiting any observable interaction between the two. Fig. 2(c) illustrates the interaction between H_2SO_4 and thiourea. The contour plot indicates that lower concentrations of both H_2SO_4 and thiourea lead to better Au leaching. This suggests a synergistic relationship between these two variables. Thiourea becomes much stable by protonation in the presence of H_2SO_4 making it a favourable leached agent in Au extraction. This protonation process results in the formation of a positively charged thiourea molecule which facilitates the effective leaching of Au [19]. Aside from that, higher concentrations of either could lead to suboptimal conditions, as both reagents have a delicate balance in the leaching environment.

It is worth noting that the contour plots obtained in this study did not form perfect ellipses. In RSM, elliptical contour plots are generally expected when quadratic and interaction effects are balanced. However, in this case, the quadratic effect of thiourea concentration was dominant and strongly negative, while the quadratic contributions of H_2O_2 and H_2SO_4 were relatively small. Similarly, the interaction between H_2SO_4 and thiourea was far stronger than the interactions involving H_2O_2. This imbalance distorted the symmetry of the response surface, leading to circular

6

or irregular contours rather than closed ellipses. Although the plots deviate from the ideal elliptical shape, they still accurately reflect the leaching response by showing that H_2SO_4 has a more pronounced synergistic effect with thiourea, making these the main controlling variables. This observation also explains the lack of a significant interaction between H_2O_2 and either H_2SO_4 or thiourea.

The optimization of response via design expert was based on the second order polynomial model. The optimum condition for Au extraction is at the maximum point of 57.971mL H_2O_2, 0.502M Thiourea and 1.006M H_2SO_4. Under the optimum condition, 2.81 ppm of Au extraction is predicted. An experiment was carried out using the estimated optimum condition for the Au extraction and the result shows that 2.87 ppm of Au was extracted with less than 5% deviation from the predicted value, indicating the experimental data as in good agreement with the predicted values. This is consistent with the acceptable error range reported in similar RSM studies [20]. The results also indicate that RSM is capable of optimizing the extraction of Au.

Dissolution of metals in thiourea leaching solution
The dissolution of metals in the thiourea leaching solution is presented in Table 4. Meanwhile aqua regia analysis shows Au at 3.84 ppm in the WPCBs, along with other metals such as Cu 45,000 ppm, Al 6,240 ppm, Ni 320 ppm, and Fe 183.5 ppm. It is noteworthy that thiourea is not stable and easy to decompose in alkaline solution therefore the reaction should occur in acidic solutions. Meanwhile, for base metals, the percentage of leached Cu, Al, Ni, and Fe was 8%, 13%, 26% and 19%, respectively. Meanwhile, the leaching efficiency of the precious metals Au is 74%, which is higher than that of base metals when dissolved in the thiourea solution under optimized conditions. On the other hand, a significantly higher distribution ratio (2.81) was observed for Au compared to the other metals. This indicates that a much higher fraction of Au present in the WPCBs is leached into the solution. Additionally, the separation factor (β) of Au over other metals was investigated to determine how well Au can be separated from each metal in the leaching process. Higher values of β imply better separation. Au is most easily separated from copper ($\beta = 31.10$) and least easily separated from nickel ($\beta = 7.94$).

Table 4 Dissolution of metals in thiourea leaching solution.

Element	% Leaching	Distribution ratio, D	Separation factor (β) of Au
Cu	8	0.09	31.10
Al	13	0.15	18.59
Fe	19	0.24	11.87
Ni	26	0.35	7.94
Au	74	2.81	NA

Conclusion
The optimization using the RSM was applied for studying the Au extraction performance. The maximum predicted value of the extraction performance of Au was 2.81 ppm at 57.971mL H_2O_2, 0.502M Thiourea and 1.006M H_2SO_4 and the observed value was 2.87 ppm. The experimental values agreed well with the predicted values, with <5% deviation, similar to error ranges reported for comparable RSM methods. Thiourea is demonstrated to be the most significant variables. The interaction between thiourea and H_2SO_4 plays an important role in the extraction process. Additionally, the separation factors show that Au can be efficiently separated from other metals, especially copper ($\beta=31.10$) and aluminum ($\beta=18.59$), due to the high separation factors. This study therefore highlights the novelty of applying RSM for optimizing thiourea leaching and establishes a foundation for selective recovery strategies such as liquid–liquid extraction.

Separation Technology - ICoST 2025
Materials Research Proceedings 59 (2026) 1-9

Materials Research Forum LLC
https://doi.org/10.21741/9781644903957-1

Acknowledgement

This research has received funding from the from the Universiti Teknologi Malaysia under UTM Professional Development Research University (PDRU) (Q.J130000.21A2.06E68).

References

[1] S. Jeon, M. Ito, C.B. Tabelin, R. Pongsumrankul, N. Kitajima, Ilhwan Park, N. Hiroyoshi, Gold recovery from shredder light fraction of E-waste recycling plant by flotation-ammonium thiosulfate leaching, Waste Management 77 (2018) 195–202. https://doi.org/10.1016/J.WASMAN.2018.04.039

[2] M. Huy Do, G. Tien Nguyen, U. Dong Thach, Y. Lee, T. Huu Bui, Advances in hydrometallurgical approaches for gold recovery from E-waste: A comprehensive review and perspectives, Miner Eng 191 (2023) 107977. https://doi.org/10.1016/J.MINENG.2022.107977

[3] C. Carelse, M. Manuel, D. Chetty, A. Corfield, Au and Ag distribution in alloys produced from the smelting of printed circuit boards-an assessment using SEM-EDS, EPMA, and LA-ICP-MS analysis, The Journal of the Southern African Institute of Mining and Metallurgy (n.d.). https://doi.org/10.17159/2411

[4] M. Baniasadi, F. Vakilchap, N. Bahaloo-Horeh, S.M. Mousavi, S. Farnaud, Advances in bioleaching as a sustainable method for metal recovery from e-waste: A review, Journal of Industrial and Engineering Chemistry 76 (2019) 75–90. https://doi.org/10.1016/J.JIEC.2019.03.047

[5] D.A. Ray, M. Baniasadi, J.E. Graves, A. Greenwood, S. Farnaud, Thiourea Leaching: An Update on a Sustainable Approach for Gold Recovery from E-waste, Journal of Sustainable Metallurgy 8 (2022) 597–612. https://doi.org/10.1007/S40831-022-00499-8/TABLES/4

[6] I. Birloaga, I. De Michelis, F. Ferella, M. Buzatu, F. Vegliò, Study on the influence of various factors in the hydrometallurgical processing of waste printed circuit boards for copper and gold recovery, Waste Management 33 (2013) 935–941. https://doi.org/10.1016/j.wasman.2013.01.003

[7] J. Li, J.D. Miller, Reaction kinetics of gold dissolution in acid thiourea solution using ferric sulfate as oxidant, Hydrometallurgy 89 (2007) 279–288. https://doi.org/10.1016/j.hydromet.2007.07.015

[8] T.H. Bui, S. Jeon, Y. Lee, Facile recovery of gold from e-waste by integrating chlorate leaching and selective adsorption using chitosan-based bioadsorbent, J Environ Chem Eng 9 (2021) 104661. https://doi.org/10.1016/J.JECE.2020.104661

[9] Z. wei Liu, X. yi Guo, Q. hua Tian, L. Zhang, A systematic review of gold extraction: Fundamentals, advancements, and challenges toward alternative lixiviants, J Hazard Mater 440 (2022) 129778. https://doi.org/10.1016/J.JHAZMAT.2022.129778

[10] E. Hsu, K. Barmak, A.C. West, A.H.A. Park, Advancements in the treatment and processing of electronic waste with sustainability: a review of metal extraction and recovery technologies, Green Chemistry 21 (2019) 919–936. https://doi.org/10.1039/C8GC03688H

[11] K. Li, Z. Xu, A review of current progress of supercritical fluid technologies for e-waste treatment, J Clean Prod 227 (2019) 794–809. https://doi.org/10.1016/J.JCLEPRO.2019.04.104

[12] P. Altinkaya, Z. Wang, I. Korolev, J. Hamuyuni, M. Haapalainen, E. Kolehmainen, K. Yliniemi, M. Lundström, Leaching and recovery of gold from ore in cyanide-free glycine media, Miner Eng 158 (2020) 106610. https://doi.org/10.1016/J.MINENG.2020.106610

[13] A. Hubau, A. Chagnes, M. Minier, S. Touzé, S. Chapron, A.G. Guezennec, Recycling-oriented methodology to sample and characterize the metal composition of waste Printed Circuit Boards, Waste Management 91 (2019) 62–71. https://doi.org/10.1016/j.wasman.2019.04.041

[14] J. Rajahalme, S. Perämäki, R. Budhathoki, A. Väisänen, Effective Recovery Process of Copper from Waste Printed Circuit Boards Utilizing Recycling of Leachate, JOM 73 (2021) 980–987. https://doi.org/10.1007/s11837-020-04510-z

[15] M. Zhao, J. Wan, W. Qin, X. Huang, G. Chen, X. Zhao, A machine learning-based diagnosis modelling of type 2 diabetes mellitus with environmental metal exposure, Comput Methods Programs Biomed 235 (2023) 107537. https://doi.org/10.1016/J.CMPB.2023.107537

[16] K. Li, Q. Li, Y. Zhang, X. Liu, Y. Yang, T. Jiang, Thiourea leaching of gold: Elucidating the mechanism of arsenopyrite catalyzed thiourea oxidation by Fe3+ and the beneficial role of oxalate through experimental and density functional theory (DFT) investigations, Miner Eng 207 (2024) 108550. https://doi.org/10.1016/J.MINENG.2023.108550

[17] O. Herreros, J. Viñals, Leaching of sulfide copper ore in a NaCl–H2SO4–O2 media with acid pre-treatment, Hydrometallurgy 89 (2007) 260–268. https://doi.org/10.1016/J.HYDROMET.2007.07.011

[18] G. Senanayake, The role of ligands and oxidants in thiosulfate leaching of gold, Gold Bulletin 2005 38:4 38 (2005) 170–179. https://doi.org/10.1007/BF03215257

[19] N.F.M. Noah, R.N.R. Sulaiman, N. Othman, N. Jusoh, M.B. Rosly, Extractive continuous extractor for chromium recovery: Chromium (VI) reduction to chromium (III) in sustainable emulsion liquid membrane process, J Clean Prod (2020). https://doi.org/10.1016/j.jclepro.2019.119167

[20] R.N. Raja Sulaiman, N. Othman, N.A. Saidina Amin, Recovery of ionized nanosilver by emulsion liquid membrane process and parameters optimization using response surface methodology, Desalination Water Treat 57 (2016) 3339–3349. https://doi.org/10.1080/19443994.2014.985724

Separation Technology - ICoST 2025
Materials Research Proceedings 59 (2026) 10-16

Materials Research Forum LLC
https://doi.org/10.21741/9781644903957-2

Effect of Temperature and Time on the Conversion of Defatted Microalgae Biomass into Bio-Oil through Hydrothermal Liquefaction

Muhammad Khairul Hilmi MOHD ZAKI[1,a], Shafirah SAMSURI[1,2,b*],
Usman BELLO[2,3,c], Nor Adilla RASHIDI[1,2,d]

[1]Chemical Engineering Department, Universiti Teknologi PETRONAS, 32610 Seri Iskandar, Perak, Malaysia

[2]HICoE—Centre for Biofuel and Biochemical Research (CBBR), Institute of Sustainable Energy and Resources, Universiti Teknologi PETRONAS, 32610 Seri Iskandar, Perak, Malaysia

[3]Department of Chemistry, Abubakar Tafawa Balewa University, Gubi Campus, 740102, Bauchi, Nigeria

[a]muhammad_20001244@utp.edu.my, [b]shafirah.samsuri@utp.edu.my,
[c]usman.bello@utp.edu.my, [d]adilla.rashidi@utp.edu.my

Keywords: Bio-Oil, Biorefining, Microalgae, Thermochemical Process, Defatted Microalgae Biomass

Abstract. The growing demand for renewable energy has increased interest in bio-oil production from microalgae biomass using hydrothermal liquefaction (HTL). However, limited research has focused on the utilization of defatted microalgae biomass (DMB), the residual biomass remaining after lipid extraction. Therefore, this study investigated the effect of varying operating conditions on the conversion of DMB derived from Chlorella sp. into bio-oil via HTL. Experiments were conducted at reaction temperatures ranging from 210–270 °C and reaction times between 30 and 90 min. The resulting bio-oils contained a range of compounds, including aliphatic acids, amines, amides, pyrazine, pyridine, and phenolic compounds. The highest yields of desirable aliphatic acids were obtained at 250–260 °C and 45 min, indicating these as optimal conditions for maximizing fuel-relevant components. While higher temperatures and longer reaction times promoted the formation of nitrogenous and aromatic compounds, they also introduced instability in the product composition. Hence, this work highlights the importance of optimizing HTL parameters to enhance bio-oil recovery from DMB.

Introduction

Global energy demand has increased over the years due to commercial and industrialization, as well as rapid population growth, along with limited fuel resources and the rise of environmental issues caused by the heightened combustion of fossil fuels. This has led to the exploration of renewable and sustainable resources to augment the depleting fossil reserves as alternative energy sources [1]. In response, energy derived from biomass (biofuel) is placed forward as an attractive alternative owing to its fuel quality parameters compared to fossil-based fuels, with a lower combustion profile and CO_2 emission [2]. However, the issue linked to the negative impact on global food security and environmental pollution caused by over-utilization of fertilizer and pesticide in cultivating food and oil crops (first-generation feedstocks), has desalted the interest in using them as feedstock for biofuel production. Second generation feedstock, also known as the non-edible crops such as forest, agricultural residues and waste oils, are used in producing biofuel, however, low product yield and incomplete reaction under certain conditions, especially for large, prompted their broader adoption [1,3]. Consequently, the third-generation feedstock (microalgae) became widely explored owing to their higher oil yield compared to that obtained using other feedstocks [4]. At the moment, many studies have been focused on extracting high-value chemicals

Separation Technology - ICoST 2025
Materials Research Proceedings 59 (2026) 10-16

Materials Research Forum LLC
https://doi.org/10.21741/9781644903957-2

and energy-dense lipids from microalgae for biofuel production, leaving behind a large amount of residual biomass after the lipid extraction, which is known as DMB [5]. These recovered residues (DMB) contain significant amounts of organic material, including proteins and carbohydrates, which can be utilized as animal feed and fertilizer or fermented for ethanol production [6]. Rather than disposed-off, these leftover residues can be processed via thermochemical conversion routes to generate valuable products like bio-oil, a process considered socially sustainable. Among different thermochemical methods, HTL was considered cheap, fast, and easy to operate compared to hydrothermal gasification, hydrothermal carbonization, and pyrolysis. Additionally, this is more favored due to its capability of converting wet microalgae biomass with high moisture content into bio-oil. Typically, HTL operates at moderately high temperatures between 280–370 °C and pressures within 5–25 MPa with the presence of water as the reaction medium [7], DMB feedstock from *Chlorella sp.* microalgae using solvent-based extraction, and to determine the effect of different operating conditions on the conversion of defatted microalgae biomass into bio-oil through HTL.

Materials and Methods

The sample Chlorella sp. microalgae powder was purchased from a local store along with other high-purity grade chemicals like chloroform, methanol, and dichloromethane.

About 50 g of Chlorella sp. microalgae powder was weighed and transferred into a 500 mL conical flask, then mixed with 200 mL of methanol. The mixture was left to soak overnight. Afterward, excess methanol was removed by filtration. The biomass was subsequently extracted with 300 mL of a methanol–chloroform mixture (2:1, v/v) under continuous stirring at 1000 rpm for 30 min at room temperature. The resulting mixture was filtered to remove residual solvents, and the extract was dried in an oven at 100 °C for 24 h [8].

About 40 g of dried DMB was mixed with 160 mL of deionized water to form a microalgae slurry with 80 wt.% moisture content. The wet biomass was then transferred into the HTL autoclave reactor, which was sealed tightly. The reactor was placed in a pre-heated oven at 210 °C and held for 1 h. After the reaction, the reactor was removed from the oven, allowed to cool to room temperature, and the contents were filtered using vacuum filtration to separate the solid residues. The resulting filtrate was transferred to a 500 mL conical flask and mixed with 200 mL of dichloromethane. The mixture was stirred for 2 min and then transferred into a separatory funnel to separate the aqueous by-product from the bio-oil [9]. This process was repeated at different operating temperatures ranging from 210 to 270 °C in 10 °C increments. Additionally, a separate set of experiments was conducted with varying reaction times from 30 to 90 min in 5-minute intervals. Fig. 1 illustrates the step-by-step procedure for the lipid extraction through HTL processing to bio-oil recovery.

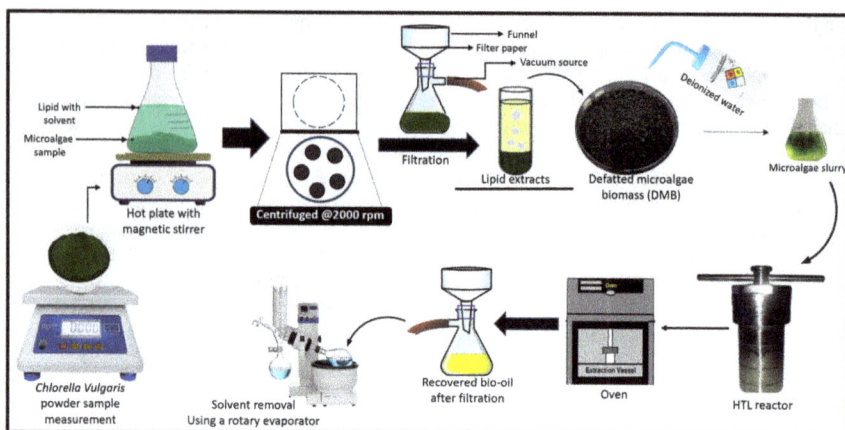

Figure 1 *The experimental steps involved in the recovery of bio-oil from DMB.*

Results and Discussion

Effect of reaction temperature

The percentage area versus reaction temperature for the respective compounds present in the bio-oil is shown in Fig. 2A–D. From the graphs, it is evident that temperature variation affects the composition of the bio-oil produced from *Chlorella sp.* (DMB) via hydrothermal liquefaction (HTL). For aliphatic acids (Fig. 2A), the data show that both hexadecanoic and octadecenoic acids follow a similar trend. The highest percentage areas of hexadecanoic acid and octadecenoic acid are observed at 250 °C and 260 °C, respectively. In the case of amines and amides (Fig. 2B), both compounds exhibit a similar trend, with their highest percentage areas recorded at 270 °C. For nitrogen-containing compounds (Fig. 2C), pyrazine is not detected at lower temperatures (210–230 °C). However, its percentage area increases significantly as the temperature rises from 240 °C to 270 °C, with the highest concentration observed at 270 °C. Pyridine is detected in small amounts at lower temperatures, but its percentage area drops below 1% at temperatures above 230 °C, indicating its near absence at higher temperatures [10]. For phenolic compounds (Fig. 2D), the data show an inconsistent trend, with the highest percentage area observed at the lowest temperature of 210 °C. Regarding overall bio-oil yield, it increases significantly with rising temperature, however, it begins to decrease beyond a certain point [11]. To determine the optimum reaction temperature, aliphatic acids were selected as reference compounds due to their relevance as precursors for biofuels and their role in improving bio-oil quality [12]. According to the study/s data, the optimum temperature range is determined to be between 250 °C and 260 °C, where the highest percentage areas of aliphatic acids were obtained.

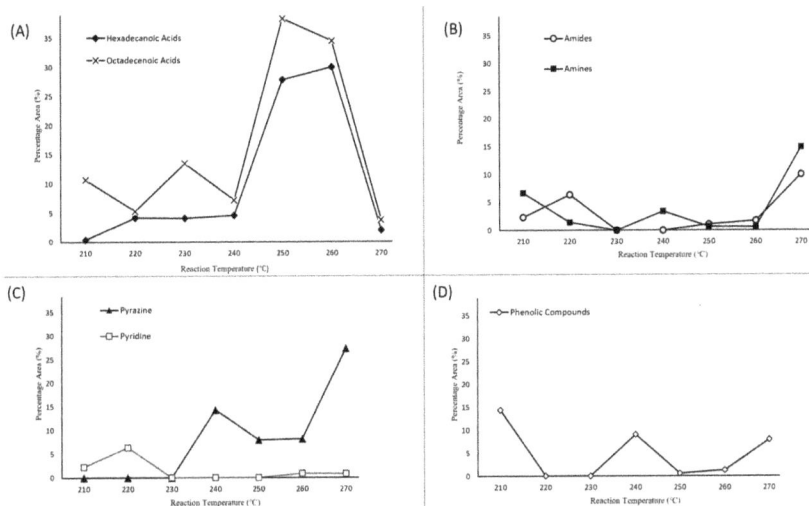

Figure 2 *Plots of percentage area vs reaction temperatures for (A) aliphatic acids, (B) amides and amines (C) pyrazine and pyridine, and (D) phenolic compounds.*

Effect of reaction time

The percentage area versus reaction time, as shown in Fig. 3A–D, indicates that variations in reaction time affect the composition of the bio-oil produced for the selected components. In the plot for aliphatic acids (Fig. 3A), both hexadecanoic and octadecenoic acids follow a similar trend, with the highest percentage areas observed at a reaction time of 45 min. Conversely, amines and amides are absent at shorter reaction times (Fig. 3B), with the highest percentage areas recorded at longer reaction times, corresponding to 75 and 90 min for amides and amines, respectively. However, the data for amines show an unstable trend at reaction times beyond 60 min. In Fig. 3C, both pyrazine and pyridine are absent at reaction times between 30 and 60 min. The percentage area of pyridine increases significantly at 75 min, followed by a slight decrease at 90 min. Similarly, the percentage area of pyrazine increases with reaction time, peaking at 75 min before slightly declining at 90 min. Thus, the highest percentage areas for both pyrazine and pyridine occur at 75 min. For phenolic compounds (Fig. 3D), the data display an unstable pattern, but with an overall increasing trend. The highest percentage area is observed at the longest reaction time of 90 min. According to Huang et., al. [11], bio-oil yield is expected to increase significantly with longer reaction times. However, the authors also reported that temperature plays a more significant role than reaction time in enhancing bio-oil yield recovery using HTL. Therefore, in determining the optimum reaction time for the process, a similar method used for evaluating reaction temperature was applied, selecting aliphatic acids as the reference components due to their desirability for biofuel production [13,14]. Based on this criterion, the optimum reaction time is determined to be 45 min, as this condition yields the highest percentage area of aliphatic acids.

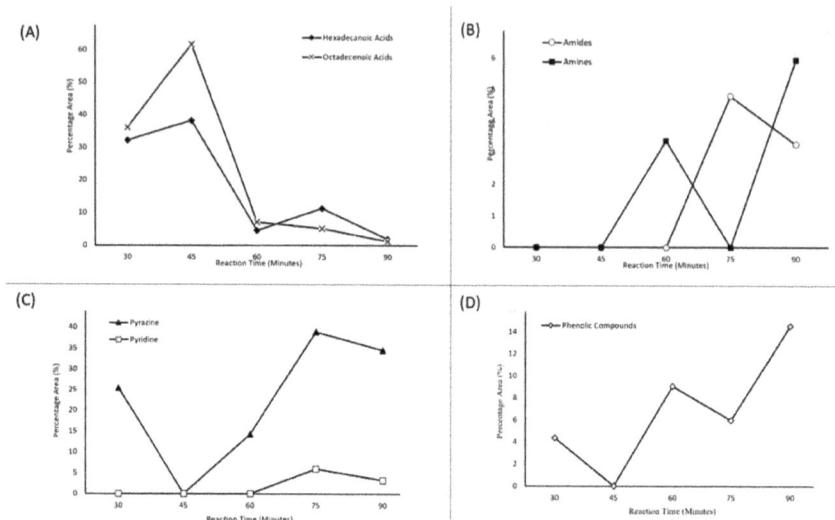

Figure 3 Plots of percentage area vs reaction times for aliphatic acids (A), for amides and amines (B), for pyrazine and pyridine (C) for phenolic compounds (D).

Conclusion

In conclusion, this work demonstrates that both reaction temperature and time significantly influence the composition and yield of bio-oil produced from Chlorella sp. via HTL. Aliphatic acids, which are key precursors for biofuel applications, were most abundant at 250–260 °C and a reaction time of 45 min, indicating these as optimal conditions for maximizing desirable bio-oil components. While longer reaction times favored the formation of nitrogen-containing and phenolic compounds, over-stretching of the reaction times led to unstable trends and diminished yields for certain components. Finally, it was established that temperature had a more pronounced effect on bio-oil quality and yield than reaction time, which is consistent with findings from previous studies.

Acknowledgment

The authors would like to acknowledge the financial assistance from PETRONAS via YUTP-FRG (Cost Centre: 015LC0-588), and the facilities support from HICoE - Centre for Biofuel and Biochemical Research (CBBR), and Chemical Engineering Department, Universiti Teknologi PETRONAS. Support from the Ministry of Education Malaysia through the HICoE award to CBBR is duly and warmly acknowledged.

References

[1] C. Xia, A. Pathy, B. Paramasivan, P. Ganeshan, K. Dhamodharan, A. Juneja, D. Kumar, K. Brindhadevi, S. Kim, and K. Rajendran, Comparative study of pyrolysis and hydrothermal liquefaction of microalgal species: Analysis of product yields with reaction temperature. *Fuel*, *311*, (2021) 121932. https://doi.org/10.1016/j.fuel.2021.121932

[2] Y. Guo, T. Yeh, W. Song, D. Xu, and S. Wang, A review of bio-oil production from hydrothermal liquefaction of algae. *Renewable and Sustainable Energy Reviews, 48*, (2015) 776–790. https://doi.org/10.1016/j.rser.2015.04.049

[3] K. Brindhadevi, S. Anto, E. R. Rene, M. Sekar, T. Mathimani, N. T. L. Chi, and A. Pugazhendhi, Effect of reaction temperature on the conversion of algal biomass to bio-oil and biochar through pyrolysis and hydrothermal liquefaction. *Fuel, 285*, (2021) 119106. https://doi.org/10.1016/j.fuel.2020.119106

[4] L. Brennan, and P. Owende, Biofuels from microalgae—A review of technologies for production, processing, and extractions of biofuels and co-products. *Renewable and Sustainable Energy Reviews, 14*(2), (2009) 557–577. https://doi.org/10.1016/j.rser.2009.10.009

[5] D. R. Vardon, B. K. Sharma, G. V. Blazina, K. Rajagopalan, and T. J. Strathmann, Thermochemical conversion of raw and defatted algal biomass via hydrothermal liquefaction and slow pyrolysis. *Bioresource Technology, 109*, (2012) 178–187. https://doi.org/10.1016/j.biortech.2012.01.008

[6] J. Yang, H. Shin, Y. Ryu, and C. Lee, Hydrothermal liquefaction of *Chlorella vulgaris*: Effect of reaction temperature and time on energy recovery and nutrient recovery. Journal of Industrial and Engineering Chemistry, 68, (2018) 267–273. https://doi.org/10.1016/j.jiec.2018.07.053

[7] S. I. Mustapha, U. A. Mohammed, F. Bux, and Y. M. Isa, Catalytic hydrothermal liquefaction of nutrient-stressed microalgae for production of high-quality bio-oil over Zr-doped HZSM-5 catalyst. *Biomass and Bioenergy, 163*, (2022) 106497. https://doi.org/10.1016/j.biombioe.2022.106497

[8] J. Zhou, M. Wang, J. A. Saraiva, A. P. Martins, C. A. Pinto, M. A. Prieto, J. Simal-Gandara, H. Cao, J. Xiao, and F. J. Barba, Extraction of lipids from microalgae using classical and innovative approaches. *Food Chemistry, 384*, (2022) 132236. https://doi.org/10.1016/j.foodchem.2022.132236

[9] C. Yuan, S. Zhao, J. Ni, Y. He, B. Cao, Y. Hu, S. Wang, L. Qian, and A. Abomohra, Integrated route of fast hydrothermal liquefaction of microalgae and sludge by recycling the waste aqueous phase for microalgal growth. *Fuel, 334*, (2022) 126488. https://doi.org/10.1016/j.fuel.2022.126488

[10] J. Arun, P. Varshini, P. K. Prithvinath, V. Priyadarshini, and K. P. Gopinath, Enrichment of bio-oil after hydrothermal liquefaction (HTL) of microalgae C. vulgaris grown in wastewater: Bio-char and post-HTL wastewater utilization studies. *Bioresource Technology, 261*, (2018) 182–187. https://doi.org/10.1016/j.biortech.2018.04.029

[11] Y. Huang, Y. Chen, J. Xie, H. Liu, X. Yin, and C. Wu, Bio-oil production from hydrothermal liquefaction of high-protein high-ash microalgae including wild Cyanobacteria sp. and cultivated Bacillariophyta sp. *Fuel, 183*, (2016) 9–19. https://doi.org/10.1016/j.fuel.2016.06.013

[12] I. Nava-Bravo, S. Velasquez-Orta, I. Monje-Ramírez, L. Güereca, A. Harvey, A. Cuevas-García, L. Yáñez-Noguez, and M. Orta-Ledesma, Catalytic hydrothermal liquefaction of microalgae cultivated in wastewater: Influence of ozone-air flotation on products, energy balance, and carbon footprint. *Energy Conversion and Management, 249*, (2021) 114806. https://doi.org/10.1016/j.enconman.2021.114806

[13] S. Chaudry, P. A. Bahri, N. R., and Moheimani, Pathways of processing of wet microalgae for liquid fuel production: A critical review. *Renewable and Sustainable Energy Reviews, 52,* (2015) 1240–1250. https://doi.org/10.1016/j.rser.2015.08.005

[14] S. Kandasamy, B. Zhang, Z. He, H. Chen, H. Feng, Q. Wang, B. Wang, V. Ashokkumar, S. Siva, N. Bhuvanendran, and M. Krishnamoorthi, Effect of low-temperature catalytic hydrothermal liquefaction of Spirulina platensis. *Energy, 190,* (2019)116236. https://doi.org/10.1016/j.energy.2019.116236

Separation Technology - ICoST 2025
Materials Research Proceedings 59 (2026) 17-24

Materials Research Forum LLC
https://doi.org/10.21741/9781644903957-3

Eco-Friendly Activation of Palm Kernel Shell Carbon using Calcium Chloride from Cockle Shells

Mohamad Razlan MD RADZI[1,a], Nor Adilla RASHIDI[1,b] *

[1]HICoE—Centre for Biofuel and Biochemical Research, Institute of Sustainable Energy and Resources (ISER), Department of Chemical Engineering, Universiti Teknologi PETRONAS, Seri Iskandar, Perak 32610, Malaysia

[a]razlan.radzi@utp.edu.my, [b]adilla.rashidi@utp.edu.my

Keywords: Biomass Utilization, Heavy Metal Adsorption

Abstract. The study aimed to develop a sustainable, low-cost method for removing heavy metals such as iron (Fe) from wastewater by producing activated carbon from palm kernel shells, activated with calcium chloride derived from waste cockle shells. Using a Taguchi L_9 orthogonal array, nine samples were synthesized by varying temperature (500-700°C), heating time (60-120 min), and calcium chloride concentration (0.25-1M). The carbon yield was affected mainly by temperature and time, with excessive values causing ash formation. The best-performing sample was tested for Pb adsorption using atomic absorption spectrometry (AAS). Results showed that higher adsorbent dosage increased removal efficiency, while higher metal concentrations reduced it. Longer contact times initially improved adsorption but eventually reached equilibrium.

Introduction

Cockle is a popular seafood in Malaysia and a key aquaculture product. However, the discarded shells, which constitute nearly two-thirds of the biomass, pose environmental concerns due to slow biodegradation and pollution risks. To mitigate this, studies have explored reusing cockle shells as low-cost adsorbents and other products [1]. Rich in calcium carbonate ($CaCO_3$) up to 95%, cockle shells offer a sustainable alternative to mined limestone, reducing the ecological damage linked to conventional extraction [2].

Activated carbon is a highly porous material derived from carbon-rich sources such as bamboo, coconut husks, and palm kernel shells (PKS). It is widely used in environmental and industrial applications due to its exceptional adsorption capacity, high surface area, and chemical stability [3]. Common applications include water and air purification, wastewater treatment, and serving as catalyst supports. Activated carbon's effectiveness stems from its extensive pore structure and the presence of surface functional groups such as carboxyl, carbonyl, phenol, lactone, and quinone, which play key roles in binding and adsorbing various contaminants [4]. It is particularly valued for its ability to withstand harsh chemical environments and its high affinity for a wide range of pollutants [5].

The production of activated carbon from biomass can be enhanced through chemical activation techniques. In this method, calcium chloride derived from cockle shells is used to impregnate the palm kernel shell biomass, promoting pore development and enhancing the adsorptive properties of the final product [6]. Utilizing calcium chloride from waste cockle shells not only adds value to a previously discarded by-product but also reduces dependence on conventional, often hazardous and costly activating agents such as phosphoric acid or zinc chloride. This approach offers a more economical, environmentally friendly, and sustainable pathway for producing high-performance activated carbon. Furthermore, the synthesized activated carbon was later evaluated for its ability to remove iron (Fe) from aqueous solutions, highlighting its potential in treating Fe-contaminated wastewater.

Materials and Methods

Cockle shells were collected from local stores in Seri Iskandar, Perak, cleaned, dried at 110°C, ground, and sieved to 100 μm. Palm kernel shells from FELCRA Nasaruddin Belia in Bota, Perak, were similarly cleaned, dried, ground, and sieved. Calcium chloride ($CaCl_2$) was synthesized by reacting cockle shell powder with hydrochloric acid (0.5–2 M), producing $CaCl_2$ solutions of 0.25-1 M after filtration.

Five grams of palm kernel shell were impregnated overnight with 100 mL of the $CaCl_2$ solution, dried, and calcined under conditions defined by a Taguchi L_9 orthogonal array. The resulting activated carbon was washed with 10% HCl and deionized water until neutral pH, then its yield was calculated using formula in Eq. 1.

$$\text{Yield (\%)} = \frac{\text{weight of AC after washing (g)}}{\text{inital PKS weight (g)}} \times 100\%$$

(1)

The activated carbon's adsorption performance for iron (Fe) removal was tested using Fe(III) nitrate solutions. Experiments varied three parameters: adsorbent dosage (0.005-0.1 g), initial Fe concentration (20-120 ppm), and contact time (30-180 min) at 30°C. After adsorption, samples were filtered and analysed via Atomic Absorption Spectrometry (AAS, Agilent 240FS). The adsorption capacity (q_e) was determined using Eq. 2 based on Fe concentrations before and after treatment, where q_e is the adsorption capacity in mg/g, C_o is the initial Fe concentration in mg/L, C_e is the equilibrium Fe concentration in mg/L, V is the volume of solution in L and m is the mass of adsorbent in g.

$$q_e = \frac{(C_0 - C_e) \times V}{m}$$

(2)

Results and Discussion

The section was subdivided into multiple subsections discussing about optimization of activated carbon synthesis using Taguchi L_9 orthogonal array based on pyrolysis temperature, pyrolysis time and concentration of calcium chloride, followed by characterization of activated carbon and lastly effect of initial Fe concentration, adsorbent dosage and contact time on Fe adsorption. Three key parameters, calcium chloride molarity (0.25-1 M), calcination temperature (500-700°C), and calcination time (60-120 min), were optimized using the Taguchi L_9 orthogonal array to maximize activated carbon yield from palm kernel shells (PKS). The yield was determined by comparing the final product weight to the initial biomass weight.

Table 1 Yield calculation of activated carbon.

Run	M	Temp. (°C)	Time (min)	PKS	Pyro	Wash	Yield (%)
		Parameters			Weight (g)		
1	0.25	500	60	5.012	0.31	0.276	5.51
2	0.25	600	90	5.009	0.07	0.031	0.62
3	0.25	700	120	5.005	0.06	0.024	0.48
4	0.5	500	90	5.006	0.55	0.392	7.83
5	0.5	600	120	5.016	0.07	0.059	1.18
6	0.5	700	60	5.013	0.15	0.097	1.93
7	1	500	120	5.009	0.20	0.19	3.79
8	1	600	60	5.011	0.06	0.036	0.72
9	1	700	90	5.006	0.04	0.032	0.64

Separation Technology - ICoST 2025
Materials Research Proceedings 59 (2026) 17-24

Materials Research Forum LLC
https://doi.org/10.21741/9781644903957-3

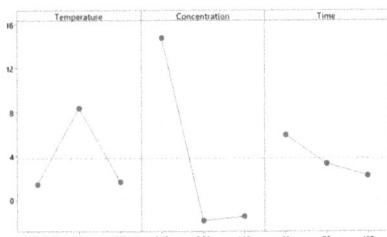

Figure 1 Signal to noise ratio (SNR) plot.

Table 1 shows that yield decreases with increasing temperature and time, due to greater decomposition of volatile components. The highest yield of 7.83% occurred at moderate conditions (M = 0.5, 500°C and 90 min), while the lowest yield of 0.48% was at harsher conditions but lower concentrations (M = 0.25, 700 °C and 120 min). This indicates that milder activation favours higher carbon recovery, whereas harsher conditions cause more material loss. From the yield, Taguchi analysis using Minitab and signal-to-noise (S/N) ratios identified the optimal conditions as 600 °C, 0.25 M calcium chloride, and 60 min as observed in Fig. 1. These settings provided the best balance between thermal decomposition and chemical activation, resulting in efficient pore formation, minimal carbon loss, and the highest overall yield. [7].

Characteristics of activated carbon
The FTIR spectrum of the synthesized activated carbon of 600°C, 0.25M calcium chloride, 60 min in Fig. 2 confirms the presence of several characteristic functional groups typically associated with carbonaceous materials. Comparison has shown that the spectrum is almost a duplicate of FTIR spectrum of activated carbon in [8]. A broad absorption band around 3397 cm^{-1} is attributed to O–H stretching vibrations, indicating the presence of hydroxyl groups, which may originate from residual moisture or phenolic compounds commonly retained on the surface of activated carbon due to its high porosity and surface area [9].

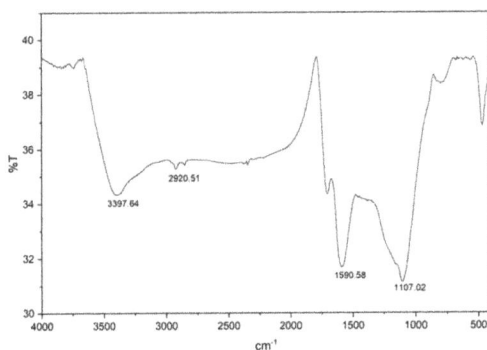

Figure 2 FTIR spectra of synthesized activated carbon.

The absorption peak near 2920 cm^{-1} corresponds to C–H stretching vibrations of aliphatic hydrocarbons, suggesting that remnants of lignin or cellulose from the original palm kernel shell precursor may still be present despite pyrolysis [8]. A distinct band near 1590 cm^{-1} is assigned to C=C stretching vibrations within aromatic ring structures, indicating the formation of aromatic domains during thermal treatment. Additionally, a peak around 1105 cm^{-1} is associated with C–O stretching vibrations, reflective of the presence of ether, alcohol, or phenolic groups on the carbon surface. Finally, absorption bands in the 670-870 cm^{-1} region are characteristic of aromatic C–H out-of-plane bending, further confirming the development of aromatic ring structures during the carbonization process.

SEM analysis in Fig. 3a revealed that the synthesized activated carbon exhibited compact, spherical, and sub-spherical particles with well-defined boundaries and a high degree of structural uniformity. At higher magnification in Fig. 3b, a dense distribution of circular micropores was clearly visible, although the pore sizes appeared slightly smaller compared to those reported in [10]. This observation confirms the successful development of a well-formed porous network, which is essential for enhancing surface area and improving adsorption efficiency. The particles were closely packed with limited large voids, while the presence of small inter-particle gaps suggested additional mesoporosity that could facilitate better diffusion of adsorbates within the material. Minor surface irregularities and debris were observed, possibly originating from residual ash, unreacted precursor material, or mechanical handling during processing.

The EDX analysis in Table 2 further supported the successful synthesis of carbon-rich material, showing that carbon (C) was the predominant element at 73.13%, followed by oxygen (O) at 25.23%, with smaller amounts of silicon (Si) at 1.52% and sulphur (S) at 0.12%.

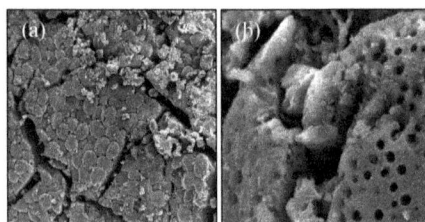

Figure 3 *FESEM micrograph of activated carbon at (a) 500x and (b) 50Kx.*

The high carbon content reflects an effective carbonization process and the efficient removal of non-carbon elements during activation. The presence of oxygen is attributed to surface functional groups such as hydroxyl, carbonyl, and carboxyl, which are known to enhance surface reactivity and adsorption performance. Trace amounts of Si and S may have originated from the raw biomass or minor impurities introduced during activation.

Table 2 *Elemental composition, BET area, pore volume and pore size of activated carbon.*

Elements	Weight %	Parameters	Value
Carbon	73.13	Surface Area (m^2/g)	435.64
Oxygen	25.23	Pore volume (cm^3/g)	0.27
Silicon	1.52	Pore size (nm)	2.48
Sulphur	0.12		

Overall, the elemental composition confirms that the synthesized activated carbon met the SNI 06-3730-1995 standard, with a carbon content exceeding the minimum requirement of 65%, thereby validating its suitability for adsorption and other environmental applications [11].

BET analysis in Table 2 revealed that the synthesized activated carbon exhibited a high specific surface area of 435.64 m^2/g, a total pore volume of 0.27 cm^3/g, and an average pore diameter of 2.48 nm, indicating a predominantly mesoporous structure suitable for adsorption applications

Separation Technology - ICoST 2025
Materials Research Proceedings 59 (2026) 17-24

Materials Research Forum LLC
https://doi.org/10.21741/9781644903957-3

[12]. Such characteristics are advantageous for accommodating larger adsorbate molecules and support potential applications in wastewater treatment, gas separation, and catalysis.

Adsorption of Fe

After optimizing the activated carbon synthesis parameters and conducting its characterization, adsorption experiments were performed to examine the effects of adsorbent dosage, initial Fe concentration, and contact time. The results, summarized in Table 3, revealed distinct trends in Fe removal efficiency for each parameter.

Table 3 *Effects of activated carbon dosage, initial Fe concentration and contact time on q_e.*

Parameters	C_o (mg/L)	C_e (mg/L)	% Removal	q_e (mg/g)
0.005 g	100	77.48	22.52	450.4
0.01 g	100	68.33	31.67	317
0.02 g	100	62.18	37.82	189.1
0.1 g	100	56.09	43.91	43.9
20 ppm	20	4.87	75.65	151.3
40 ppm	40	23.65	40.88	163.5
60 ppm	60	37.49	37.52	225.1
80 ppm	80	50.09	37.39	299.1
100 ppm	100	68.45	31.55	315.5
120 ppm	120	90.13	24.89	299
30 min	100	69.01	30.99	309.9
60 min	100	68.27	31.73	317.3
90 min	100	67.08	32.92	329.2
120 min	100	66.65	33.35	333.5
150 min	100	66.57	33.43	334.3
180 min	100	66.48	33.52	335.2

In Fig. 4, increasing adsorbent dosage enhanced overall Fe removal, with equilibrium concentration (C_e) decreasing from 77.48 mg/L at 0.005 g to 56.09 mg/L at 0.1 g. However, adsorption capacity (q_e) declined from 450.4 mg/g to 43.9 mg/g, reflecting reduced efficiency at higher dosages due to particle aggregation and site overlap [13].

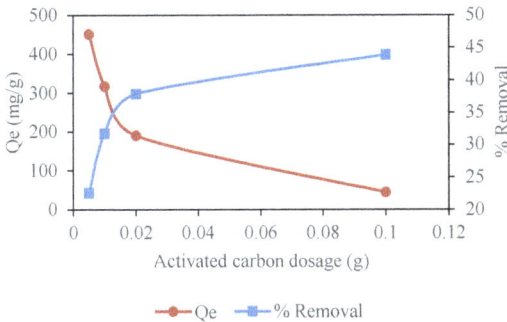

Figure 4 *Effect of activated carbon dosage on adsorption capacity and % removal of Fe.*

The percentage of Fe removal increased from 22.52% to 43.91% as dosage rose, indicating that additional surface area and active sites improved overall metal removal [14]. These results

Separation Technology - ICoST 2025   Materials Research Forum LLC
Materials Research Proceedings 59 (2026) 17-24   https://doi.org/10.21741/9781644903957-3

highlight the need for dosage optimization to achieve an effective balance between Fe removal efficiency and adsorbent utilization.

Fig. 5 shows the effect of initial Fe concentration on the adsorption performance of the activated carbon. The results show that increasing the initial Fe concentration improves q_e, particularly between 20 ppm and 100 ppm. This trend can be attributed to the increased driving force for mass transfer between the aqueous phase and the surface of the adsorbent, which facilitates more effective diffusion of Fe ions to the available active sites. The higher availability of Fe ions at elevated concentrations allows for more active sites to be occupied, resulting in a higher q_e [15]. However, at 120 ppm, a slight drop in q_e is observed, signifying the onset of adsorbent site saturation. This indicates that most of the active sites are already occupied and further increases in Fe concentration do not lead to higher adsorption. In fact, excessive concentrations might lead to pore blockage or competition among Fe ions for limited active sites, thus reducing adsorption efficiency. On the other hand, the percentage removal of Fe decreases as the initial concentration increases. This inverse relationship is likely due to the fixed 0.01g adsorbent dosage being inadequate to handle excess Fe ions at higher concentrations, resulting in a total Fe being adsorbed [16]. These findings show the importance of balancing initial metal concentration with appropriate adsorbent dosage to maximize both removal efficiency and adsorption capacity.

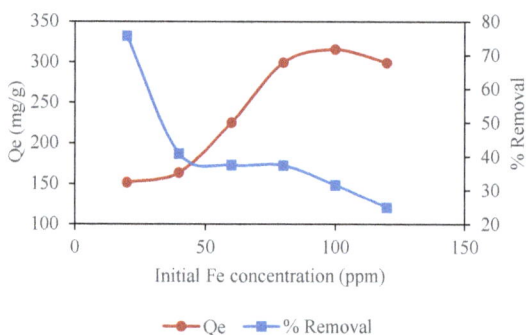

Figure 5 *Effect of initial Fe concentration on adsorption capacity and % removal of Fe.*

As shown in Fig. 6, the effect of contact time on Fe adsorption using activated carbon was evaluated. The results indicate that increasing contact time leads to a corresponding increase in both q_e and percentage removal. However, the rate of increase begins to taper after 120 min, suggesting that the adsorption process is nearing equilibrium. The minimal difference in Fe removal observed between 120 and 180 min further supports this, indicating that extending the contact time beyond 120 min provides only marginal improvement in adsorption performance [17]. Therefore, 120 min can be considered the optimal contact time for efficient Fe adsorption under the specified experimental conditions.

Separation Technology - ICoST 2025

Materials Research Proceedings 59 (2026) 17-24

Materials Research Forum LLC

https://doi.org/10.21741/9781644903957-3

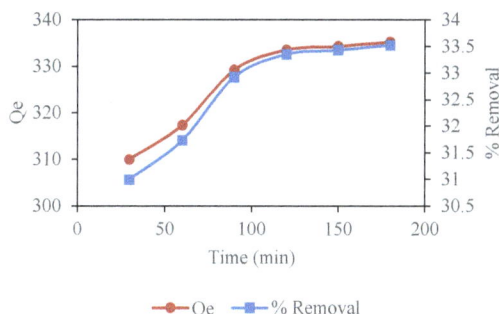

Figure 6 *Effect of contact time on adsorption capacity and % removal of Fe.*

Conclusion

Activated carbon was successfully synthesized from palm kernel shell using calcium chloride derived from cockle shell through an eco-friendly activation process. Optimization using the Taguchi L_9 orthogonal array identified the best synthesis conditions as 600°C, 60 min, and 0.25 M calcium chloride. The resulting activated carbon exhibited a high surface area of 435.64 m^2/g, good pore volume, and small pore size, with FESEM images revealing compact sub-spherical particles containing well-arranged circular pores. The Fe adsorption performance depended on adsorbent dosage, initial concentration, and contact time. Lower adsorbent dosage increased adsorption capacity (q_e), reaching 450.2 mg/g at 0.005 g, while higher dosages achieved greater overall removal efficiency (up to 43.91%). Lower initial Fe concentrations resulted in higher removal efficiency (75.65% at 20 ppm), whereas higher concentrations yielded greater q_e values. Longer contact times improved adsorption until equilibrium was reached between 120-180 min, with a maximum q_e of 335.2 mg/g. Overall, maximum Fe removal occurred at low initial concentrations and high adsorbent dosages, while maximum adsorption capacity was achieved at low adsorbent dosages and high Fe concentrations, with 120 min sufficient to reach near equilibrium.

Acknowledgement

This research was supported in full with Kurita Asia Research Grant (24Pmy227) provided by Kurita Water and Environment Foundation.

References

[1] A. Nayeem, F. Mizi, M.F. Ali, J.H. Shariffuddin, Utilization of cockle shell powder as an adsorbent to remove phosphorus-containing wastewater. Environ. Res. (2023). 216. 114514. https://doi.org/10.1016/j.envres.2022.114514

[2] N.S. Mohd Isha, F. Mohd Kusin, N.M. Ahmad Kamal, S.N.M. Syed Hasan, V.L.M. Molahid, Geochemical and mineralogical assessment of sedimentary limestone mine waste and potential for mineral carbonation. Environ. Geochem. Health (2021). 43.5. 2065-2080. https://doi.org/10.1007/s10653-020-00784-z

[3] Z. Heidarinejad, M.H. Dehghani, M. Heidari, G. Javedan, I. Ali, M. Sillanpää, Methods for preparation and activation of activated carbon: a review. Environ. Chem. Lett. (2020). 18.2. 393-415. https://doi.org/10.1007/s10311-019-00955-0

[4] M.A. Islam, M.V. Jacob, E. Antunes, A critical review on silver nanoparticles: From synthesis and applications to its mitigation through low-cost adsorption by biochar. J. Environ. Manag. (2021). 281. 111918. https://doi.org/10.1016/j.jenvman.2020.111918

[5] A. Bhatnagar, W. Hogland, M. Marques, M. Sillanpää, An overview of the modification methods of activated carbon for its water treatment applications. Chem. Eng. J. (2013). 219. 499-511. https://doi.org/10.1016/j.cej.2012.12.038

[6] K. Udyani, D.Y. Purwaningsih, Chemical and Physical Activation Using a Microwave to Increase the Ability of Activated Carbon to Adsorb Dye Waste. J. Phys. Conf. Ser. (2021). 2117.1. 012030. https://doi.org/10.1088/1742-6596/2117/1/012030

[7] X.l. Long, P.y. Chen, X.y. Jin, Effect of modification with hydrobromic acid on the performance of activated carbon in the removal of hexavalent chromium from aqueous solution. Environmental Progress & Sustainable Energy (2023). 43. https://doi.org/10.1002/ep.14245

[8] T.V. Nagalakshmi, K.A. Emmanuel, C. Babu, C. Challari, P. Divakar, Preparation of Mesoporous Activated Carbon from Jackfruit PPI-1 Waste and Development of Different Surface Functional Groups. ILCPA (2015). 54. 189-200. https://doi.org/10.18052/

[9] H. Kristianto, A.A. Arie, R.F. Susanti, M. Halim, J.K. Lee, The effect of activated carbon support surface modification on characteristics of carbon nanospheres prepared by deposition precipitation of Fe-catalyst. IOP Conf. Ser.: Mater. Sci. Eng. (2016). 162.1. 012034. https://doi.org/10.1088/1757-899X/162/1/012034

[10] M. Wei Tze, M. Aroua, M. Szlachta, Palm Shell-based Activated Carbon for Removing Reactive Black 5 Dye: Equilibrium and Kinetics Studies. Bioresour. (2016). 11. 1432-1447. https://doi.org/10.15376/biores.11.1.1432-1447

[11] M.P. Divya, S. Krishnamoorthi, R. Ravi, V.G. Jenner, K. Baranidharan, M. Raveendran, P. Hemalatha, Preparation and characterization of activated carbon from commercially important bamboo species in north eastern India. Adv. Bamboo Sci. (2025). 11. 100148. https://doi.org/10.1016/j.bamboo.2025.100148

[12] A.R. Hidayu, N. Muda, Preparation and Characterization of Impregnated Activated Carbon from Palm Kernel Shell and Coconut Shell for CO_2 Capture. Procedia Eng. (2016). 148. 106-113. https://doi.org/10.1016/j.proeng.2016.06.463

[13] U. Kouakou, A.S. Ello, J.A. Yapo, A. Trokourey, Adsorption of iron and zinc on commercial activated carbon. J. Environ. Chem. Ecotoxicol (2013). 5.6. 168-171

[14] T.M. Alslaibi, I. Abustan, M.A. Ahmad, A.A. Foul, Kinetics and equilibrium adsorption of iron (II), lead (II), and copper (II) onto activated carbon prepared from olive stone waste. Desalination Water Treat. (2014). 52.40-42. 7887-7897. https://doi.org/10.1080/19443994.2013.833875

[15] R.S. Brishti, R. Kundu, M.A. Habib, M.H. Ara, Adsorption of iron(III) from aqueous solution onto activated carbon of a natural source: Bombax ceiba fruit shell. Results Chem. (2023). 5. 100727. https://doi.org/10.1016/j.rechem.2022.100727

[16] S.O. Okonji, L. Yu, J.A. Dominic, D. Pernitsky, G. Achari, Adsorption by Granular Activated Carbon and Nano Zerovalent Iron from Wastewater: A Study on Removal of Selenomethionine and Selenocysteine. (2021). 13.1. 23

[17] M.E. Wahyuhadi, R.A. Kusumadewi, R. Hadisoebroto, Effect of Contact Time on The Adsorption Process of Activated Carbon from Banana Peel in Reducing Heavy Metal Cd and Dyes Using a Stirring Tub (Pilot Scale). IOP Conf. Ser.: Earth Environ. Sci. (2023). 1203.1. 012035. https://doi.org/10.1088/1755-1315/1203/1/012035

Separation Technology - ICoST 2025
Materials Research Proceedings 59 (2026) 25-32

Materials Research Forum LLC
https://doi.org/10.21741/9781644903957-4

Pectin-Functionalized Iron Oxide Nanoparticles for Enhanced Removal of Cationic Dyes: Influence of Pectin Methoxylation on Surface Charge and Adsorption Efficiency

Yen Yan NG[1,a], Peck Loo KIEW[1,b *], Lian See TAN[1,c], Man Kee LAM[2,d], Wei Ming YEOH[3,e], Muhamad Ali MUHAMMAD YUZIR[1,f]

[1]Department of Chemical and Environmental Engineering (ChEE), Malaysia-Japan International Institute of Technology, Universiti Teknologi Malaysia, Jalan Sultan Yahya Petra, 54100 Kuala Lumpur, Malaysia

[2]Department of Chemical Engineering, HICoE-Centre for Biofuel and Biochemical Research, Institute of Self-Sustainable Building, University Teknologi PETRONAS, Seri Iskandar, 32610 Perak, Malaysia

[3]Department of Petrochemical Engineering, Universiti Tunku Abdul Rahman, Jalan Universiti, Bandar Barat Kampar, 31900 Perak, Malaysia

[a]ngyan@graduate.utm.my, [b]plkiew@utm.my, [c]tan.liansee@utm.my, [d]lam.mankee@utp.edu.my, [e]yeohwm@utar.edu.my, [f]muhdaliyuzir@utm.my

Keywords: Iron Oxide Nanoparticles, Pectin Functionalization, Methylene Blue, Adsorption, Surface Charge

Abstract. The discharge of industrial dye effluents, particularly those containing persistent cationic dyes such as methylene blue, poses significant threats to aquatic ecosystems and public health due to their high stability and resistance to conventional treatment methods. Iron oxide nanoparticles (IONPs) have emerged as effective adsorbents owing to their large surface area and strong dye affinity; however, their tendency to agglomerate reduces dispersion and adsorption efficiency. This study presents a sustainable approach to enhance IONP performance by functionalizing them with pectin, a biodegradable polysaccharide derived from agricultural waste. The effects of pectin methoxylation, particularly comparing high-methoxyl (HMP) and low-methoxyl (LMP) pectin, on surface charge, colloidal stability, and dye adsorption were systematically examined. Zeta potential, FESEM–EDX, and adsorption analyses confirmed successful functionalization and improved performance. Increasing the precipitation pH enhanced the negative surface charge (-39.4 mV for HMP and -39.6 mV for LMP; versus -17.3 mV for bare IONPs) and methylene blue removal (HMP: 84.3 to 89.0%; LMP: 94.4 to 95.4%). These findings demonstrate the critical roles of synthesis pH and pectin structure in tailoring nanoparticle properties for efficient and eco-friendly wastewater treatment.

Introduction

The widespread use of synthetic dyes in the textile industry has made it a significant contributor to global water pollution. Among the various types of dyes employed, cationic dyes, often referred to as basic dyes are particularly concerning due to their chemical stability, solubility in water, vibrant coloration, and tendency to persist in wastewater [1]. These dyes, which contain functional groups such as $-NR_3^+$ or $=NR_2^+$ [2], are not only visually polluting but can also elevate the biochemical and chemical oxygen demand of natural water bodies [3], disrupt photosynthetic processes, restrict aquatic plant growth, and introduce toxic, mutagenic, or even carcinogenic substances into food chains [4]. Their resilience against conventional biological degradation allows them to remain in the environment over long periods, posing significant risks to both ecosystems and human health.

To counter this problem, numerous methods for dye removal from wastewater have been explored, with adsorption emerging as one of the most effective and practical techniques [5]. Adsorption has long been widely recognized for its simplicity, low operational cost, and effective performance in removing dye pollutants [5,6]. Iron oxide nanoparticles (IONPs) have attracted particular interest as adsorbents due to their high surface-area-to-volume ratio, non-toxic nature, biocompatibility, chemical stability, and ease of production [7,8]. Compared to bulk materials, these nanoparticles offer markedly superior adsorption capacities due to their large surface area and high particle number per unit mass, which increase the number of binding sites for target molecules [7,9]. While IONPs are also widely applied in medical technologies such as drug delivery, tissue engineering, diagnostic imaging, and hyperthermia therapy [7], their use in water treatment is often hindered by agglomeration. This phenomenon, in which nanoparticles cluster together to form larger aggregates, significantly reduces the available surface area for adsorption and impairs performance [10].

In response, recent research has focused on modifying the nanoparticle surface to prevent agglomeration and enhance performance through coating, stabilization and functionalization [8], which involve altering the surface chemistry of the adsorbent to improve its adsorption capacity and selectivity for specific target molecules [11]. Among them, natural polymers such as pectin, an abundant and plant-based polysaccharide commonly extracted from fruit peels and other agricultural by-products, has shown potential in stabilizing IONPs and preventing agglomeration during synthesis [12]. Functionalizing IONPs with pectin not only improves particle dispersion and adsorption efficiency but also supports sustainability by utilizing biodegradable and renewable resources [13]. Importantly, introducing pectin-functionalized IONPs into aquatic systems poses minimal environmental risk, as pectin is naturally derived, non-toxic, and easily biodegradable. This eco-friendly approach offers a dual benefit: effective dye removal and valorization of agricultural waste.

Despite these benefits, there are still gaps in understanding the factors that govern the effectiveness of pectin-functionalized IONPs for dye adsorption. One underexplored aspect is the degree of methoxylation in pectin, which significantly alters its physicochemical properties, such as charge density and gelation behavior [14]. Pectin is generally categorized based on its degree of esterification (DE) into high-methoxyl pectin (HMP), with a DE greater than 50%, and low-methoxyl pectin (LMP), with a DE less than 50%; these two types also differ in their charge density and gel-forming abilities. They may interact differently with IONPs, influencing the surface charge, colloidal stability, and adsorption capacity of the functionalized IONPs. In addition, synthesis parameters, particularly temperature and pH, are known to affect nanoparticle morphology, crystallinity, and surface characteristics, yet their influence in conjunction with pectin functionalization remains underexplored. Studies systematically investigating how these factors jointly affect the adsorption efficiency of pectin-functionalized IONPs are limited in the literature.

The major problem addressed in this study is the lack of comprehensive understanding regarding the influence of pectin methoxylation and precipitation pH on the structure, surface properties, and cationic dye adsorption performance of pectin-functionalized IONPs. Without such knowledge, it is difficult to optimize these nanocomposites for practical use in wastewater treatment applications. Therefore, the primary objective of this research to develop a green and sustainable nanocomposite adsorbent by functionalizing IONPs with pectin derived from agricultural waste and to systematically evaluate how the degree of pectin methoxylation (HMP vs. LMP) and pH of IONPs precipitation affect the surface charge, agglomeration and adsorption efficiency of the resulting nanocomposites. Methylene blue is used as the model cationic dye to assess the adsorption performance of these nanocomposites. Overall, this work aims to deepen the understanding of the structure-function relationship in pectin-functionalized nanoparticle

Separation Technology - ICoST 2025
Materials Research Proceedings 59 (2026) 25-32

Materials Research Forum LLC
https://doi.org/10.21741/9781644903957-4

adsorbents and provide practical guidance for the development of advanced, eco-friendly materials for the efficient removal of dyes from wastewater.

Materials and Methods

Iron (II) chloride tetrahydrate (FeCl$_2$·4H$_2$O), Iron (III) chloride hexahydrate (FeCl$_3$·6H$_2$O), ammonium hydroxide (NH$_4$OH), and methylene blue (MB) dye were obtained from R&M Chemicals. Commercial HMP (DE= 81.4%) was purchased from Sigma-Aldrich while LMP (DE = 44.3%) was procured from Pomona's Universal Pectin. All chemicals were used as received, without further purification. Millipore ultrapure deionized water was used throughout all synthesis and experimental procedures.

Based on the synthesis method described by Nsom et al. [5], with slight modifications, 0.994 g of FeCl$_2$·4H$_2$O and 2.701 g of FeCl$_3$·6H$_2$O were dissolved in 100 mL of deionized water, maintaining a constant molar ratio of Fe^{2+}: Fe^{3+} = 1:2 throughout the experiment. The mixture was stirred at 550 rpm and heated to 50 °C for 3 hours using an overhead stirrer. Under continuous stirring, NH$_4$OH was added dropwise to the mixture until the pH reached 11, initiating the precipitation of IONPs. After a further 30 minutes of reaction time, the suspension was centrifuged at 9000 rpm for 15 minutes to separate the precipitate. The collected solid was washed repeatedly with deionized water until the supernatant reached a neutral pH. The washed precipitate was then dried overnight in an oven at 60 °C, ground into fine powder using a pestle and mortar, and stored in a sealed zip-lock bag for further use. The chemical reaction involved is illustrated in Eq. 1.

$$2FeCl_3 + FeCl_2 + 8NH_3 + 4H_2O \rightarrow Fe_3O_4 + 8NH_4Cl \tag{1}$$

To prepare pectin-functionalized IONPs, 0.5 g of HMP was first dissolved in 100 mL of deionized water. The solution was stirred and heated to 40 °C until fully dissolved. Separately, 0.994 g of FeCl$_2$·4H$_2$O and 2.701 g of FeCl$_3$·6H$_2$O were dissolved in 100 mL of deionized water, maintaining a consistent Fe^{2+}: Fe^{3+} molar ratio of 1:2. This iron salt solution was then added dropwise into the pectin solution under continuous stirring. The resulting mixture was stirred at 550 rpm and maintained at 50 °C for 3 hours (including the addition time). Subsequently, NH$_4$OH was added dropwise under continuous stirring (heating was stopped at this stage) to adjust the pH of the mixture, to target values of pH 9, 10 and 11. After 30 minutes of reaction at the set pH, the solution was centrifuged at 9000 rpm for 15 minutes to isolate the precipitate. The collected precipitate was subsequently rinsed multiple times with deionized water until the supernatant achieved a neutral pH. The final product was oven-dried at 60 °C overnight, then ground finely using a pestle and mortar, and stored in a zip-lock bag for subsequent characterization and experiments. The IONPs functionalized with HMP were labelled as HMP-iron oxide. The same procedure was repeated using low-methoxyl pectin (LMP) to produce LMP-IONPs.

Batch dye adsorption experiments were conducted. 90 mL of 10 mg/L MB solution mixed with 0.15 g of the synthesized IONPs adsorbent. The mixtures were placed in an incubator shaker at 200 rpm for 2 hours at room temperature to ensure sufficient contact between the MB dye and the adsorbent. After the adsorption process, the final concentration of MB dye in the solution was measured using a PerkinElmer Lambda 365 UV–Vis spectrophotometer at wavelengths ranging from 663 nm to 668 nm, corresponding to the maximum absorbance of MB. The dye removal percentage (%) was calculated using Eq. 2.

$$\text{Dye Removal Percentage (\%)} = \frac{(C_0 - C_i)}{C_0} \times 100\% \tag{2}$$

C_0 and C_i are the initial and final concentration of MB solution respectively.

The zeta potential of both bare and pectin-functionalized IONPs was determined using a Malvern Zetasizer Nano ZSP to analyze their surface charge. The adsorbents were redispersed in deionized water at a concentration of 5 mg/mL, and the samples were maintained at room temperature. Each sample was measured in triplicate. The morphology and particle size of the

Separation Technology - ICoST 2025
Materials Research Proceedings 59 (2026) 25-32

Materials Research Forum LLC
https://doi.org/10.21741/9781644903957-4

synthesized bare and HMP-iron oxide were examined using a Hitachi Regulus 8220 Field Emission Scanning Electron Microscope (FESEM). This high-resolution imaging technique enabled detailed visualization of the surface structures and estimation of particle size distribution. FESEM images also offered valuable insights into how pectin functionalization helps mitigate nanoparticle agglomeration. To further investigate the elemental composition, energy-dispersive X-ray spectroscopy (EDS) was performed using an Oxford EDX Windowless 100 mm detector integrated with the FESEM system. The EDS analysis confirmed the presence of key elements and provided evidence for the successful surface functionalization of IONPs with pectin.

Results and Discussion
Dye removal performance and corresponding surface charge
Fig. 1 presents a combined view of the MB dye removal percentage and zeta potential results for both HMP- and LMP-functionalized IONPs, enabling a direct visual comparison of their surface charge behaviour and dye adsorption performance across the tested pH values. For comparison, the performance of bare (unmodified) IONPs is also included to highlight the enhancement achieved through pectin functionalization.

Based on the results presented in Fig. 1, the surface charge behaviour, as indicated by zeta potential measurements, plays a significant role in determining the dye removal performance of IONPs synthesized under various precipitation pH conditions. The zeta potential is a well-known indicator of colloidal stability, with values exceeding $+30$ mV or falling below -30 mV generally indicating a stable dispersion due to strong electrostatic repulsion between particles [15]. In this study, bare (unmodified) IONPs exhibited a zeta potential of -17.3 mV, suggesting poor electrostatic stability. This relatively low surface charge implies a high tendency for agglomeration, which reduces the effective surface area available for dye adsorption. In contrast, pectin-functionalized IONPs displayed more negative zeta potential values, ranging from -30 mV to -40 mV, depending on the type of pectin and pH. This enhancement in surface charge can be attributed to the presence of negatively charged functional groups from the pectin layer [14], which promotes repulsion and prevents particle aggregation in relative to the unmodified IONPs. As a result, the functionalized IONPs exhibited good dispersion in suspension, preserving a high surface area and providing greater accessibility to active sites for MB dye adsorption. This was further supported by the FESEM images presented in the following characterization section.

Interestingly, it was observed that the degree of this surface charge enhancement was found to be dependent on the type of pectin used. LMP-functionalized IONPs exhibited more negative zeta potentials and superior adsorption performance compared to HMP-functionalized ones. This may be due to the lower degree of esterification (DE) in LMP, which results in a greater number of free carboxylate groups ($-COO^-$) available on the IONPs surface [16]. These carboxylate groups contribute to both stronger surface charge and enhanced electrostatic attraction toward cationic MB dye molecules.

On the other note, the precipitation pH during synthesis also significantly influenced the surface properties of the IONPs. At higher pH values, increased deprotonation of carboxyl groups in pectin can occur, exposing more anionic molecules [16,17] and strengthening coordination interactions between pectin and iron ions. This results in a more effective functionalization process, yielding nanoparticles with enhanced surface charge and improved MB dye adsorption capability. Consequently, LMP-functionalized IONPs demonstrated the highest dye removal percentage of 95.4 % at pH 11, owing to their high surface charge, excellent dispersion, and strong electrostatic interaction with the cationic MB dye. HMP-functionalized IONPs also showed improved performance compared to bare IONPs, but to a lesser extent due to the higher DE and fewer available carboxylate groups.

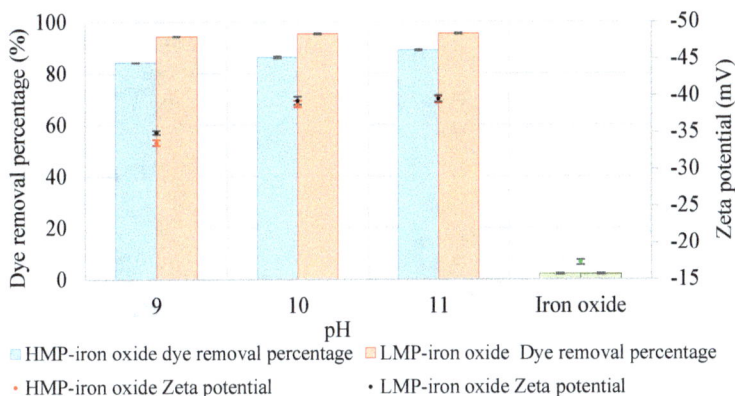

Figure 1 *Comparison of dye removal percentage and zeta potential of HMP- and LMP-iron oxide synthesized at different precipitation pH values.*

Bare IONPs showed significantly lower dye removal performance, likely due to their lower zeta potential, which led to agglomeration and in solution, and subsequently reduced the available surface area for adsorption. Moreover, the pH-dependent nature of iron oxide's surface charge [18] further limited its effectiveness: at lower pH values, the surface becomes more positively charged, which repels the cationic dye molecules, whereas at higher pH values, partial negative charge develops but still lacks the colloidal stability. Since MB is a cationic dye and naturally in slightly acidic conditions, the positively charged surface of bare IONPs at low pH leads to electrostatic repulsion, further inhibiting dye adsorption.

FESEM and EDS analysis
Fig. 2 presents the FESEM images of bare and HMP-IONPs, respectively. Both samples exhibit similar particle size distributions, ranging approximately from 12 to 20 nm. However, clear differences in surface morphology are observed between the two. The bare IONPs appear densely packed and severely agglomerated, reflecting their lower zeta potential and limited colloidal stability. In contrast, the HMP-iron oxide displays a looser and more porous structure with visibly reduced agglomeration. This less compact morphology in the functionalized nanoparticles suggests that the pectin functionalization provides improved dispersion, likely due to enhanced electrostatic repulsion resulting from a more negative surface charge (presented in Fig. 1). The improved colloidal stability of the HMP-iron oxide leads to better separation between particles, increasing the effective surface area and the number of accessible active sites for dye interaction. This structural difference aligns with the zeta potential results and is further supported by the MB dye adsorption results, which demonstrate significantly higher removal efficiency for the HMP-iron oxide. The reduced severity of agglomeration, coupled with enhanced surface charge, plays a key role in facilitating more effective adsorption of MB cationic dye molecules.

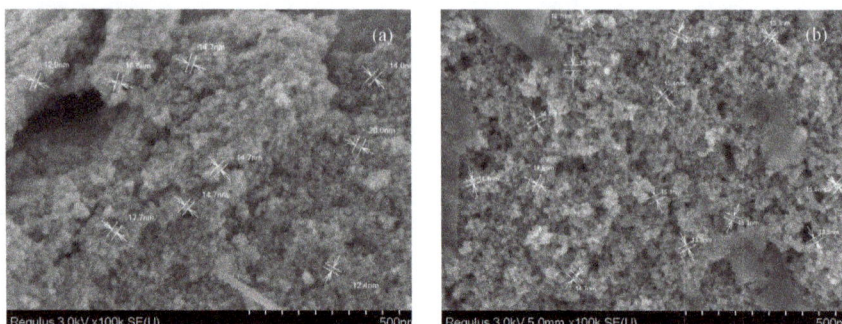

Figure 2 *FESEM images of (a) bare IONPs and (b) HMP-iron oxide precipitated at pH 11.*

Fig. 3 shows the EDS spectra of both bare IONPs and HMP-iron oxide synthesized in this study. The EDS analysis of the bare IONPs confirmed the presence of iron (Fe) and oxygen (O) only, consistent with the expected stoichiometry of Fe_3O_4, indicating successful synthesis of magnetite nanoparticles without any surface modification. In contrast, the EDS spectrum of the HMP-iron oxide synthesized at pH 11 revealed a distinct carbon (C) peak, in addition to Fe and O, confirming the successful surface functionalization with pectin.

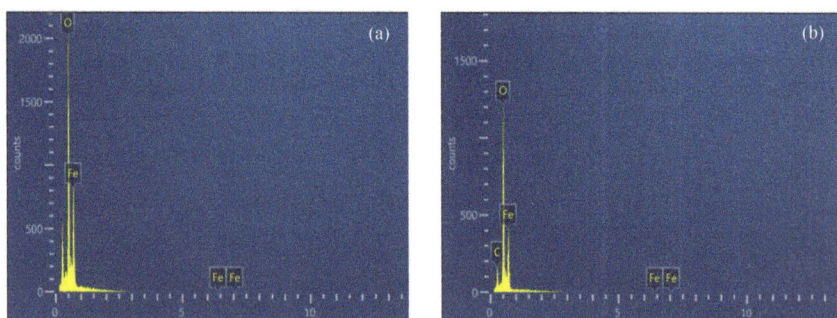

Figure 3 *EDS images of (a) bare IONPs and (b) HMP-iron oxide precipitated at pH 11.*

Table 1 tabulates the elemental composition data for bare IONPs, as well as HMP-iron oxide functionalized at pH 9 and pH 11.

Table 1 *Elemental composition of all IONPs synthesized in this study.*

Element	Weight Percentage [%]		
	Bare IONPs	HMP-IONPs (pH 9)	HMP-IONPs (pH 11)
C	-	3.54	6.35
O	27.65	26.16	30.62
Fe	72.35	70.30	63.03

It is clear that compared to the functionalized IONPs synthesized at pH 9, the pH 11 sample exhibited a higher carbon content, indicating greater pectin incorporation under more alkaline conditions. This finding is consistent with the zeta potential and MB dye removal results, where a

Separation Technology - ICoST 2025
Materials Research Proceedings 59 (2026) 25-32

Materials Research Forum LLC
https://doi.org/10.21741/9781644903957-4

higher degree of pectin functionalization at pH 11 contributed to more negative surface charge and improved adsorption performance.

Conclusion

This study explored the synthesis and performance of pectin-functionalized IONPs for cationic dye removal, with a particular focus on the type of pectin and precipitation pH. Results showed that among the two pectin types studied, LMP-functionalized IONPs consistently outperformed their HMP counterparts across all precipitating pH investigated, especially under higher pH conditions. This performance is attributed to LMP's greater number of free carboxyl groups and higher dissociation ability, which enhanced the surface charge and created more active binding sites for dye molecules. The more negative zeta potentials observed in LMP-based samples further support this mechanism, as they promote stronger electrostatic attraction and reduce nanoparticle agglomeration. The precipitation pH during synthesis was also found to significantly influence nanoparticle performance. Higher (alkaline) pH levels promoted better polymer–metal coordination, leading to stronger pectin functionalization, increased carbon content (as confirmed by EDS analysis), and more negative zeta potentials. These conditions further improved colloidal stability and enhanced dye removal efficiencies. Morphological analysis revealed that pectin-functionalized IONPs exhibit a more porous and less compact structure than bare IONPs, providing greater accessible surface area for dye interaction. These results show the importance of selecting the appropriate pectin type and optimizing synthesis pH to tailor the surface characteristics and functional performance of iron oxide-based nanoadsorbents. These insights support the development of eco-friendly and efficient materials for wastewater treatment applications.

Acknowledgement

This work was financially supported by the Ministry of Higher Education (MOHE) Malaysia under the Fundamental Research Grant Scheme (FRGS/1/2023/TK08/UTM/02/45).

References

[1] M.A. Minhas, A. Rauf, S. Rauf, F.T. Minhas, N. Memon, A. Jabbar, M.I. Bhanger, M.I. Malik, Selective and Efficient Extraction of Cationic Dyes from Industrial Effluents through Polymer Inclusion Membrane, Sep. Purif. Technol. 272 (2021). https://doi.org/10.1016/j.seppur.2021.118883

[2] N.P. Raval, P.U. Shah, N. K. Shah, Malachite Green "A Cationic Dye" and Its Removal from Aqueous Solution by Adsorption, Appl. Water Sci. 7 (2016) 3407-3445. https://doi.org/10.1007/s13201-016-0512-2

[3] B. Lellis, C.Z, Fávaro-Polonio, J.A. Pamphile, J.C. Polonio, Effects of Textile Dyes on Health and The Environment and Bioremediation Potential of Living Organisms, Biotechnol. Res. Innov. 3 (2019) 275-290. https://doi.org/10.1016/j.biori.2019.09.001

[4] V. Chandanshive, S. Kadam, N. Rane, B.H. Jeon, J. Jadhav, S. Govindwar, In Situ Textile Wastewater Treatment in High Rate Transpiration System Furrows Planted with Aquatic Macrophytes and Floating Phytobeds, Chemosphere. 252 (2020) 126513. https://doi.org/10.1016/j.chemosphere.2020.126513

[5] M.V. Nsom, E.P. Etape, J.F. Tendo, B.V. Namond, P.T. Chongwain, M.D. Yufanyi, N. William, A Green and Facile Approach for Synthesis of Starch-Pectin Magnetite Nanoparticles and Application by Removal of Methylene Blue from Textile Effluent, J. Nanomater. (2019) 1-12. https://doi.org/10.1155/2019/4576135

[6] E.J.E. Abdelrazek, A.A. Gahlan, G.A. Gouda, A.S.A Ahmed, Cost-effective Adsorption of Cationic Dyes Using ZnO Nanorods Supported by Orange Peel-derived Carbon, Sci. Rep. 15 (2025), 4123. https://doi.org/10.1038/s41598-025-86209-2

[7] A.V. Samrot, H.H. Ali, A.J. Selvarani, E. Faradjeva, S.S. Kumar, Adsorption Efficiency of Chemically Synthesized Superparamagnetic Iron Oxide Nanoparticles (SPIONs) on Crystal Violet Dye, Curr. Res. Green Sustain. Chem. 4 (2021). https://doi.org/10.1016/j.crgsc.2021.100066

[8] M. Jain, M. Yadav, T. Kohout, M. Lahtinen, V.K. Garg, M. Sillanpää, Development of Iron Oxide/Activated Carbon Nanoparticle Composite for The Removal Of Cr(VI), Cu(II) And Cd(II) Ions from Aqueous Solution, Water Resour. Ind. 20 (2018) 54-74. https://doi.org/10.1016/j.wri.2018.10.001

[9] N. Joudeh, D. Linke, Nanoparticle Classification, Physicochemical Properties, Characterization, and Applications: A Comprehensive Review for Biologists, J. Nanobiotechnol. 20 (2022) 262. https://doi.org/10.1186/s12951-022-01477-8

[10] W.R. Diephuis, A.L. Molloy, L.L. Boltz, T.B. Porter, A. Aragon Orozco, R. Duron, D. Crespo, L.J. George, A.D. Reiffer, G. Escalera, A. Bohloul, C. Avendano, V.L. Colvin, N.I. Gonzalez-Pech, The Effect of Agglomeration on Arsenic Adsorption Using Iron Oxide Nanoparticles, Nanomaterials. 12 (2022). https://doi.org/10.3390/nano12091598

[11] A. Chinedu, Role of Surface Functional Groups on the Adsorption Capacity of Carbon Nanomaterials in Nigeria, J. Chem. 3 (2024) 1-11. https://doi.org/10.47672/jchem.1972

[12] M. Nemiwal, T. Zhang, D. Kumar, Pectin Modified Metal Nanoparticles and Their Application in Property Modification of Biosensors, Carbohydr. Polym. Technol. Appl. 2 (2021). https://doi.org/10.1016/j.carpta.2021.100164

[13] F.T.J. Ngenefeme, N.J. Eko, Y.D. Mbom, N.D. Tantoh, K.W.M. Rui, A One Pot Green Synthesis and Characterisation of Iron Oxide-Pectin Hybrid Nanocomposite, Open J. Compos. Mater. 03 (2013) 30-37. https://doi.org/10.4236/ojcm.2013.32005

[14] O.D. Akin-Ajani, A.O. Okunlola, Pharmaceutical Applications of Pectin, in: M.A. Masuelli (Ed.), Pectins - The New-Old Polysaccharides, IntechOpen, 2022. https://doi.org/10.5772/intechopen.100152

[15] S. Mansur, A. Rai, R.A. Holler, T. Mewes, Y. Bao, Synthesis and Characterization of Iron Oxide Superparticles with Various Polymers, J. Magn. Magn. Mater. 515 (2020) 167265. https://doi.org/10.1016/j.jmmm.2020.167265

[16] Y. Han, X. Jin, Z. Zhang, Q. Sun, Characterization of Pectin with Different Structural Features and Its Effects on Yogurt Quality, LWT. 229 (2025) 118236. https://doi.org/10.1016/j.lwt.2025.118236

[17] B. Bindereif, H.P. Karbstein, K. Zahn, U.S. van der Schaaf, Effect of Conformation of Sugar Beet Pectin on the Interfacial and Emulsifying Properties, Foods. 11 (2022) 214. https://doi.org/10.3390/foods11020214

[18] Y.W. Choi, H. Lee, Y. Song, D. Sohn, Colloidal Stability of Iron Oxide Nanoparticles with Multivalent Polymer Surfactants, J. Colloid Interface Sci. 443 (2015) 8-12. https://doi.org/10.1016/j.jcis.2014.11.068

Separation Technology - ICoST 2025
Materials Research Proceedings 59 (2026) 33-40

Materials Research Forum LLC
https://doi.org/10.21741/9781644903957-5

One-Step Phosphoric Acid Activated Pomegranate Peel Powder for Adsorptive Removal of Tetracycline in Aqueous Solution: Synthesis, Adsorption and Mechanism

Mohammed Awwal SULEIMAN[1,a], Muhammad Abbas AHMAD ZAINI[1,2,b*], Nuhu Dalhat MU'AZU[3,c]

[1]Faculty of Chemical and Energy Engineering, Universiti Teknologi Malaysia, 81310, Skudai, Johor, Malaysia

[2]Centre of Lipids Engineering and Applied Research (CLEAR), Ibnu-Sina Institute for Scientific and Industrial Research, Universiti Teknologi Malaysia, 81310, Skudai, Johor, Malaysia

[3]Department of Environmental Engineering, College of Engineering, Imam Abdulrahman Bin Faisal University, P.O. Box 1982, 31451 Dammam, Saudi Arabia

[a]masuleiman@graduate.utm.my, [b]abbas@cheme.utm.my, [c]nmdalhat@iau.edu.sa

Keywords: Adsorption, Biochar, Pomegranate Peel, Phosphoric Acid, Tetracycline

Abstract. In this study, a high surface area biochar was synthesized from pomegranate peel powder via one-step phosphoric acid activation and pyrolysis (at 600°C for 1 h), and evaluated for the adsorption of tetracycline (TCL) from aqueous solution. Characterization using BET, FESEM and FTIR confirmed the development of mesoporous architecture and the presence of functional groups favorable for TCL interaction. The activated biochar exhibited high specific surface area of 1882 m^2/g, total pore volume of 1.67 cm^3/g, and average pore radius of 3.5 nm. Batch adsorption experiments were performed to assess the effects of adsorption parameters. The biochar achieved a maximum TCL adsorption capacity of 435.4 mg/g. Equilibrium data were fitted to Langmuir, Freundlich, Redlich–Peterson, and Sips isotherm models, with the Langmuir model providing the best fit, indicating monolayer adsorption. Kinetic data fitted the pseudo second-order model, suggesting chemisorption as the dominant mechanism. Mechanistic analysis revealed that TCL removal was driven by electrostatic interactions, π–π electron donor–acceptor interactions, and hydrogen bonding. These results demonstrate the efficacy of phosphoric acid-activated pomegranate peel biochar as a low-cost, sustainable adsorbent for removing pharmaceutical contaminants from water.

Introduction

The widespread utilization of antibiotics such as TCLs especially in human and animal health have been attributed to the problem of TCL contamination in surface water bodies [1]. It is reported that more than 70% of TCL groups (tetracycline, chlorotetracycline, doxycycline, and oxytetracycline) that are consumed by humans or used in livestock treatments are excreted or discharged in their original form into aquatic environment [2]. TCL exhibit degradation resistance in natural envirionment. As a result, tetracycline remain persistent in aqueous medium in their original forms long after they are discharged from effluent sources. Investigations have shown that efforts to remove TCL using conventional wastewater treatment processes have not succeeded in completely eliminating the pollutant in effluent sources [3]. Thus the need to explore and develop efficient remediation options for teracycline removal in aqueous solution is of great research importance.

Pomegrante peel is a fruit waste that is generated from the consumption or processing of pomegranate fruit. The peel represents between 40-50 % of the total fruit weight, equivalent to around 1.62 million tons of wastes generated annually [4]. They are often disposed in municipal waste bins or garbage, thereby constituting environmental pollution and aesthetic problems. Considering that these peels are available in abundance and offer rich sources of cellulose, hemicellulose and phenolic compounds compared to other types of biomass precursors, they can be explored as sustainable low-cost precursor for the production of biochar for application in separation technologies.

Studies have shown that pristine biochar exhibit low adsorption capacity for TCL in aqueous solution when compared to chemically modified biochar [5,6]. This is because modification using chemical agents such as acids, bases and salts improve the structural and chemical properties of biochar. Chen et al. [7] prepared tannic-acid modified rice straw biochar at 700°C to enhance the oxygen-containing functional groups present in the biochar. The modified biochar exhibited higher adsorption capacity for TCL adsorption compared to the unmodified biochar.

Based on literature data and to the best of our knowledge, studies related to the synthesis of phosphoric acid-activated pomegranate peel waste biochar for TCL adsorption in aqueous medium is scarce. As a result, this study aims to prepare activated pomegranate peel biochar using one-step phosphoric activation of the peel waste followed by pyrolysis at 600°C. The prepared biochar was characterized and applied in batch adsorption of TCL in aqueous solution.

Materials and Methods

Pomegranate peels were retrieved from the fresh fruit purchased in Jubail Industrial City, Saudi Arabia. TCL, sodium hydroxide (NaOH), sodium chloride (NaCl), phosphoric acid (H_3PO_4) and hydrochloric acid (HCl) were analytical grade chemicals purchased from Sigma-Aldrich Co., USA and used as-received without further purification. 1000 mL stock solution of TCL was prepared by dissolving 250 mg of TCL ($C_{22}H_{24}N_2O_8$) in deionized water.

Preparation of activated biochar

The peel was dried in oven (Thermo Scientific Heratherm) at 110°C for 2 h. The dried peel was ground in a high-speed multi-function electric grinder (RRH-200) to make pomegranate powder (PPP) of particle size < 0.6 mm and stored in air tight glass jar. Activated PPP biochar was prepared following these steps: (i) 10 g of the sample was mixed with 1 M H_3PO_4 in ratio of 3:1 w/w (acid : precursor), stirred for 3 h at 25°C and dried in oven at 110°C; (ii) the impregnated precursor was pyrolyzed at 600°C for 1 h and the resulting biochar was soaked in 1 M HCl for 24 h, washed severally with distilled water to remove excess acids and minerals, and then dried in oven at 110°C for 24 h. The sample was labelled as PAC (phosphoric acid activated carbon). The sample was characterized and utilized in batch adsorption of TCL.

Characterization of adsorbent material

The structural and chemical properties of PAC were determined using various characterization techniques. BET specific surface area and pore parameters were estimated based on the results of physical adsorption-desorption isotherm of N_2 at 77 K (Micromeritics, ASAP 2020) and analyzed using Micrometrics TriStar II 3020 software. Surface functional groups were determined within the range of 4000-650 cm^{-1} using FTIR spectrometer (Thermo Scientific, Nicolet iS 10). Morphology was obtained using a field emission scanning electron microscope (FESEM) (Hitachi SU8020).

Adsorption experiments

Experimental adsorption runs were performed in a constant temperature water bath shaker (Thermo Scientific, MaxQ 7000) maintained at 25°C and agitated at 230 rpm. TCL solution of different initial concentrations (10, 80 and 150 mg/L) were prepared from the stock solution. For the adsorption experiments, 40 ml of each solution was measured into a plastic vial containing predetermined mass of adsorbent (5, 10 or 15 mg) and the mixture pH was adjusted using 0.1 M nitric acid or 0.1 M sodium hydroxide to achieve the desired solution pH (3, 6 or 9). The solution

Separation Technology - ICoST 2025   Materials Research Forum LLC
Materials Research Proceedings 59 (2026) 33-40   https://doi.org/10.21741/9781644903957-5

was agitated continuously for 3 h. Afterwards, the samples were centrifuged to obtain clear supernatant. The supernatant was analyzed for residual TCL concentration using a Hach DR6000 UV-vis spectrophotometer at wavelength of 370 nm. The removal percentage and adsorption capacity of TCL were calculated.

Results and Discussion
Properties of activated biochar
Presented in Table 1 are the textural properties of PBC (pomegranate peel biochar without activation) and PAC. It can be seen that PAC is characterized by significantly high specific surface area of 1882 m^2/g as compared to that of PBC (195.32 m^2/g). Results of the average pore radius and total pore volume suggests that these materials belong to mesoporous category based on IUPAC classification. In general, the high surface properties could be due to the ability of phosphoric acid to enhance the formation of meso- and micropores at temperature range of 400–700°C [8].

Table 1 Physical properties of PBC and PAC.

Material	Specific surface area S_{BET} (m^2/g)	Total pore volume V_t (cm^3/g)	Average pore width (nm)
PBC	195	0.102	3.340
PAC	1882	1.668	3.543

As shown in Fig. 1, FTIR spectra of PBC and PAC reveal distinct peaks. PAC is characterized by several sharp peaks that could be attributed to structural and chemical modifications induced by phosphoric acid activation. The broad band observed at 3158 cm^{-1} in the spectra of PBC corresponds to O–H stretching vibrations of hydroxyl groups from alcohols, phenols, and carboxylic acids, which are typical of lignocellulosic-derived biochars [9]. The peak at 1571 cm^{-1} represents aromatic C=C vibrations, confirming the development of condensed aromatic structures characteristic of biochar [10]. In contrast, PAC spectrum shows significant changes, with the band at 2336 cm^{-1} assigned to P–H stretching vibrations and that at 2111 cm^{-1} linked to P–C or C≡C stretching, confirming chemical modification by phosphoric acid [11]. The strong peak at 1046 cm^{-1} corresponds to P–O–C and P=O stretching, verifying the incorporation of phosphate groups into the carbon matrix [12].

FESEM images for the surface morphology of PBC and PAC biochars are presented in Fig. 2. Fig. 2a revealed the surface of PBC biochar to comprise of fragmented irregular structures forming overlapping layers characterized by few cavities of various sizes. Fig. 2b shows that the surface of PAC is distinguished by coarsely agglomerated particles of various sizes forming a porous structure which could significantly enhance the surface area and effectively contribute to high adsorption capacity. The effect of chemical treatment is the partial degradation of the lignin and hemicellulose components of the biomass thus enriching the carbon structure. During carbonization, these chemicals prevent excessive shrinkage and promote the formation of a stable, porous structure characterized by high surface area.

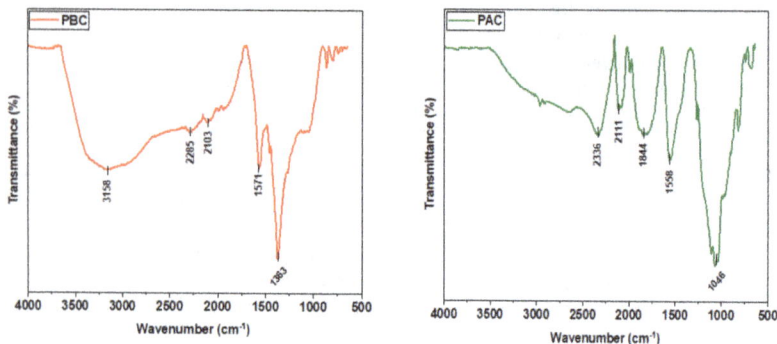

Figure 1 *FTIR spectra of (a) PBC and (b) PAC.*

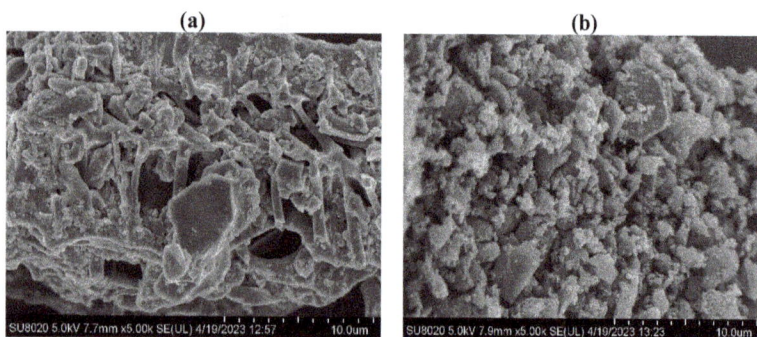

Figure 2 *FESEM images of (a) PBC and (b) PAC.*

Adsorption studies

Presented in Fig. 3 is the effect of solution pH on TCL adsorption by PBC and PAC. The result shows that PAC exhibited significantly higher removal efficiency (37-57 %) across pH range of 2-10 compared to PBC with maximum efficiency <6%. The superior performance of PAC can be attributed to the significant enhancement in surface area and pore volume as reported in the BET analysis, and the abundance of oxygen-containing functional groups as a result of phosphoric acid activation. These improvements facilitate multiple interactions such as electrostatic attraction, hydrogen bonding, and π–π electron donor–acceptor interactions with TCL molecules [11]. In contrast, the low removal efficiencies achieved for PBC highlights the limited porosity and functional group availability of pristine biochar. It can be seen that the removal efficiency trend for PAC was relatively stable between pH 3 and 8 with maximum removal achieved at pH 4 (~57 %). The observed pH-dependent behavior is consistent with the amphoteric nature of TCL with three pKa values of 3.3, 7.6 and 9.7 [13].

Separation Technology - ICoST 2025
Materials Research Proceedings 59 (2026) 33-40

Materials Research Forum LLC
https://doi.org/10.21741/9781644903957-5

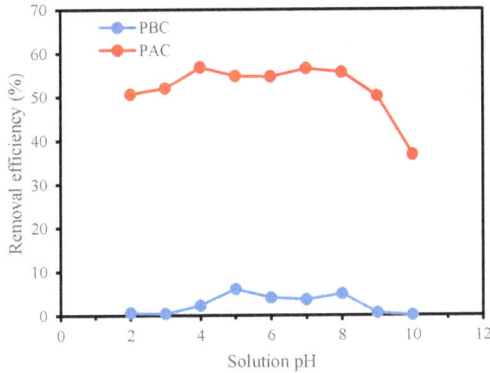

Figure 3 *Effect of initial pH on TCL adsorption onto PBC and PAC.*

Fig. 4 shows the effect of contact time on the adsorption of TCL (C_o = 150 mg/l, pH = 4 and T = 25°C) onto PAC. The adsorption dynamics reveals a rapid initial increase in TCL adsorption within the first 50 min followed by a slow approach to equilibrium until 280 min. This phenomenon indicates quick occupation of readily available surface sites in the early phase of the adsorption process and subsequent diffusion-limited adsorption as equilibrium is reached. The kinetic data were fitted to pseudo-first order (PFO) and pseudo-second order (PSO) models. Analysis of the kinetic parameters as presented in Table 2 shows that PSO model exhibited the best fit with coefficients of determination (R^2) value of 0.988 and estimated equilibrium adsorption capacity, $q_{e,cal}$ = 429.9 mg/g. Thus, the adsorption process can be considered to be predominantly governed by chemical process involving valence forces through electron sharing or exchange between TCL molecules and the oxygen-containing functional groups present on PAC surfaces [13].

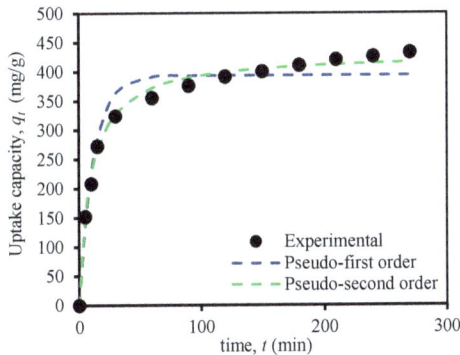

Figure 4 *Effect of contact time on TCL adsorption using PAC (C_o = 150 mg/L, adsorbent dosage = 5 mg, pH 4.0 and 25°C).*

Table 2 *Kinetic model parameters for TCL adsorption.*

Material	Pseudo-first order			Pseudo-second order		
	q_e (mg/g)	k_1 (min^{-1})	R^2	q_e (mg/g)	k_2 (g mg^{-1} min^{-1})	R^2
PAC	393.1	0.0863	0.933	429.9	0.0002	0.988

Adsorption isotherm curves for two (Langmuir and Freundlich) and three (Redlich-Peterson) parameter models were fitted to the experimental data as shown in Fig. 5. The model parameters for the non-linear adsorption isotherms are presented in Table 3. Langmuir model exhibited the highest R^2 (0.999) with corresponding high adsorption capacity (q_m = 435.4 mg/g) and strong adsorption energy (K_L = 1.069 L/mg), which indicates monolayer adsorption of TCL on the surface of PAC biochar. Although the R^2 for Freundlich model is lower than that of Langmuir model, the heterogeneity parameter value ($1/n$ = 0.139) suggests that the adsorption process proceeds favorably based on the criteria of favorable adsorption (0< $1/n$ <1). The Redlich-Peterson exponent parameter (n_{RP}) is estimated to be 0.862 as reported in Table 3. Considering that this value is close to 1, it can be inferred that the behavior of the model approaches Langmuir isotherm.

Presented in Table 4 is the comparison of maximum adsorption capacity of activated biochars derived from fruit and crop wastes for the removal of TCL in aqueous solution. Result from the this study shows that PAC exhibited higher adsorption capacity compared to the other adsorbents. Thus, PAC possess the potential to serve as an excellent adsorbent for the removal of TCL in aqueous solution.

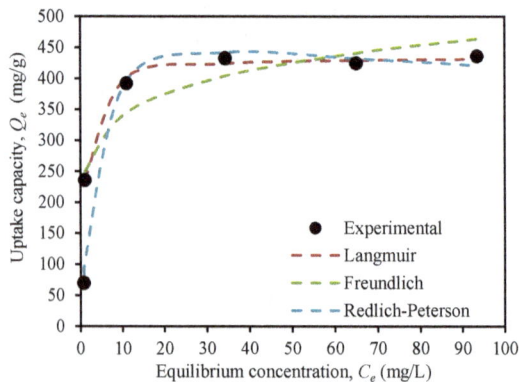

Figure 5 *Fitting of isotherm models with equilibrium data.*

Table 3 *Constants of adsorption models for TCL removal by PAC.*

Langmuir		Freundlich		Redlich-Peterson	
Q_m (mg/g)	435.4	K_F (mg^{1-n} ·L^n ·g^{-1})	246.6	α_{RP}	0.148
K_L (L/mg)	1.069	$1/n$	0.139	K_R	14.97
R^2	0.999	R^2	0.892	n_{RP}	0.862
				R^2	0.928

Table 4 Comparison of TCL capacity (Q_m) by some biochar materials.

Materials	Activator	Capacity (mg/g)	Reference
Rice straw activated biochar	Tannic acid	308.0	[7]
Corn straw activated biochar	Phosphoric acid	227.3	[14]
Pomegranate peel activated biochar	Phosphoric acid	435.4	This study

Conclusion

In this study, activated pomegranate peel biochar (PAC) with high surface area (1882 m^2/g) was synthesized via one-step phosphoric acid activation and used for the adsorption study of TCL in aqueous solution. PAC exhibited high adsorption capacity (Q_m = 435.4 mg/g) for TCL. PSO kinetic model and Langmuir, Redlich-Peterson and Sips isotherm models provided good fit to the experimental data. TCL removal is governed by electrostatic interactions, π–π EDA interactions, and hydrogen bonding. These results demonstrate the efficacy of phosphoric acid-activated pomegranate peel biochar as a low-cost, sustainable adsorbent for removing TCL contaminants from water.

Acknowledgement

This work was supported by UTM Fundamental Research Grant No. 23H09.

References

[1] R. Vinayagam, T. Varadavenkatesan, R. Selvaraj, Tetracycline Adsorption Research (2015–2025): A Bibliometric Analysis of Trends, Challenges, and Future Directions, Results Eng. (2025) 106383. https://doi.org/https://doi.org/10.1016/j.rineng.2025.106383

[2] J. Antos, M. Piosik, D. Ginter-Kramarczyk, J. Zembrzuska, I. Kruszelnicka, Tetracyclines contamination in European aquatic environments: A comprehensive review of occurrence, fate, and removal techniques, Chemosphere. 353 (2024) 141519. https://doi.org/https://doi.org/10.1016/j.chemosphere.2024.141519

[3] B. Wang, Z. Xu, B. Dong, Occurrence, fate, and ecological risk of antibiotics in wastewater treatment plants in China: A review, J. Hazard. Mater. 469 (2024) 133925. https://doi.org/https://doi.org/10.1016/j.jhazmat.2024.133925

[4] M. Cano-Lamadrid, L. Martínez-Zamora, N. Castillejo, F. Artés-Hernández, From pomegranate byproducts waste to worth: a review of extraction techniques and potential applications for their revalorization, Foods. 11 (2022) 2596.

[5] D. Naghipour, L. Hoseinzadeh, K. Taghavi, J. Jaafari, A. Amouei, Effective removal of tetracycline from aqueous solution using biochar prepared from pine bark: isotherms, kinetics and thermodynamic analyses, Int. J. Environ. Anal. Chem. 103 (2023) 5706–5719. https://doi.org/https://doi.org/10.1080/03067319.2021.1942462

[6] V.T. Nguyen, T.B. Nguyen, C.P. Huang, C.W. Chen, X.T. Bui, C. Di Dong, Alkaline modified biochar derived from spent coffee ground for removal of tetracycline from aqueous solutions, J. Water Process Eng. 40 (2021) 101908. https://doi.org/10.1016/j.jwpe.2020.101908

[7] J. Chen, H. Li, J. Li, F. Chen, J. Lan, H. Hou, Efficient removal of tetracycline from water by tannic acid-modified rice straw-derived biochar: Kinetics and mechanisms, J. Mol. Liq. 340 (2021) 117237. https://doi.org/https://doi.org/10.1016/j.molliq.2021.117237

[8] I. Neme, G. Gonfa, C. Masi, Activated carbon from biomass precursors using phosphoric acid: A review, Heliyon. 8 (2022). https://doi.org/10.1016/j.heliyon.2022.e11940

[9] J. Zhang, J. Liu, R. Liu, Effects of pyrolysis temperature and heating time on biochar obtained from the pyrolysis of straw and lignosulfonate, Bioresour. Technol. 176 (2015) 288–291. https://doi.org/https://doi.org/10.1016/j.biortech.2014.11.011

[10] M. Nuhanović, N. Smječanin, N. Mulahusić, J. Sulejmanović, Pomegranate peel waste biomass modified with H3PO4 as a promising sorbent for uranium(VI) removal, J. Radioanal. Nucl. Chem. 328 (2021) 617–626. https://doi.org/https://doi.org/10.1007/s10967-021-07664-5

[11] Z. Lu, H. Zhang, A. Shahab, K. Zhang, H. Zeng, A.U.R. Bacha, I. Nabi, H. Ullah, Comparative study on characterization and adsorption properties of phosphoric acid activated biochar and nitrogen-containing modified biochar employing Eucalyptus as a precursor, J. Clean. Prod. 303 (2021) 127046. https://doi.org/https://doi.org/10.1016/j.jclepro.2021.127046

[12] M.A.H. Shouman, S.A.A. Khedr, Removal of cationic dye from aqueous solutions by modified acid-treated pomegranate peels (Punica granatum): Equilibrium and kinetic studies, Asian J. Appl. Sci. 3 (2015).

[13] L. Yan, X. Song, J. Miao, Y. Ma, T. Zhao, M. Yin, Removal of tetracycline from water by adsorption with biochar: A review, J. Water Process Eng. 60 (2024) 105215. https://doi.org/https://doi.org/10.1016/j.jwpe.2024.105215

[14] Q. Yang, P. Wu, J. Liu, S. Rehman, Z. Ahmed, B. Ruan, N. Zhu, Batch interaction of emerging tetracycline contaminant with novel phosphoric acid activated corn straw porous carbon: Adsorption rate and nature of mechanism, Environ. Res. 181 (2020) 108899. https://doi.org/https://doi.org/10.1016/j.envres.2019.108899

Separation Technology - ICoST 2025
Materials Research Proceedings 59 (2026) 41-47

Materials Research Forum LLC
https://doi.org/10.21741/9781644903957-6

Biogas Production Potential from Animal Farm Waste: A Preliminary Study at Sirukam Dairy Farm, Indonesia

Erda Rahmilaila DESFITRI[1,a*], Adillah Rahmi PUTRI[2,b], Bunga Karuni PUTRI[1,c], Ellyta SARI[2,d], Reni DESMIARTI[2,e]

[1]Renewable Energy Engineering Technology, Faculty of Industry, Universitas Bung Hatta, West Sumatera, Indonesia

[2]Chemical Engineering Faculty of Industry, Universitas Bung Hatta, West Sumatera, Indonesia

[a]rahmilaila@bunghatta.ac.id, [b]adilahrahmiputri9@gmail.com, [c]bungakaruniputri2018@gmail.com, [d]ellytasari@bunghatta.ac.id, [e]renitk@bunghatta.ac.id

Keywords: Biogas, Anaerobic Digestion, C/N Ratio, Livestock Waste, Renewable Energy, Sirukam Dairy Farm

Abstract. The increasing demand for renewable energy highlights the importance of utilizing livestock waste as a sustainable energy source. This study examines the biogas production potential from dairy cattle manure at Sirukam Dairy Farm in West Sumatra, Indonesia. Cattle manure and organic waste were co-digested at four mixing ratios with water (1:1:1, 2:1:1, 1:2:1, and 2:0:1, w/w). Substrate characteristics, including moisture content, total solids (TS), and carbon-to-nitrogen (C/N) ratio, were analyzed to evaluate their influence on biogas yield. Results showed that the 1:1:1 ratio achieved the highest cumulative biogas production (276 mL) with optimal substrate characteristics (moisture 89.2%, TS 10.8%, C/N 21). Lower performance was observed in the manure-only mixture (2:0:1), which yielded only 107.5 mL due to an imbalanced C/N ratio (7) and excessive nitrogen, leading to ammonia inhibition. The study confirms that optimal conditions for biogas production occur at moisture levels of 88–89%, TS of 9–11%, and C/N ratios between 20–30. These findings demonstrate the technical feasibility of utilizing livestock waste for renewable energy generation in Indonesia, providing a scientific basis for scaling up farm-based anaerobic digestion systems.

Introduction

Energy demand continues to increase in line with population growth and economic activities, while dependence on fossil fuels remains high [1]. To date, most of the energy consumed worldwide still comes from fossil sources such as petroleum, natural gas, and coal, whose availability is declining and which have negative impacts on the environment [2–5]. For instance, coal utilization can generate fly ash containing heavy metals that may be released into the environment [6–10]. In addition, the combustion of fossil fuels contributes significantly to greenhouse gas (GHG) emissions, which are the primary drivers of global climate change [11,12].

In this context, renewable energy becomes a strategic solution to address the dual challenges of energy security and environmental sustainability. Among the various renewable energy resources, biogas is considered promising due to its dual benefits: energy production and waste management. Biogas is produced through the anaerobic fermentation of organic waste such as livestock manure, agricultural residues, or municipal solid waste [13–16]. This process not only generates a renewable source of energy but also reduces methane (CH_4) emissions, a potent greenhouse gas with a global warming potential 25 times greater than that of carbon dioxide. Furthermore, the by-product of anaerobic digestion, digestate, can be used as an organic fertilizer, thereby supporting sustainable agriculture and circular economy practices [17].

Separation Technology - ICoST 2025   Materials Research Forum LLC
Materials Research Proceedings 59 (2026) 41-47   https://doi.org/10.21741/9781644903957-6

Livestock waste, particularly dairy cow manure, has been extensively studied as a feedstock for biogas production. On average, a dairy cow produces between 1.2 and 1.33 m^3 of biogas per day, equivalent to approximately 6.67–8.52 kWh of energy [21,22]. However, the actual yield depends on factors such as manure composition, volatile solids content, carbon-to-nitrogen (C/N) ratio, and digestion conditions. Previous studies have reported significant variability in manure output depending on breed, body weight, feed intake, and milk productivity. The chemical composition of manure, particularly its proportions of fiber, protein, lipids, and moisture, strongly affects substrate biodegradability and methane yield. Manure with high lignocellulosic content tends to degrade more slowly, while excessive protein or fat levels can cause ammonia or long-chain fatty acid inhibition if not properly balanced [18]. For instance, the FAO reported that a dairy cow weighing around 500 kg and producing 15 liters of milk per day yields approximately 35 kg of fresh manure daily with about 88% moisture content [19]. Similarly, lactating Holstein cows with an average body weight of 630 kg and a dry matter intake (DMI) of ~21.7 kg/day were observed to excrete 66.3 kg of manure per day [21].

Sirukam Dairy Farm, the largest dairy farm in West Sumatra, represents a strategic case study for the implementation of biogas technology. Based on a field survey, the farm manages approximately 100 dairy cows (increasing to 150 in recent years), each producing an average of 20 kg of manure per day, yielding a total of 2,000–3,000 kg/day of manure. This estimation, although conservative, is reasonable for local cattle breeds with moderate productivity and feeding regimes. If managed optimally, this manure could generate 350–525 m^3 of biogas per day, equivalent to replacing 200–350 liters of LPG per month, sufficient to meet local household energy demand and support the farm's green ecotourism initiatives.

Despite the promising potential, there is limited scientific data on the biogas yield characteristics of livestock waste from Sirukam Dairy Farm, including manure composition, C/N ratio, and theoretical methane generation. Therefore, this study aims to analyze the biogas production potential from dairy cow manure at Sirukam Dairy Farm, West Sumatra, focusing on estimating manure generation, characterizing waste properties, and theoretically calculating methane yield. The findings are expected to provide a basis for scaling up renewable energy applications in rural livestock systems, contributing to Indonesia's energy transition and sustainable waste management strategies.

Materials and Methods

This study was conducted using cattle manure collected from Sirukam Dairy Farm, Solok Regency, West Sumatra, Indonesia. Organic waste obtained from traditional markets in Nanggalo, Padang City. Both materials were selected as the main substrates for biogas production, with water added as a diluent to adjust the consistency of the mixtures. Fresh samples were transported in sealed containers and processed within 24 hours to minimize degradation before analysis.

The experimental setup utilized several laboratory instruments, including gas-tight glass syringes (100 mL, Hamilton®, USA) for periodic biogas volume measurements, an electric drying oven (Memmert UN55, Germany) for determining total solids, and a muffle furnace (Nabertherm L9/11, Germany) for volatile solids analysis. Samples were cooled in a vacuum desiccator (Duran®, Germany) before weighing on an analytical balance (Shimadzu AY220, ±0.0001 g, Japan). Standard laboratory glassware such as Erlenmeyer flasks, evaporation dishes, and stainless-steel spatulas (Iwaki®, Japan) were used throughout the analyses. The UV–vis spectrophotometer (Shimadzu UV-1900, Japan) was employed for total carbon determination, while nitrogen content was analyzed using the Kjeldahl digestion and distillation apparatus (Gerhardt Vapodest 45s, Germany) following standard procedures (APHA 2005).

The study considered three categories of variables. The controlled variables included the types of substrates used, namely organic waste, cattle manure, and water. The independent variable was the proportion of these components, with four substrate mixing ratios tested: 1:1:1, 2:1:1, 1:2:1,

Separation Technology - ICoST 2025
Materials Research Proceedings 59 (2026) 41-47

Materials Research Forum LLC
https://doi.org/10.21741/9781644903957-6

and 2:0:1 (manure: organic waste: water, w/w). The dependent variables consisted of the C/N ratio and the cumulative volume of biogas produced.

Before biogas production tests, feedstock samples underwent several analytical procedures. The moisture content was determined gravimetrically by drying approximately 25 g of the sample at 105°C until a constant weight was achieved. The total solids (TS) were calculated from the dried mass. The carbon-to-nitrogen (C/N) ratio was evaluated based on the total organic carbon (TOC) and total nitrogen (TN) contents of the samples. The Walkley method analyzed carbon content using the Black dichromate oxidation method, in which organic carbon is oxidized using 1 N potassium dichromate ($K_2Cr_2O_7$) and concentrated sulfuric acid (H_2SO_4, 98%) at 150°C. The remaining dichromate was titrated with 0.5 N ferrous ammonium sulfate ($Fe(NH_4)_2(SO_4)_2$) using ferroin as an indicator. The percentage of organic carbon was calculated using the standard equation from AOAC (2019). For nitrogen analysis, the Kjeldahl method was employed using a Gerhardt Vapodest 45s (Germany) system. Samples were digested with concentrated H_2SO_4 and a catalyst mixture (K_2SO_4 + $CuSO_4$) at 380°C until transparent, followed by distillation with 40% NaOH and titration of the released ammonia with standardized 0.1 N HCl.

The moisture content, total solids (TS), and volatile solids (VS) of the feedstock were analyzed according to the Standard Methods for the Examination of Water and Wastewater (APHA, 2005). Approximately 25 g of a well-mixed sample was weighed and dried in an electric oven (Memmert UN55, Germany) at 105°C for 24 h to a constant weight, to determine the moisture content. The TS (%) was calculated as the ratio of dry mass to initial wet mass. The dried residue was then placed in a muffle furnace (Nabertherm L9/11, Germany) and heated to 550°C for 2 h to remove organic matter. The VS (%) was calculated as the fraction of mass lost during ignition relative to the total solids. All analyses were performed in triplicate, and average values were reported.

Results and Discussion
Biogas production
Anaerobic digestion performance is significantly influenced by the composition of feedstock mixtures and their physicochemical characteristics, including the C/N ratio and moisture content. To evaluate the effect of different substrate combinations on biogas generation, cattle manure, organic waste, and water were mixed in several proportions. The cumulative biogas production for each mixture was monitored over a 17-day digestion period, and the results are presented in Figure 1. This figure illustrates the dynamic pattern of daily biogas accumulation across the four tested ratios (1:1:1, 2:1:1, 1:2:1, and 2:0:1), providing insight into microbial adaptation, degradation efficiency, and the overall stability of the anaerobic process.

Fig. 1 shows the cumulative biogas production from different substrate ratios of cattle manure, organic waste, and water (1:1:1, 2:1:1, 1:2:1, and 2:0:1). All treatments exhibited a short lag phase during the first 2–3 days, followed by a rapid increase until day 7–8, after which production gradually stabilized. This trend represents the hydrolysis, acidogenesis, and methanogenesis phases typically observed in anaerobic digestion [23]. The 1:1:1 mixture achieved the highest cumulative biogas yield (276 mL), consistent with the data shown in Figure 1. This performance is attributed to its balanced nutrient composition and favourable moisture content ($\approx$approximately 89%), with total solids (TS) of around 11% and a C/N ratio of roughly 22, which supports microbial activity and stable methanogenesis. The 1:2:1 ratio also produced a relatively high yield (243 mL), although slightly lower, likely due to partial acidification caused by the excess carbon-rich substrate. In contrast, the 2:1:1 mixture (190 mL) and the 2:0:1 mixture (107 mL) demonstrated reduced performance due to the limited availability of degradable organics and lower moisture levels ($\approx$75–78%), resulting in slower hydrolysis and restricted microbial activity.

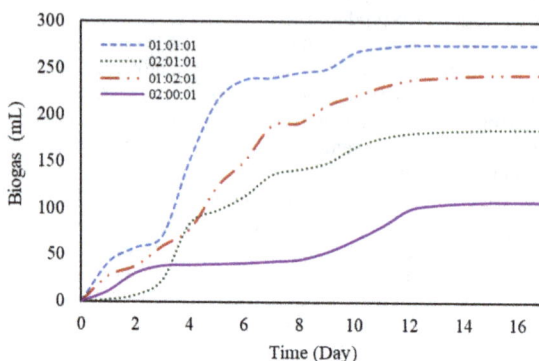

Figure 1 *Daily biogas yield during anaerobic digestion of dairy manure and co-digestion mixtures.*

Effect of total solids on biogas production

The influence of total solids (TS) on cumulative biogas yield is shown in Fig. 2. The data indicate that biogas production is strongly dependent on the concentration of TS in the substrate mixture. The highest gas production (276 mL) was obtained at a TS concentration of 10.8% (ratio 1:1:1), followed by 243 mL at 9.8% TS (ratio 2:1:1). Lower gas yields were observed at higher TS concentrations: 185 mL at 11.4% TS (ratio 1:2:1) and only 107.5 mL at 12.2% TS (ratio 2:0:1).

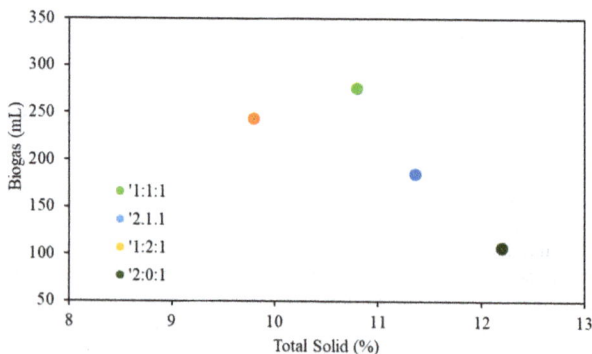

Figure 2 *Effect of total solids on biogas production.*

Our results are consistent with previous studies. Other research reported that anaerobic digestion of manure and crop residues is most efficient at TS levels between 8–12%, where hydrolysis and methanogenesis remain balanced [24]. Similarly, earlier studies found that TS above 12% can cause process instability due to restricted diffusion and localized accumulation of volatile fatty acids, which inhibit methane yield [25]. On the other hand, very low TS (<8%) may decrease methane productivity due to substrate dilution [26].

In this study, the most favourable biogas yield (276 mL) was obtained at TS ≈ 10%, within the range that supports stable digestion. This observation indicates an empirically optimal TS

Separation Technology - ICoST 2025   Materials Research Forum LLC
Materials Research Proceedings 59 (2026) 41-47   https://doi.org/10.21741/9781644903957-6

condition under the tested setup, though no statistical or multi-parameter optimization analysis was conducted. For practical application at Sirukam Dairy Farm, maintaining TS between 9–11% through controlled water addition and mixing is recommended to sustain stable and efficient biogas generation. This approach enhances methane yield and minimizes process instability, making the system more reliable for farm-scale renewable energy utilization.

Effect of C/N ratio on biogas production

The effect of the carbon-to-nitrogen (C/N) ratio on biogas production is shown in Fig. 3. The results demonstrate a clear relationship, where the highest biogas yield (276 mL) was obtained at a C/N ratio of ~24 (1:1:1 mixture. These results indicate that an optimal C/N ratio for anaerobic digestion lies between 20–30, consistent with theoretical expectations. A balanced ratio ensures sufficient carbon as an energy source and adequate nitrogen for microbial growth, thereby supporting efficient methane production. Comparable findings have been reported in previous studies.

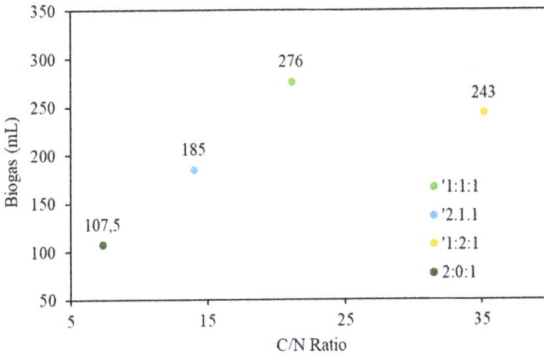

Figure 3 *Effect of C/N ratio on biogas production.*

A C/N ratio between 25–30 provides the most favourable conditions for methane fermentation, as excess carbon slows microbial growth while excess nitrogen leads to ammonia toxicity [27]. Similarly, co-digestion of cattle manure with crop residues achieved maximum methane yield when the C/N ratio was maintained around 25 [28]. Moreover, C/N ratios outside the range of 20–30 significantly destabilize digestion processes due to either nutrient imbalance or acidification [24]. The findings underscore the importance of substrate balancing for optimal anaerobic digestion. At Sirukam Dairy Farm, mixtures that provide a C/N ratio 21 are expected to maximize methane output while ensuring process stability. This suggests that proper co-digestion strategies combining cattle manure with organic waste of higher carbon content can improve the sustainability of farm-based biogas systems

Conclusion

This study highlights the biogas production potential of livestock waste from Sirukam Dairy Farm in West Sumatra, achieved through the anaerobic co-digestion of cattle manure and organic waste. The results indicate that substrate composition and physicochemical properties have a significant influence on digestion performance. The 1:1:1 ratio (manure: organic waste: water) provided the highest yield of 276 mL, corresponding to optimal moisture content (89%), TS (10.8%), and C/N ratio (21). In contrast, the manure-only mixture (2:0:1) produced the lowest yield (107.5 mL), confirming that nutrient imbalance, particularly excess nitrogen, suppresses methane production. For practical applications, this implies that co-digestion strategies and proper substrate preparation can enhance renewable energy generation at Sirukam Dairy Farm, supporting Indonesia's energy transition and sustainable livestock waste management.

References

[1]   IEA. World Energy Outlook. 2024. 23–28 p.

[2]   Carvalho H. Coal: in a burning world, the dark side of energy still rules. J Glob Health. 2025;15:03007.

[3]   Awan TI, Afsheen S, Mushtaq A. Introduction to Sustainable Energy and Climate Change BT - Influence of Noble Metal Nanoparticles in Sustainable Energy Technologies. In: Awan TI, Afsheen S, Mushtaq A, editors. Cham: Springer Nature Switzerland; 2025. p. 1–18. Available from: https://doi.org/10.1007/978-3-031-80983-5_1

[4]   Desfitri ER, Sutopo UM, Hayakawa Y, Kambara S. Effect of Additive Minerals on Controlling Chromium (Cr) Leaching from Coal Fly Ash. Minerlas [Internet]. 2020;10(6):563. Available from: doi:10.3390/min10060563

[5]   Hapsauqi I, Desfitri ER, Farrah D, Hanum F. Development of Aloe Vera-Based Desulfurization Method to Improve The Quality Of Sumatra's Coal. 2024;14:73–80. https://doi.org/10.31938/jsn.v14i2.699

[6]   Desfitri ER, Hanum FF, Hayakawa Y, Kambara S. Calcium Performance in Paper Sludge Ash as Suppressing Material. IOP Conf Ser Mater Sci Eng. 2019;543(1). DOI 10.1088/1757-899X/543/1/012092

[7]   Desfitri ER, Sutopo UM, Hayakawa Y, Kambara S. Effect of additive material on controlling chromium (Cr) leaching from coal fly ash. Minerals. 2020;10(6):1–12. DOI:10.3775/jie.100.102

[8]   Sutopo UM, Desfitri ER, Hanum FF, Hayakawa Y, Kambara S. An experimental and thermodynamic equilibrium analysis on the leaching process of arsenic (As) from coal fly ash. Nihon Enerugi Gakkaishi/Journal Japan Inst Energy. 2021;100(8):102–9. https://doi.org/10.3775/jie.100.102

[9]   Hanum FF, Desfitri ER, Hayakawa Y, Kambara S. The Role of Calcium Compound on Fluorine Leaching Concentration. IOP Conf Ser Mater Sci Eng. 2019;543(1). DOI 10.1088/1757-899X/543/1/012091

[10]   Sutopo UM, Desfitri ER, Hayakawa Y, Kambara S. A role of mineral oxides on trace elements behavior during pulverized coal combustion. Minerals. 2021;11(11):1–13. https://doi.org/10.3390/min11111270

[11]   Kong F, Ren H. Advances in Green Energy, Environment and Carbon Neutralization. Energies. 2025;18(5):8–11. DOI:10.3390/en18051016

[12]   Omemen KM, Aldbbah MO. International Journal of Electrical Engineering Climate Change : Key Contributors and Sustainable Solutions. 2025;https://ijees.org/index.php/ijees/article/view/105

[13]   Ankathi SK, Chaudhari US, Handler RM, Shonnard DR. Sustainability of Biogas Production from Anaerobic Digestion of Food Waste and Animal Manure. Appl Microbiol. 2024;4(1):418–38. https://doi.org/10.3390/applmicrobiol4010029

[14]   Hoyos-Sebá JJ, Arias NP, Salcedo-Mendoza J, Aristizábal-Marulanda V. Animal manure in the context of renewable energy and value-added products: A review. Chem Eng Process - Process Intensif. 2024;196(December 2023). https://doi.org/10.1016/j.cep.2023.109660

[15]   Alengebawy A, Ran Y, Osman AI, Jin K, Samer M, Ai P. Anaerobic digestion of

agricultural waste for biogas production and sustainable bioenergy recovery: a review. Environ Chem Lett [Internet]. 2024;22(6):2641–68. Available from: https://link.springer.com/article/10.1007/s10311-024-01789-1

[16] Moltames R, Noorollahi Y, Yousefi H, Azizimehr B. Assessment of potential sites for biogas production plants from domestic , agricultural , and livestock waste. Fuel Commun [Internet]. 2025;22(September 2024):100132. Available from: https://doi.org/10.1016/j.jfueco.2024.100132

[17] Müller, Z. O., Feed from animal wastes: state of knowledge. Fao Animal Production And Health Paper 18, FAO, Rome. [Internet]. 2025. Availabe at: http://www.fao.org/docrep/004/x6518e/X6518E00.htm#TOC

[18] P. Li et al., "Evaluation of lignin inhibition in anaerobic digestion from the perspective of reducing the hydrolysis rate of holocellulose," Bioresour. Technol., vol. 333, p. 125204, 2021, doi: https://doi.org/10.1016/j.biortech.2021.125204

[19] P. Singh, H. T. Behera, S. Mishra, and L. Ray, "Chapter 10 - Biofertilization of biogas digestates: An insight on nutrient management, soil microbial diversity and greenhouse gas emission," H. B. Singh and A. B. T.-N. and F. D. in M. B. and B. Vaishnav, Eds. Elsevier, 2022, pp. 199–21

[20] T. D. Nennich et al., "Prediction of Manure and Nutrient Excretion from Dairy Cattle," J. Dairy Sci., vol. 88, no. 10, pp. 3721–3733, Oct. 2005, T. D. Nennich et al., "Prediction of Manure and Nutrient Excretion from Dairy Cattle," J. Dairy Sci., vol. 88, no. 10, pp. 3721–3733, Oct. 2005, doi: 10.3168/jds.S0022-0302(05)73058-7

[21] Nleya Y, Young B, Nooraee E, Baroutian S. Anaerobic digestion of dairy cow and goat manure: Comparative assessment of biodegradability and greenhouse gas mitigation. Fuel [Internet]. 2025;381(PB):133458. Available from: https://doi.org/10.1016/j.fuel.2024.133458

[22] Arshad M, Ansari AR, Qadir R, Tahir MH, Nadeem A, Mehmood T, et al. Green electricity generation from biogas of cattle manure: An assessment of potential and feasibility in Pakistan. Front Energy Res. 2022;10(August):1–10. DOI:10.3389/fenrg.2022.911485

[23] F. Hanum, Y. Atsuta, and H. Daimon, "Methane production characteristics of anaerobic co-digestion of pig manure and fermented liquid feed," Biomass Convers. Biorefin., vol. 13, pp. 245–254, 2023, https://doi.org/10.3390/molecules27196509

[24] L. Liu, S. Wang, X. Guo, T. Zhao, and B. Zhang, "Succession and diversity of microorganisms and their association with physicochemical properties during green waste thermophilic composting," Waste Manag., vol. 73, pp. 101–112, 2018, doi: https://doi.org/10.1016/j.wasman.2017.12.026.

[25] Mao, C., Feng, Y., Wang, X., & Ren, G. (2015). Review on research achievements of biogas from anaerobic digestion. Renewable and Sustainable Energy Reviews, 45, 540–555. https://doi.org/10.1016/j.rser.2015.02.030

[26] Li, Y., Park, S. Y., & Zhu, J. (2018). Solid-state anaerobic digestion: A review of current developments and future perspectives. Waste Management, 78, 483–498.

[27] Yen, H. W., & Brune, D. E. (2007). Anaerobic co-digestion of algal sludge and waste paper to produce methane. Bioresource Technology, 98(1), 130–134. https://doi.org/10.1016/j.biortech.2006.12.043

[28] Yadvika, Santosh, T. R., Sreekrishnan, K. R., Kohli, S., & Rana, V. (2004). Enhancement of biogas production from solid substrates using different techniques—a review. Bioresource Technology, 95(1), 1–10. https://doi.org/10.1016/j.biortech.2004.02.049

Separation Technology - ICoST 2025
Materials Research Proceedings 59 (2026) 48-55

Materials Research Forum LLC
https://doi.org/10.21741/9781644903957-7

Electrocoagulation Process of Palm Oil Mill Effluent: Effect of Applied Voltage on Removal of Organic Content

Nofri NALDI[1,2,a], Ariadi HAZMI[1,b,*], Reni DESMIARTI[3,c], Primas EMERALDI[1,d], Maulana Yusup ROSADI[4,e], Nofrizon RAHMAN[5,f], Erda Rahmilaila DESFITRI[2,g]

[1]Department of Electrical Engineering, Universitas Andalas, Padang 25166, Indonesia

[2]Department of Renewable Energy Engineering Technology, Universitas Bung Hatta, Padang 25147, Indonesia

[3]Department of Chemical Engineering, Universitas Bung Hatta, Padang 25147, Indonesia

[4]Department of Civil Engineering, Borobudur University, Jakarta 13620, Indonesia

[5]Kohken Watertech Indonesia, Bekasi 17530, Indonesia

[a]nofrinaldi@bunghatta.ac.id, [b]ariadi@eng.unand.ac.id, [c]renitk@bunghatta.ac.id, [d]primas.emeraldi@eng.unand.ac.id, [e]mrosadi@borobudur.ac.id, [f]nof@kohkenwatertech.com, [g]rahmilaila@bunghatta.ac.id

Keywords: Palm Oil Mill Effluent (POME), Electrocoagulation, Applied Voltage, Aluminium Electrode, Organic Content

Abstract. Palm oil mill effluent (POME) has high organic content or extremely polluted waste water. Electrocoagulation is advanced alternative technology to treat POME. This study aims to investigate the effects of applied voltage and residence time on chemical oxygen demand (COD), biological oxygen demand (BOD), and total dissolved solid (TDS) removal. Aluminum electrodes were used and applied voltage was varied between 5, 7, and 9 volts. The sampling time was conducted at 30, 60, 90, and 120 min. The results showed that the removal of COD, BOD, and TDS increased with the increasing applied voltage. The optimum applied voltage at 9 volts and pH 7.7, achieving the highest removal efficiencies: 96.7% for COD, 98.8% for BOD, and 99.8% for TDS at 120 min of processing time. This research found that electrocoagulation process is very effective to treat highly polluted wastewater such as palm oil mill effluent.

Introduction

Palm oil mill effluent (POME) is a highly polluted wastewater generated from palm oil processing, containing high concentrations of BOD, COD, suspended solids, oil and grease, and various organic and inorganic compounds that pose serious environmental risks [1,2,3]. Conventional anaerobic pond treatment can reduce biodegradable matter but still leaves significant amounts of colour, phenolic, tannin, and ammonium compounds [4], resulting in COD levels that exceed regulatory limits (BOD ≤ 100 mg/L, COD ≤ 350 mg/L, pH 6–9) as specified in Indonesian Ministry of Environment Regulation No. 5 of 2014.

Electrocoagulation (EC) has emerged as a promising alternative due to its ability to generate coagulants *in situ* (from Al/Fe anodes), reduce chemical additive requirements, and minimize sludge production [5,6,7]. EC performance depends on operational parameters such as voltage, current density, pH, electrode spacing, and reaction time. Studies have demonstrated that EC can effectively reduce COD, BOD, and colour in POME, with optimal COD removal (>70%) typically achieved at 20–25 V [8,9,10,11].

Balancing treatment efficiency, energy consumption, and electrode passivation remains a major challenge. Therefore, optimizing operating conditions is essential to meet discharge standards at a reasonable cost. This study aims to evaluate the effect of voltage on the reduction of organic

Separation Technology - ICoST 2025
Materials Research Proceedings 59 (2026) 48-55

Materials Research Forum LLC
https://doi.org/10.21741/9781644903957-7

content (particularly COD) in POME to enhance the efficiency, sustainability, and economic feasibility of the electrocoagulation process.

Materials and Methods

Fresh POME was collected from the receiving tank of a POME wastewater facility in Dharmasraya, West Sumatra, Indonesia. The sample was stored at a temperature of 4–8°C before use. The EC system consisted of a 1 L rectangular acrylic vessel, a DC power supply, two pairs of aluminum (Al) electrodes, and a mechanical mixer, as shown in Fig. 1. The Al electrodes had a total surface area of 150 cm^2 with a thickness of 3 mm and were arranged 1 cm apart, with a total of 4 electrodes. The Al electrode was connected to a DC power source capable of delivering up to 30 A of current and 15 V of output voltage. The EC reactor was filled with 1000 mL of POME. The process was initiated by applying applied voltage at selected intensities of 5, 7 and 9 V, and contact times of 30 min, 60 min, 120 min and 180 min.

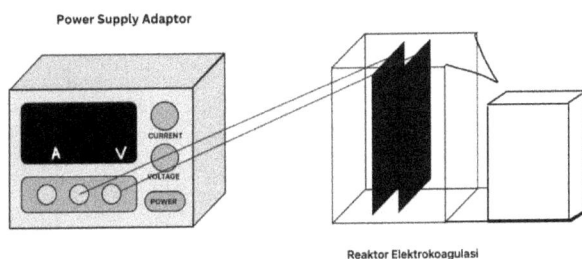

Figure 1 Electrocoagulation setup.

Results and Discussion

Effect of pH profile on electrocoagulation cell performance

The profile of changes in pH value on the performance of electrocoagulation cells based on the variables of contact time and electrical voltage in palm oil mill liquid waste can be seen in Fig. 2. The pH variation of POME during electrocoagulation using aluminum electrodes at 5, 7 and 9 V for 30–120 min. The initial acidic pH of 4.9 increased with both voltage and contact time. After 30 min, pH values reached 6.6 (5 V), 7.0 (7 V), and 7.3 (9 V). Higher voltages produced greater pH increases, with the maximum value of 7.7 achieved at 9 V after 120 min, indicating a shift toward neutral conditions.

The increase in pH during electrocoagulation is generally attributed to the electrolysis of water, which produces hydroxide ions (OH$^-$) at the cathode [4,9]. The pH increase during electrocoagulation is attributed to reduction reactions at the aluminum cathode that generate H$^+$ and OH$^-$ ions. Higher applied voltages enhance the formation of these ions, causing the pH to shift from acidic toward neutral or slightly alkaline conditions. This pH change plays an important role in improving pollutant removal efficiency during the electrocoagulation process [9,10].

The shift of pH toward neutral values (pH 6–8) benefits POME treatment by improving coagulation efficiency and minimizing electrode corrosion. Voltage is identified as the main factor controlling pH during electrocoagulation, while contact time serves as a secondary influence.

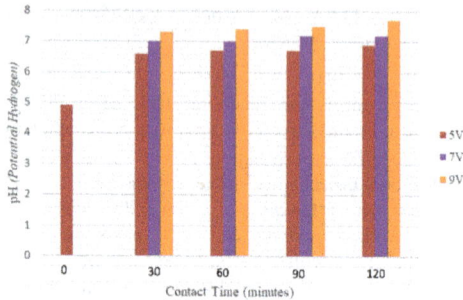

Figure 2 *The pH profile against electrical voltage and contact time.*

These results align with previous studies highlighting voltage and reaction time as key parameters in optimizing industrial wastewater treatment [8,9]. The experimental results indicate that the pH values of POME after electrocoagulation at various voltages (5–9 V) and contact times (30–120 min) meet the effluent quality standard specified by the Indonesian Ministry of Environment Regulation No. 5 of 2014, which requires a pH range of 6–9.

Effect of COD profile on electrocoagulation cell performance
The profile of COD value changes on electrocoagulation cell performance based on contact time and electrical voltage variables in palm oil mill liquid waste can be seen in Fig. 3. COD levels in POME decreased significantly after electrocoagulation, with the highest removal at 9 V and 120 min, reducing COD from 10,560 to 346.7 mg/L (96.7% efficiency). Higher voltage and longer contact time enhanced pollutant removal through $Al(OH)_3$ floc formation, confirming that optimized electrocoagulation conditions greatly improve COD reduction efficiency [8].

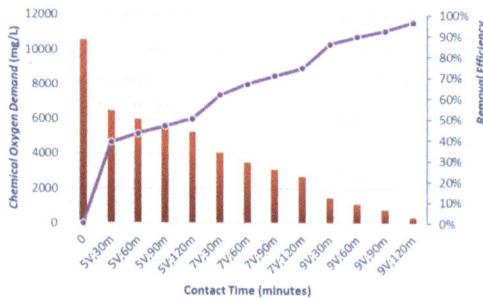

Figure 3 *The COD profile against electrical voltage and contact time.*

These observations are consistent with theoretical principles and recent literature, which state that higher voltages accelerate anodic oxidation and increase the dissolution of metallic ions from the electrodes, acting as effective coagulants [12,13]. The metal hydroxide flocs generated serve as adsorption sites for negatively charged organic molecules, thereby enhancing COD removal. Moreover, extended contact time improves particle collision frequency and flocculation, which further enhances the process efficiency [14,15]. The COD concentrations after electrocoagulation met the Indonesian discharge standard (≤ 350 mg/L), confirming that electrocoagulation, especially at higher voltages and longer contact times, is an effective method for POME treatment.

BOD profile on electrocoagulation cell performance

The profile of changes in BOD value on electrocoagulation cell performance based on contact time and electrical voltage variables in Palm Oil Mill Liquid Waste can be seen in Fig. 4.

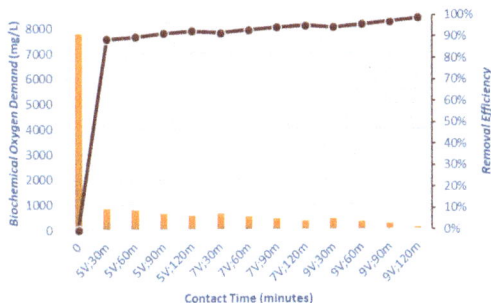

Figure 4 BOD profile against variations in electric voltage and contact time.

The initial BOD of 7,782 mg/L decreased sharply after electrocoagulation, reaching 94 mg/L at 9 V and 120 min. Higher voltage enhanced Al^{3+} ion release, promoting floc formation, pollutant removal, and easier biodegradation of remaining organics [13]. The BOD concentration of POME before treatment was 7,782 mg/L. After treatment using the electrocoagulation method with aluminium electrodes at voltage variations of 5, 7 and 9 V for 120 min, with sampling at 30, 60, 90 and 120 min, the following removal efficiencies were obtained: for 5 V, 89%, 90.1%, 91.7% and 92.7%; for 7 V, 91.9%, 93.3%, 94.4% and 95.3%; and for 9 V, 94.5%, 95.9%, 97% and 98.8%, respectively. When compared to previous studies, such as the one conducted, the BOD removal efficiency using Al-Al electrodes was only 74% [16].

Higher voltages and longer contact times enhance electrocoagulation efficiency by increasing coagulant ion release, which destabilizes colloidal particles and promotes organic floc formation. At optimal conditions (9 V, 120 min), BOD reduction reached 98.8%, outperforming 5 V (92.7%) and 7 V (95.3%). These results align with previous studies, which reported BOD reduction efficiencies of 84–96% depending on voltage, electrode configuration, and reaction time [17,18]. Electrocoagulation effectively reduces organic matter in wastewater, with the optimal condition found at 9 V and 120 min. The BOD values of POME after electrocoagulation at various voltages and contact times met the discharge standard for palm oil mill effluent (BOD ≤100 mg/L) as regulated by Ministry of Environment Decree No. 5 of 2014.

TDS profile on electrocoagulation cell performance

The profile of changes in TDS values on the performance of electrocoagulation cells based on the variables of contact time and electrical voltage in palm oil mill liquid waste can be seen in Fig. 5. The initial TDS concentration of the wastewater was 4,050 mg/L, and electrocoagulation significantly reduced it at all voltage and contact time variations. The highest reduction occurred at 9 V and 120 min, where TDS decreased to 290 mg/L. This reduction is attributed to higher applied voltage and longer contact time, which both contribute to more floc formation and lower TDS values [19,20].

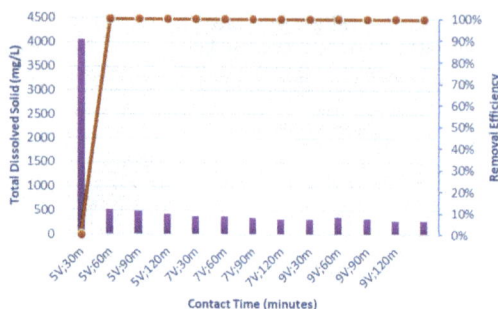

Figure 5 *TDS profile against variations in electrical voltage and contact time.*

The TDS of POME before treatment was 4050 mg/L. After being treated using the electrocoagulation method with aluminium electrodes under voltage variations of 5, 7 and 9 V for 120 min, with sampling times at 30, 60, 90, and 120 min, the following removal efficiencies were obtained: for 5 V the efficiencies were 99.9%, 99.8%, 99.8% and 99.8%; for 7 V the efficiencies were 99.9%, 99.8%, 99.8% and 99.8%; and for 9 V the efficiencies were 99.9%, 99.8%, 99.8% and 99.8%. These results show that the higher the applied voltage, the greater the decrease in TDS concentration. This is attributed to the increased release of Al^{3+} ions from the aluminium electrode at higher voltages, which accelerates the coagulation and flocculation of dissolved ions in wastewater [21]. In addition, longer contact time also plays an important role, as extended duration allows for more floc formation and sedimentation of dissolved particles [22]. Thus, it can be concluded that the combination of higher voltage (9 V) and longer contact time (120 min) is the optimum condition for reducing TDS concentration. This significant reduction demonstrates that electrocoagulation is effective as a treatment method for wastewater with high TDS content, including industrial wastewater and palm oil mill effluent [23,24]. The experimental results show that TDS values of POME after electrocoagulation at various voltages and contact times met the Indonesian wastewater standard ($\leq$2000 mg/L) as specified in Ministry of Environment Decree No. 5/2014.

Characterization of POME organic compounds using FTIR

The FTIR spectrum in Fig.6 compares functional groups in POME before and after electrocoagulation with aluminum electrodes at 5 V, 7 V and 9 V for 120 min. A significant reduction in the O–H stretching vibration band (3331–3333 cm^{-1}) was observed, particularly at 9 V, indicating a decrease in polar organic compounds like fatty acids and phenols. This supports previous studies that show electrocoagulation reduces COD and phenol content through adsorption and floc precipitation [4,25]. Before treatment, absorption bands were observed at 2182.92 cm^{-1} and 2107.12 cm^{-1}, indicating the presence of triple bonds C≡C or C≡N from complex organic compounds. After electrocoagulation, the intensity of these bands decreased significantly, particularly at higher voltages (7–9 V), indicating the degradation or coagulation of complex organic compounds in POME.

The band at 1637 cm^{-1}, linked to aromatic or amide groups, decreased after treatment, indicating reduced lignin and tannins. Bands at 451–567 cm^{-1} also weakened, suggesting the formation and precipitation of organic–metal complexes between Al^{3+} ions and organic compounds [26].

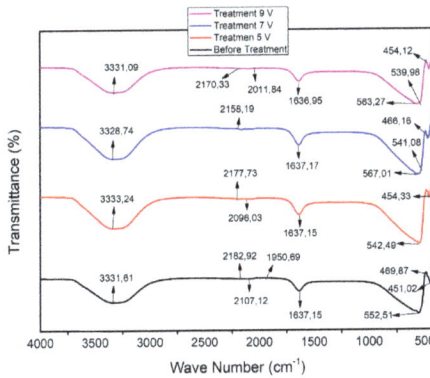

Figure 6 *Characterization of POME organic compounds at voltages of 5V, 7V and 9V for 120 min using FTIR.*

At 5 V, functional group changes were minimal, while 7 V showed better organic degradation, and 9 V achieved the greatest reduction in complex organics. FTIR results confirm that electrocoagulation at 9 V for 120 min effectively degraded or coagulated alcohols, carboxylic acids, and aromatic compounds, though higher voltage may increase energy use [27].

Conclusion

This study examined the effects of applied voltage (5, 7 and 9 V) and reaction time (30–120 min) on the reduction of organic pollutants in palm oil mill effluent (POME) using electrocoagulation with aluminum electrodes. The POME's initial pH of 4.9 increased to 6–8 after treatment. Under optimal conditions (9 V, 120 min), COD decreased from 10,560 to 346.7 mg/L (96.7% removal), BOD from 7,782 to 94.1 mg/L (98.8% removal, meeting Indonesian effluent standards), and TDS from 4,050 to 290 mg/L (99.8% removal). The POME color changed from dark brown to clear white. FTIR analysis revealed degradation of organic compounds such as alcohols, carboxylic acids and aromatics, with the most significant reduction at 9 V. Overall, electrocoagulation proved highly effective for POME treatment, with potential for sludge reuse in adsorbent, construction or pigment applications.

Acknowledgments

The authors acknowledge the financial support provided by the Ministry of Higher Education, Science, and Technology of the Republic of Indonesia under Grant No. 060/C3/DT.05.00/PL/2025. The authors also extend their sincere appreciation to the students of the Department of Electrical Engineering, Universitas Andalas, and the Department of Chemical Engineering, Universitas Bung Hatta, for their valuable contributions to this work.

References

[1] S. Shehu, S. Sani, M. Mujahid, and R. Adnan, "Valuable resources recovery from palm oil mill effluent (POME): A short review on sustainable wealth reclamation," vol. 3, no. December 2024, pp. 1–16, 2025.

[2] R. Desmiarti, M. Y. Rosadi, A. Hazmi, M. M. Rahman, N. Naldi, and J. A. Fajri, "Biogas Production from Palm Oil Mill Effluent Using Dielectric Barrier Discharge Integrated with the Aerated Condition," *Water (Switzerland)*, vol. 14, no. 22, pp. 1–15, 2022.

https://doi.org/10.3390/w14223774

[3] A. Hazmi, R. Desmiarti, P. Emeraldi, M. I. Hamid, Edwardo, and E. P. Waldi, "Preliminary study on biogas production from POME by DBD plasma," *Telkomnika (Telecommunication Comput. Electron. Control.*, vol. 15, no. 2, pp. 554–559, 2017. https://doi.org/10.12928/TELKOMNIKA.v15i2.5574

[4] P. Khongkliang, K. Chalearmkul, K. Boonloh, and N. Kanjanasombun, "Performance of combined organic precipitation , electrocoagulation , and electrooxidation in treating anaerobically treated palm oil mill effluents," *Appl. Water Sci.*, vol. 14, no. 10, pp. 1–16, 2024. https://doi.org/10.1007/s13201-024-02288-y

[5] S. Jo, R. Kadam, H. Jang, and D. Seo, "Recent Advances in Wastewater Electrocoagulation Technologies : Beyond Chemical Coagulation," *energies MDPI*, vol. 17, no. 5863, 2024. https://doi.org/https://doi.org/10.3390/en17235863

[6] M. Daudsyah, S. Mulyati, C. M. Rosnelly, and A. C. Ambarita, "Improving the efficiency of the electrocoagulation process for Palm Oil Mill effluent : a case study in POME processing in a palm oil processing factory in Meulaboh , West Aceh Regency," *BIO Web Conf.*, vol. 02002, 2024. https://doi.org/https://doi.org/10.1051/bioconf/202414802002

[7] E. Windiastuti, N. S. Indrasti, U. Hasanudin, Y. Bindar, W. Java, and W. Java, "Effect of Electrocoagulation on Improving the Quality of Palm Oil Liquid Waste," *EnvironmentAsia*, vol. 17, no. 3, pp. 174–184, 2024. https://doi.org/10.14456/ea.2024.45

[8] M. Aiyd, F. Yasir, and A. Dawood, "Studying the effect of reactor design on the electrocoagulation treatment performance of oily wastewater," *Heliyon*, vol. 9, no. July, 2023. https://doi.org/10.1016/j.heliyon.2023.e17794

[9] D. T. Moussa, M. H. El-Naas, M. Nasser, and M. J. Al-Marri, "A comprehensive review of electrocoagulation for water treatment: Potentials and challenges," *J. Environ. Manage.*, vol. 186, pp. 24–41, 2017. https://doi.org/10.1016/j.jenvman.2016.10.032

[10] M. E. Bote, "Studies on electrode combination for COD removal from domestic wastewater using electrocoagulation," *HLY*, vol. 7, no. 12, p. e08614, 2021. https://doi.org/10.1016/j.heliyon.2021.e08614

[11] E. Science, "Reactor design optimization on the electrocoagulation treatment of sugarcane molasses-based distillery wastewater Reactor design optimization on the electrocoagulation treatment of sugarcane molasses-based distillery wastewater," *IOP Conf. Ser. Earth Environ. Sci.*, 2019. https://doi.org/10.1088/1755-1315/344/1/012023

[12] M. A. Nasution, Z. Yaakob, E. Ali, and S. M. Tasirin, "Electrocoagulation of Palm Oil Mill Effl uent as Wastewater Treatment," *Tech. REPORTS WASTE Manag. Tech. REPORTS Electrocoagulation*, vol. 40, no. December 2015, pp. 2–10, 2011. https://doi.org/10.2134/jeq2011.0002

[13] K. L. Terrones-díaz, S. S. Segura-vera, and G. L. Huerta-chombo, "Favourable Conditions for the Removal of BOD and COD in Municipal Wastewater by Electrocoagulation," *MDPI*, vol. 17, no. 7803, pp. 1–22, 2025.

[14] M. Syaamil *et al.*, "Journal of King Saud University – Science Techno-economic analysis of an integrated electrocoagulation- membrane system in treatment of palm oil mill effluent," *J. King Saud Univ. - Sci.*, vol. 34, no. 4, p. 102015, 2022. https://doi.org/10.1016/j.jksus.2022.102015

[15] C. O. Demand and T. Improvement, "Chemical Oxygen Demand and Turbidity Improvement of Deinked Tissue Wastewater using Electrocoagulation Techniques,"

Separation Technology - ICoST 2025
Materials Research Proceedings 59 (2026) 48-55

Materials Research Forum LLC
https://doi.org/10.21741/9781644903957-7

bioresources, vol. 12, pp. 4327–4341, 2017.

[16] S. U. Khan *et al.*, "Efficacy of Electrocoagulation Treatment for the Abatement of Heavy Metals : An Overview of Critical Processing Factors , Kinetic Models and Cost Analysis," vol. 15, no. 2, 2023. https://doi.org/https://doi.org/10.3390/su15021708

[17] M. Javad, A. Takdastan, S. Jor, A. Neisi, and M. Farhadi, "Data in Brief Electrocoagulation process to Chemical and Biological Oxygen Demand treatment from carwash grey water in Ahvaz megacity , Iran," *Data Br.*, vol. 11, pp. 634–639, 2017. https://doi.org/10.1016/j.dib.2017.03.006

[18] N. Muhammad *et al.*, "Journal of Water Process Engineering Performance of batch electrocoagulation with vibration-induced electrode plates for land fi ll leachate treatment," *J. Water Process Eng.*, vol. 36, no. April, 2020. https://doi.org/10.1016/j.jwpe.2020.101282

[19] S. A. Bakry, M. E. Matta, and K. Zaher, "Electrocoagulation process performance in removal of TOC , TDS , and turbidity from surface water," *Desalin. Water Treat.*, vol. 129, pp. 127–138, 2018. https://doi.org/10.5004/dwt.2018.23070

[20] A. A. Al-raad and M. M. Hanafiah, "Environmental Technology & Innovation SO 4 2 − , Cl − , Br − , and TDS removal by semi continuous electrocoagulation reactor using rotating anode," *Environ. Technol. Innov.*, vol. 28, p. 102917, 2022. https://doi.org/10.1016/j.eti.2022.102917

[21] M. Kobya and E. Demirbas, "Evaluations of operating parameters on treatment of can manufacturing wastewater by electrocoagulation," *J. Water Process Eng.*, vol. 8, pp. 64–74, 2015. https://doi.org/10.1016/j.jwpe.2015.09.006

[22] E. Bazrafshan, L. Mohammadi, A. Ansari-Moghaddam, and A. H. Mahvi, "Heavy metals removal from aqueous environments by electrocoagulation process - A systematic review," *J. Environ. Heal. Sci. Eng.*, vol. 13, no. 1, 2015. https://doi.org/10.1186/s40201-015-0233-8

[23] O. Sahu, B. Mazumdar, and P. K. Chaudhari, "Treatment of wastewater by electrocoagulation: A review," *Environ. Sci. Pollut. Res.*, vol. 21, no. 4, pp. 2397–2413, 2014. https://doi.org/10.1007/s11356-013-2208-6

[24] F. Ilhan, K. Ulucan-Altuntas, Y. Avsar, U. Kurt, and A. Saral, "Electrocoagulation process for the treatment of metal-plating wastewater: Kinetic modeling and energy consumption," *Front. Environ. Sci. Eng.*, vol. 13, no. 5, pp. 1–8, 2019. https://doi.org/10.1007/s11783-019-1152-1

[25] O. Mill *et al.*, "Post treatment of Palm Oil Mill Effluent Using Electro-coagulation-peroxidation (ECP) Technique," *J. Clean. Prod.*, 2018. https://doi.org/10.1016/j.jclepro.2018.10.073

[26] A. Brian, L. Choong, and A. P. Peter, "Treatment of palm oil mill effluent (POME) using chickpea (Cicer arietinum) as a natural coagulant and flocculant: Evaluation, process optimization and characterization of chickpea powder," *Biochem. Pharmacol.*, 2018. https://doi.org/10.1016/j.jece.2018.09.038

[27] A. W. Zularisam, "Electrode design for electrochemical cell to treat palm oil mill effluent by electrocoagulation process," *Environ. Technol. Innov.*, 2017. https://doi.org/10.1016/j.eti.2017.10.001

Separation Technology - ICoST 2025
Materials Research Proceedings 59 (2026) 56-63

Materials Research Forum LLC
https://doi.org/10.21741/9781644903957-8

CBD-Grown MoS₂Thin Films on Plastic Optical Fiber for NH₃ Sensing: Fabrication and Performance

Nor Akmar MOHD YAHYA*[1,a], Mohd Rashid YUSOF HAMID[2],
Abdul Hadi ISMAIL[3,4], Nurul Atiqah Izzati MD ISHAK[5], Mohd Hanif YAACOB[6],
Saidur RAHMAN[7]

[1]School of Engineering, Faculty of Engineering and Technology, Sunway University, No. 5, Jalan Universiti, Bandar Sunway, 47500 Selangor Darul Ehsan, Malaysia

[2]Nanotechnology and Catalysis Research Centre (NANOCAT), Institute for Advanced Studies (IAS), Universiti Malaya 50603 Kuala Lumpur, Malaysia

[3]Centre of Innovative Nanostructures & Nanodevices (COINN), Universiti Teknologi PETRONAS, 32610, Seri Iskandar, Perak Darul Ridzuan, Malaysia

[4]Faculty of Data Science and Computing, Universiti Malaysia Kelantan City Campus, Pengkalan Chepa, 16100, Kota Bharu, Kelantan, Malaysia

[5]Research Centre for Carbon Dioxide Capture and Utilisation (CCDCU), Faculty of Engineering and Technology, Sunway University, No. 5 Jalan Universiti, Bandar Sunway, 47500 Petaling Jaya, Selangor Darul Ehsan, Malaysia

[6]Wireless and Photonics Network Research Centre, Faculty of Engineering, University Putra Malaysia, 43000 UPM Serdang, Selangor, Malaysia

[7]Research Centre for Nanomaterials and Energy Technology (RCNMET), Faculty of Engineering and Technology, Sunway University, No. 5, Jalan Universiti, Bandar Sunway, 47500 Selangor Darul Ehsan, Malaysia

[a]akmarm@sunway.edu.my

Keywords: Molybdenum Disulfide, Optical Ammonia Sensing, CBD, POF

Abstract. This study presents the development of an optical ammonia (NH₃) gas sensor based on plastic optical fiber (POF) coated with molybdenum disulfide (MoS₂), synthesized via chemical bath deposition (CBD). The MoS₂ thin film was uniformly deposited along the unclad region of the fiber, allowing surface interaction with NH₃ molecules through nanoscale adsorption mechanisms. These interactions result in measurable changes in absorbance, enabling real-time detection of NH₃. The morphological and optical properties of the MoS₂ coating were characterized using field emission scanning electron microscopy (FESEM), energy-dispersive X-ray spectroscopy (EDX), and UV–vis spectroscopy. The sensor exhibited a sensitivity of 1.55 a.u./% NH₃ over a concentration range of 0.0625% to 1.00%, demonstrating reliable performance for low-level NH₃ detection. This work highlights the potential of nanomaterial-coated fiber optic sensors for environmental monitoring applications.

Introduction

Ammonia (NH₃) plays a vital role in agriculture, refrigeration and chemical industries but its toxicity at low concentrations raises significant safety and environmental concerns. Prolonged exposure can lead to respiratory issues and is subject to strict regulatory limits, necessitating accurate, and real-time monitoring solutions. Conventional ammonia sensors, such as metal oxide semiconductors, electrochemical, and colorimetric sensors, have been widely used but often face challenges like high power consumption, limited selectivity, and poor stability under ambient conditions [1].

Separation Technology - ICoST 2025   Materials Research Forum LLC
Materials Research Proceedings 59 (2026) 56-63   https://doi.org/10.21741/9781644903957-8

To overcome these limitations, researchers are increasingly incorporating nanomaterials into sensing platforms to improve sensitivity. Two-dimensional (2D) materials such as molybdenum disulfide (MoS_2) offer high surface area, tunable bandgap, and strong adsorption affinity for ammonia molecules [2]. MoS_2-based sensors have shown promising room-temperature performance, making them attractive for portable and wearable applications [3]. Composite strategies have been explored to enhance MoS_2 sensing capability: MoS_2/CeO_2 composites achieved room-temperature NH_3 detection, though with limited response and recovery times [4]; plasma-modified MoS_2 nanoflowers demonstrated excellent selectivity and rapid response, but the plasma process is complex and difficult to scale [5]. Conducting polymer composites such as MoS_2/WS_2–PANI and MoS_2–PPy improved sensitivity and response [6,7], yet stability and integration remain challenging. Similarly, MoS_2 coupled with rGO [8], SnO_2 [9], or MXenes [10] enhanced charge transfer and selectivity, though synthesis procedures are often complex and limited to electrical readouts. Flexible electrical platforms such as screen-printed PANI/MoS_2 composites achieved sub-ppm NH_3 detection with long-term stability [11], while PPy/MoS_2 nanocomposites exhibited high repeatability [12].

However, most MoS_2 sensors rely on electronic transduction, which is susceptible to electromagnetic interference, electrical noise, and wiring complexity. Optical fiber sensors provide an alternative with immunity to electromagnetic noise, remote operation, and passive functionality [13]. An all-fiber MoS_2-coated side-polished silica fiber recently achieved 86% sensitivity with 21 s and 60 s response and recovery times under ambient conditions [14]. While this demonstrates strong potential, silica side-polished fibers are fragile and costly. Plastic optical fibers (POFs) offer lower cost, higher flexibility, and mechanical robustness [1]. Integrating MoS_2 with POF enables room-temperature optical NH_3 detection that merges the high sensitivity of 2D materials with a scalable, cost-effective fiber-optic platform. This work therefore proposes a MoS_2–POF ammonia sensor operating at room temperature to bridge high-performance nanomaterial sensing and practical optical implementation.

Experimental

The work starts with plastic optical fiber (POF) preparation. For the stability testing, a multimode plastic optical fiber (POF) made of polymethyl-methacrylate (PMMA) with a core diameter of 980 μm was utilized. The POF was selected for its flexibility, ease of handling, and suitability for room-temperature gas sensing operation without risk of breakage.

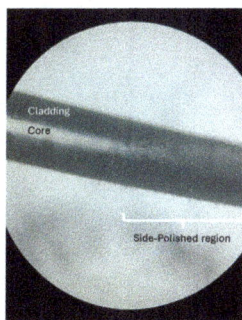

Figure 1 The side polished of plastic optical fiber (POF).

To prepare the sensor region, the protective jacket was stripped using a fiber stripper and a razor blade. The POF was then mounted onto a custom-designed polishing jig, where sandpaper was used to carefully polish one side and expose the fiber core by removing the cladding. A 2 cm polished section was produced being the maximum length compatible with the custom-built gas chamber. Finally, the polished surface was thoroughly cleaned with ethanol followed by deionized water to ensure a contaminant-free sensing area. Fig. 1 shows a microscope image taken of the side-polished POF sample.

The MoS_2 synthesis was prepared using a chemical bath approach. Initially, 10 mL of sodium sulfide, Na_2S (0.2 M) was mixed with 1 mL of sulfuric acid, H_2SO_4 (1 M) in a fume hood, resulting in a cloudy white suspension due to H_2S gas evolution. Subsequently, 10 mL of ammonium molybdate, $(NH_4)_6Mo_7O_{24}$ (1 mM) was added under temperature of 60°C in water bath, causing the solution to turn dark orange/brown, consistent with the formation of a molybdenum–sulfur complex. The side-polished POF is then immersed in this solution, which is placed in a water bath held for 30 minutes to promote the growth of the MoS_2 thin film on the POF. Upon completion of the deposition, the MoS_2 coated POF is gently rinsed with distilled water and dried at 60 °C, concluding the process. Fig. 2 illustrates the fabrication process of the MoS_2-POF.

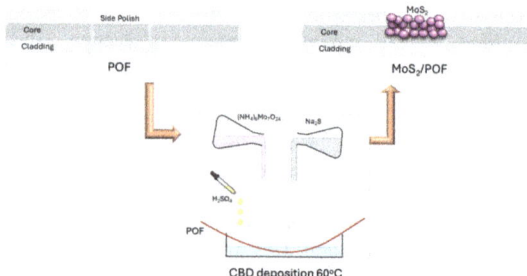

Figure 2 *Illustration of MoS$_2$-POF fabrication process.*

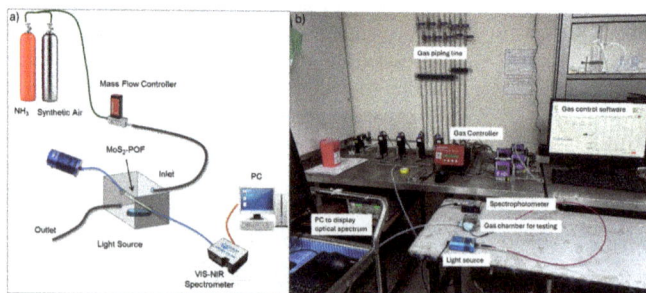

Figure 3 *Gas sensing setup: (a) Illustration diagram, (b) In-house setup.*

For testing setup, the sample MoS_2-POF was placed in the gas chamber as presented in Fig. 3. A broadband light source (HL-2000, Ocean Optics, USA; 360–2500 nm) and a spectrophotometer (USB4000 VIS-NIR, Ocean Optics, USA; 200–1100 nm) were used for the optical measurements. The sample prepared was positioned inside a sealed, custom-fabricated gas chamber. Gas flow into the chamber was controlled using a mass flow controller set to 200 sccm, with purified air used to dilute NH_3 to various concentrations (0.0625%, 0.125%, 0.25%, 0.5%, 0.75%, and 1%) for sensitivity evaluation. The response and recovery times at 1% NH_3 concentration were determined

Separation Technology - ICoST 2025
Materials Research Proceedings 59 (2026) 56-63

Materials Research Forum LLC
https://doi.org/10.21741/9781644903957-8

from the dynamic absorbance behavior, following standard procedures reported in the literature [2].

Results and Discussion

This section is divided into two parts. The first one is the result of MoS_2 characterization, and the latter is on the absorbance measurement when the samples are exposed to ammonia (NH_3) gas as per the experimental setup in Fig. 3.

Fig. 4 shows FESEM images of MoS_2-POF synthesized using the chemical bath deposition technique. The FESEM image of a sample reveals a highly porous and rough surface morphology with well-defined interconnected voids and nanoscale features. The porous structure consists of a continuous framework with irregularly shaped pores in the sub-micron to micron range. At higher magnification, the surface appears to be densely packed with fused nanoparticles or clusters, and agglomerated nanostructures, resulting in forming a network-like morphology. The high porosity and rough texture provide abundant active sites for gas adsorption, while the interconnected pathways enable efficient gas diffusion through the sensing layer. Such architecture enhances the sensitivity and response speed of the sensor toward target gases like ammonia (NH_3).

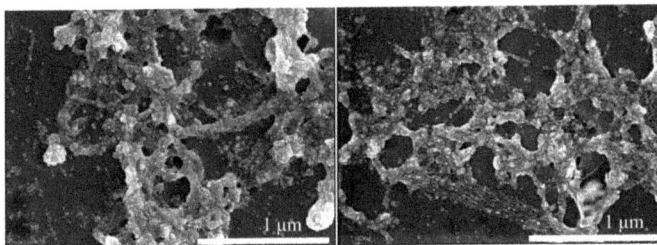

Figure 4 FESEM images of MoS₂ synthesized via chemical bath deposition method.

The elemental composition of the synthesized MoS_2 sample was examined using energy-dispersive X-ray spectroscopy (EDX), as shown in Fig. 5. Strong peaks corresponding to molybdenum (Mo) and sulfur (S) confirm the successful deposition of MoS_2. The additional presence of oxygen (O) can be attributed to surface oxidation of MoS_2 during exposure to air and possible adsorbed moisture. Peaks of sodium (Na) and calcium (Ca) are most likely derived from the soda-lime glass substrate as a place holder, while aluminum (Al) may originate from the aluminum SEM stub or crucible contamination during preparation. The barium (Ba) peaks observed are attributed to the substrate composition, as certain glass materials contain barium oxide additives, or may arise as an artifact from the EDX system. Overall, the EDX result has validated the formation of MoS_2 while the other elements detected are associated with the substrate and measurement environment rather than the active sensing layer.

The bandgap energy of the synthesized MoS_2 was determined using the Kubelka–Munk function, where the Tauc plot was extrapolated from its linear region. As shown in Fig. 6(a), the UV-Vis absorbance spectrum of MoS_2 exhibits strong absorption in the visible region, with a prominent peak around 500–600 nm, indicating the material's ability to absorb visible light. This absorption behavior is consistent with semiconducting MoS_2 but suggests deviations from typical monolayer MoS_2, which usually exhibits a non-zero bandgap with high absorption across the visible spectrum [3]. The corresponding Tauc plot in Fig. 6(b) reveals that the bandgap energy of the synthesized MoS_2 is approximately 2.4 eV, obtained from the linear fit extrapolation. This value is higher than that reported for pristine MoS_2, which may be attributed to structural defects such as oxygen substitution at sulfur sites or sulfur vacancies, both of which are known to significantly alter the electronic and optical properties of MoS_2 [4,5].

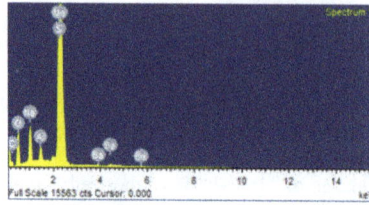

Figure 5 EDX spectrum of MoS_2.

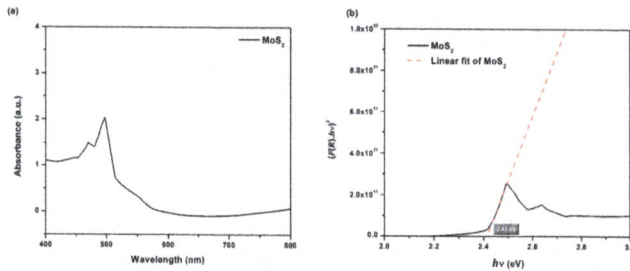

Figure 6 (a) UV-vis absorbance spectra of the synthesized MoS_2, (b) bandgap energy (eV) of the synthesized MoS_2.

The XRD pattern of the synthesized MoS_2 nanostructures in Fig. 7 exhibited distinct diffraction peaks at 2θ values of $32.03°$, $33.87°$, $38.68°$, $43.04°$, and $48.76°$, which are in good agreement with the standard hexagonal $2H–MoS_2$ phase. These reflections can be indexed to the and crystallographic planes, respectively, confirming the formation of crystalline MoS_2. The slight deviations in peak positions relative to the standard values may be attributed to strain effects, size-induced broadening, or interfacial interactions with the substrate during the deposition process.

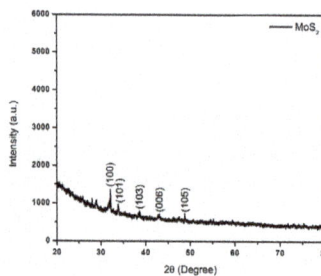

Figure 7 The XRD spectra of the synthesized MoS_2.

In optical fiber-based gas sensors, modifying the cladding is essential to enable interaction between the evanescent field from the fiber core and the surrounding medium. Side-polishing of POF provides an effective approach to expose the evanescent wave, where the polishing depth and length can be precisely controlled with minimal structural damage. When MoS_2 is coated on the polished region of the POF, the exposed evanescent field offers the most sensitive interaction zone, resulting in enhanced adsorption of NH_3 molecules. The polished length provides a large active

Separation Technology - ICoST 2025
Materials Research Proceedings 59 (2026) 56-63

Materials Research Forum LLC
https://doi.org/10.21741/9781644903957-8

area, thereby increasing the effective adsorption sites for NH_3 and improving the overall sensitivity of the sensor.

The absorbance measurements of MoS_2-POF exposed to 1% NH_3 gas at room temperature can be shown in Fig. 8(a) in which the absorbance response shift starts at 370nm up to 900nm indicating sufficient optical response to proceed with dynamic response as displayed in Fig. 8(b) and Fig. 9(a). The response time was determined based on 90% of the total magnitude of the absorbance increase at 1% NH_3 gas, meanwhile the inverse trend was determined for the recovery time of the samples. Response and recovery times measured are 60s and 120s respectively.

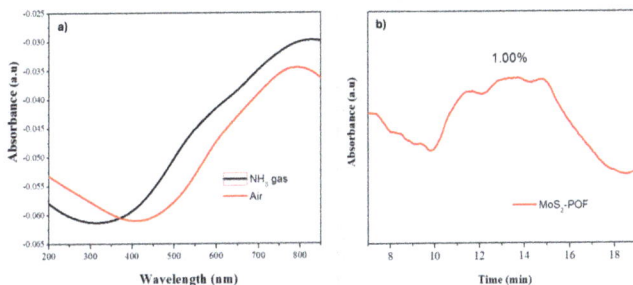

Figure 8 MoS_2-POF optical spectrum when exposed to 1% NH_3 gas at room temperature: (a) absorbance shift and (b) dynamic response.

The sensing mechanism of the MoS_2-POF optical sensor can be explained based on the interaction between NH_3 molecules and the MoS_2 surface. The MoS_2 nanostructures possess numerous active sites and surface defects, which readily adsorb atmospheric oxygen, leading to partial oxidation into MoO_x/MoO_3 species. These species introduce hole carriers, giving rise to p-type behavior in the MoS_2 sensing layer. Upon exposure to NH_3, which acts as an electron-donor molecule, the adsorbed NH_3 donates electrons to MoS_2 and reduces the hole carrier concentration [6]. This charge transfer modifies the electronic structure of MoS_2 and alters its optical properties.

As mentioned before, the side-polished region allows the evanescent field from the fiber core to penetrate into the MoS_2 sensing layer. Any change in the optical properties of MoS_2, particularly refractive index and absorption directly alter the evanescent wave. The adsorption of NH_3 increases the effective refractive index of the surrounding medium and induces a detectable change in absorbance intensity. As a result, the sensor exhibits an optical response that can be monitored as a shift in absorbance variations at certain wavelengths. Thus, the synergistic effects of NH_3 adsorption at MoS_2 active sites and the strong coupling of the evanescent field with the sensing layer form the basis of the high sensitivity of the MoS_2-POF optical ammonia sensor.

The sensitivity of the fabricated sensor was determined from the slope of the calibration curve obtained by plotting the absorbance change (Δ absorbance) against NH_3 gas concentrations in Fig. 9(b). The resulting sensitivity was 1.55 a.u./% NH_3 indicating that for every 1% increase in NH_3 concentration, the absorbance increased by approximately 1.55 arbitrary units (a.u.).

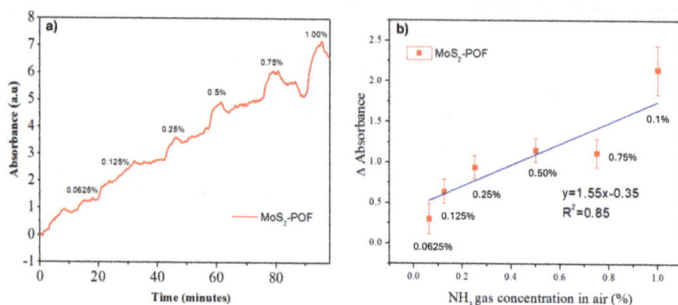

Figure 9 (a) The dynamic response of MoS_2-POF towards a range of NH_3 gas concentration, (b) absorbance change as a function of NH_3 gas concentration.

Conclusion

A MoS_2-coated plastic optical fiber (POF) gas sensor was successfully fabricated and characterized. The sensor exhibited good NH_3 detection performance with a sensitivity of 1.55 au/% NH_3, a 60s response, and a 120s recovery time. These results are attributed to the high surface area and active sites of MoS_2 that enhance NH_3 adsorption and alter the fiber's evanescent field. Further studies are required to assess selectivity toward gases such as H_2, CH_4, and VOCs, as well as long-term stability. Future work will explore surface modification and hybrid nanostructures to improve selectivity and durability. Overall, MoS_2-POF demonstrates strong potential as a low-cost, room-temperature optical gas sensing platform.

Acknowledgement

This work has been funded by Research Accelerator Grant Scheme (RAGS) Sunway University (GRTIN-RAG-DEN-08-2024).

References

[1] A.H. Ismail, M.H. Yaacob, M.A. Mahdi, Y. Sulaiman, Influence of HKUST-1 and emeraldine based on the long-term stability of emeraldine salt-coated SP-POF for room temperature optical NH_3 gas sensing, Sens. Actuators A Phys. 335 (2022) 113395. https://doi.org/10.1016/j.sna.2022.113395

[2] C.K. Ho, A. Robinson, D.R. Miller, M.J. Davis, Overview of sensors and needs for environmental monitoring, Sensors 5 (2005) 4–37. https://www.mdpi.org/sensors

[3] D. Burman, A. Sharma, P.K. Guha, Flexible large MoS_2 film based ammonia sensor, IEEE Sens. Lett. 2(2) (2018) 2817651. https://doi.org/10.1109/LSENS.2018.2817651

[4] N. Dogra, K. Dadwal, A. Kumar, S. Sharma, MoS_2/CeO_2 based composite for ammonia sensing, IOP Conf. Ser. Mater. Sci. Eng. 1225(1) (2022) 012059. https://doi.org/10.1088/1757-899X/1225/1/012059

[5] A. Kashyap, B. Chakraborty, T. Hazarika, S. Chouhan, B. Kakati, H. Kalita, Highly selective ammonia sensing at room temperature using DC plasma-modified MoS_2 nanoflowers, Mater. Adv. 6(12) (2025) 3828–3840. https://doi.org/10.1039/D5MA00232J

[6] H. Parangusan, J. Bhadra, R.A. Al-Qudah, E.C. Elhadrami, N.J. Al-Thani, Comparative study on gas-sensing properties of 2D (MoS_2, WS_2)/PANI nanocomposites-based sensor, Nanomaterials 12(24) (2022) 44423. https://doi.org/10.3390/nano12244423

[7] S. Ahmad, I. Khan, A. Husain, A. Khan, A.M. Asiri, Electrical conductivity based ammonia sensing properties of polypyrrole/MoS$_2$ nanocomposite, Polymers (Basel) 12(12) (2020) 3047. https://doi.org/10.3390/polym12123047

[8] R. Singh, R.K. Mishra, P. Yadav, M. Kumar, Room temperature ammonia (NH$_3$) gas sensor based on molybdenum disulfide and reduced graphene oxide (MoS$_2$/rGO) heterojunction, J. Phys.: Conf. Ser. 2663 (2023) 012022. https://doi.org/10.1088/1742-6596/2663/1/012022

[9] S. Singh, R.M. Sattigeri, S. Kumar, P.K. Jha, S. Sharma, Superior room-temperature ammonia sensing using a hydrothermally synthesized MoS$_2$/SnO$_2$ composite, ACS Omega 6(17) (2021) 11602–11613. https://doi.org/10.1021/acsomega.1c00805

[10] Z. Guo, L. Zhang, J. Li, X. Zhao, Y. Wang, Tailoring MoS$_2$ nanoflakes over MXenes nanobelts for efficient ammonia detection at room temperature, J. Alloys Compd. 1010 (2025) 177710. https://doi.org/10.1016/j.jallcom.2024.177710

[11] A. Jain, A.M. Parambil, S. Panda, Screen-printed PANI/MoS$_2$-based flexible gas sensor for sub-ppm NH$_3$ detection: Experimental and DFT investigations, Sens. Actuators B Chem. 432 (2025) 137453. https://doi.org/10.1016/j.snb.2025.137453

[12] Y. Sood, S.D. Lawaniya, H. Mudila, A. Katoch, K. Awasthi, A. Kumar, Ultra-high performance of PPy/MoS$_2$ 2D nanocomposites for ammonia sensing, Sens. Actuators B Chem. 417 (2024) 136165. https://doi.org/10.1016/j.snb.2024.136165

[13] N.A.M. Yahya, M.H. Yaacob, Y. Sulaiman, A.A. Bakar, H. Ahmad, H$_2$ sensor based on tapered optical fiber coated with MnO$_2$ nanostructures, Sens. Actuators B Chem. 246 (2017) 421–427. https://doi.org/10.1016/j.snb.2017.02.084

[14] J. Mohanraj, M. Valliammai, S. Addanki, N. Vinodhkumar, G. Gulothungan, R. Rishav, All fiber-optic ammonia gas sensor using few-layer MoS$_2$ coated side polished fiber, Proc. IEEE Workshop Recent Adv. Photonics (WRAP 2023), Institute of Electrical and Electronics Engineers, (2023). https://doi.org/10.1109/WRAP59682.2023.10712905

[15] Y. Qu, M. Huang, L. Chen, J. Lin, Room temperature NH$_3$ selective gas sensors based on double-shell hierarchical SnO$_2$@polyaniline composites, Sensors 24(6) (2024) 1824. https://doi.org/10.3390/s24061824

[16] X.Y. Deng, X.H. Deng, F.H. Su, N.H. Liu, J.T. Liu, Broadband ultra-high transmission of terahertz radiation through monolayer MoS$_2$, J. Appl. Phys. 118 (2015) 015501. https://doi.org/10.1063/1.4921234

[17] M. Sharma, A. Kumar, P.K. Ahluwalia, Electron transport and thermoelectric performance of defected monolayer MoS$_2$, Physica E Low-Dimens. Syst. Nanostruct. 107 (2019) 117–123. https://doi.org/10.1016/j.physe.2018.11.011

[18] V. Křínek, Y. Ruzickova, K. Mikulik, F. Roubicek, J. Vyskocil, K. Mach, M. Petrov, G. Jarosova, Spectroscopic properties of nanostructured molybdenum oxysulfide deposits fabricated by MoO$_3$ evaporation in H$_2$S, Mater. Lett. 275 (2020) 128191. https://doi.org/10.1016/j.matlet.2020.128191

Separation Technology - ICoST 2025
Materials Research Proceedings 59 (2026) 64-71

Materials Research Forum LLC
https://doi.org/10.21741/9781644903957-9

Screening and Integration of Watermelon Rind Extract (WMRE) for Functional Chitosan Thin Films

Rozaini ABDULLAH[1,2,a*], Sharifah Zati Hanani SYED ZUBER[1,2,b],
Noor Amirah ABDUL HALIM[1,3,c], Muhammad Zafri Aziman MOHD SALLEH[1,d]

[1]Faculty of Chemical Engineering & Technology Kompleks Pusat Pengajian Jejawi 3 Kawasan Perindustrian Jejawi Universiti Malaysia Perlis (UniMAP) 02600 Arau, Perlis

[2]Centre of Excellence for Frontier Materials Research, Universiti Malaysia Perlis (UniMAP), No. 64-66, Blok B, Taman Pertiwi Indah, Jalan Kangar - Alor Setar, Kampung Seriap, 01000 Kangar, Perlis

[3]Centre of Excellence for Biomass Utilization, Kompleks Pusat Pengajian Jejawi 3 Kawasan Perindustrian Jejawi Universiti Malaysia Perlis (UniMAP) 02600 Arau, Perlis

[a]rozainiabdullah@unimap.edu.my, [b]sharifahzati@unimap.edu.my,
[c]amirahhalim@unimap.edu.my, [d]zafriaziman99@gmail.com

Keywords: Watermelon Rind Extract, Thin Film, Active Packaging, Two-Level Factorial

Abstract. This study explores the development of chitosan-based thin films incorporated with watermelon rind extract (CS-WMRE) as a sustainable material for active food packaging. Initially, the WMRE was obtained by comparing two maceration techniques: hot plate and water bath shaker, to maximize total phenolic content (TPC) and antioxidant activity. The hot plate method yielded the highest TPC (5.68 ± 0.47 mg GAE/g) and antioxidant activity ($16.49 \pm 0.37\%$). Then, three different effects of the parameters were screened via a Two-level factorial design, specifically focusing on the amounts of chitosan (wt.%), glycerol (wt.%), and watermelon rind extract (WMRE) (wt.%). The respective response was antioxidant activity (%). ANOVA analysis indicates that these three parameters and their interactions were highly significantly associated with enhanced antioxidant activity and overall performance. FTIR analysis showed the presence of phenolic groups, appearing between $1405.86–1407.85$ cm^{-1}, which are linked to O–H bending. Meanwhile, water vapour permeability (WVP) analysis showed that incorporating WMRE reduces the film's WVP, thereby improving its barrier properties. These findings highlight the potential of upcycled watermelon rind as a functional additive in chitosan-based thin films, offering a sustainable approach for various applications, particularly in packaging and other fields that require moisture-barrier characteristics.

Introduction

The use of synthetic polymers in active packaging raises environmental concerns due to their inability to biodegrade, high recycling costs, and contamination of food products with packaging materials [1]. Most conventional plastics are resistant to natural degradation processes, resulting in degradation times that can span centuries [2]. Consequently, there has been increasing interest in developing active biopolymer packaging, which offers biodegradability, safety, and the ability to enhance food quality and shelf life. In recent years, growing consumer awareness of sustainability and food safety has further driven the demand for active packaging with antibacterial and antioxidant functionalities [3].

Active biopolymer packaging can be produced either by integrating active substances directly into the packaging matrix or by introducing active pads within the package [4]. Among the materials explored, chitosan-based thin films have attracted considerable attention due to their film-forming properties, non-toxicity, biodegradability, and inherent antimicrobial activity.

Chitosan, a linear cationic polysaccharide ($C_56H_103N_9O_39$) obtained from the deacetylation of chitin in shellfish waste, is composed of β-(1−4)-linked D-glucosamine and N-acetyl-D-glucosamine units [5]. Despite these advantages, pure chitosan films generally exhibit poor mechanical and functional stability, limiting their potential in industrial food packaging. To address these limitations, recent studies have focused on enhancing chitosan films by incorporating natural bioactive compounds [6].

Plant-derived extracts are rich in phenolic compounds, flavonoids, and terpenes, offering an environmentally friendly solution to enhance the antioxidant, antimicrobial, and mechanical properties of chitosan-based films [7]. A valuable, underutilised resource for this purpose is watermelon rind waste (WMR), a substantial agricultural by-product from Malaysia's high watermelon production (174,000 tonnes in 2024) [8]. The resulting watermelon rind extract (WMRE) contains diverse bioactive constituents, including phenols and flavonoids, whose antioxidant and cross-linking properties are crucial for improving the physical and functional characteristics of chitosan-based films [9].

In this study, chitosan-based thin films incorporated with WMRE were screened for active biopolymer packaging applications. The effects of key parameters, including the amounts of chitosan, glycerol, and WMRE, were evaluated using a Two-level factorial design to determine their influence on antioxidant activity and the overall performance of the thin film. The obtained data showed desirable structural and barrier properties, demonstrating its potential for sustainable food packaging or other applications that require moisture barrier performance.

Materials and Methods

The red flesh watermelon rind (*Citrullus lanatus*) was obtained from a local stall in Perlis. All chemicals and reagents used in this study were of analytical grade (99.99%) and did not require further purification.

The watermelon rind (WMR) was oven-dried at 60°C for 24 h, ground into a fine powder, and extracted with 80% methanol at 40°C for 60 min using a water bath shaker and a hot plate. The extract was then evaporated at 40°C using a rotary evaporator, filtered, and stored in a screw-cap bottle for further analysis.

Chitosan (1-3% wt.) was added to a 1% acetic acid aqueous solution. Then, glycerol (30-60% wt.) and WMRE (5-10% wt.) were added to the mixture. The solution was filtered to remove the precipitate it contained. The thin film was cast using flexible casting and then dried at room temperature under controlled humidity. All thin films were equilibrated in a desiccator at 50% relative humidity and 25°C for 2 days before testing.

DPPH radical scavenging activity for WRME and thin film

A free radical assay was performed using the method of [11] with some modifications. About 3 mL of sample extract was mixed with 1 mL of 0.025 M DPPH solution. Then, the mixtures were shaken and placed in the dark at room temperature for 1 hour. The absorbance of the mixtures and the blank was measured at 517 nm using a UV-Vis spectrometer. The antioxidant activity was measured by using Eq. 1.

$$Antioxidant\ activity\ (\%) = \frac{A_c - A_t}{A_c} \times 100 \tag{1}$$

Where A_c is the absorbance of the methanol solution with DPPH as control, and A_t is the absorbance of the sample solution at 517 nm.

Films of each suggested by experimental design via Design Expert software (Version 13, Stat-Ease Inc., Minneapolis, USA) (900 mg) were put into 18 mL of methanol at room temperature with gentle stirring at 150 rpm for 3 h to obtain the supernatant. Then, 1 mL of supernatant was mixed with 2 mL of 0.06 mM DPPH. The mixture was stirred and incubated in the dark at room

Separation Technology - ICoST 2025
Materials Research Proceedings 59 (2026) 64-71

Materials Research Forum LLC
https://doi.org/10.21741/9781644903957-9

temperature for 30 min. The control for this experiment is a methanol solution without film [12]. The antioxidant activity will be measured by using the following Eq. 2.

$$Antioxidant\ activity\ (\%) = 1 - \frac{A_s}{A_c} \times 100 \tag{2}$$

In this context, A_s denotes the absorbance of the sample solution, while A_c represents the absorbance of the methanolic DPPH solution (control) at 517 nm. The antioxidant activity was quantified as the percentage of DPPH radical scavenging activity per 100 mg of film.

Experimental design and statistical analysis

The Design-Expert software (Version 13, Stat-Ease Inc., Minneapolis, USA) was used to screen using a Two-level Factorial Design to identify experimental parameters and interactions that significantly influence the formulation of CS-WMRE thin films. All independent parameters are listed in Table 1. A 2^3 fractional factorial design, consisting of 30 runs including triplicates and six centre points, was implemented for all parameters, as outlined in Table 2.

Table 1 *Variables and levels for the Two-level factorial design.*

Parameter	Unit	Symbol	Level	
			High	Low
Chitosan	(wt.%)	A	1	3
Glycerol	(wt.%)	B	30	60
WRME	(wt.%)	C	5	10

Fourier transform infrared spectroscopy (FTIR) and water vapour permeability (WVP)

FTIR analysis was performed to investigate the structural characteristics of chitosan films containing WMRE using an FTIR spectrometer (PerkinElmer, MA, USA) equipped with Attenuated Total Reflectance (ATR) technology. The CS–WMRE thin film was placed on the ATR attachment, and spectra were recorded in transmission mode over a wavenumber range of 4000–400 cm^{-1} with 16 scans.

The cup was filled with distilled water, and the film sample was sealed on the mouth of the cup. The sealed cup was placed in the desiccator at room temperature with silica gel for 4 days. The weight of the cup will be recorded every 24 h [12]. The water vapour permeability (WVP) was calculated according to the following equation.

$$WVP = \frac{(\Delta w).x}{A.(\Delta t).(p_2 - p_1)} \times 100 \tag{3}$$

Where Δw is the weight of water absorbed in the cup, x is the film thickness (mm), A is the area of the exposed film (m^2), Δt is the time for weight change (days), and p_2-p_1 is the differential of water vapour pressure (kPa).

Results and Discussion

WMR extraction and analysis

The optimal extraction of bioactive compounds from watermelon rind (WMR) was achieved by comparing two distinct maceration techniques: a hot plate stirrer and a water bath shaker. The hot plate stirrer technique produced a higher Total Phenolic Content (5.68 ± 0.47 mg GAE/g) than the water bath shaker (3.56 ± 0.59 mg GAE/g). Similarly, DPPH analysis showed greater antioxidant activity with the hot plate stirrer (16.49 ± 0.37%) compared to the water bath shaker (15.17 ± 0.61%). The results demonstrated that the hot plate stirrer significantly enhanced extraction efficiency, attributed to the synergistic effect of heat combined with dynamic, continuous agitation. The superior results acknowledge that the combination of heat and vigorous stirring effectively facilitates cell wall breakdown and mass transfer, confirming the hot plate stirrer as the optimal technique for producing WMRE.

Design matrix of CS-WRME thin films

A Two-level factorial design was employed to identify significant parameters and evaluate their effects on the formulation of CS-WMRE thin films. Table 2 displays the results from the 30 iterations proposed by Design-Expert Software version 13. The screening results from the Two-level factorial design revealed distinct optimal conditions for each film property. The highest antioxidant activity (~50%) was achieved with a combination of 1% wt. chitosan, 60% wt. glycerol, and 10% wt. WRME. The minimum antioxidant activity (~41.24%), was observed with a film containing 3% wt. chitosan, 30% wt. of glycerol, and 10% wt. of WRME.

Table 2 *Experimental results of Two-level factorial designs for CS-WMRE thin films.*

No. of Experiment	Chitosan A	Glycerol B	WRME C	Antioxidant Activity (%) Y_1
1	3	30	5	45.43
2	3	60	5	45.33
3	1	30	10	47.09
4	3	60	5	44.02
5	1	60	10	50.27
6	3	30	5	45.55
7	2	45	7.5	45.86
8	2	45	7.5	45.09
9	1	60	5	48.08
10	1	30	5	44.57
11	3	30	10	41.18
12	1	30	5	43.58
13	1	30	5	44.08
14	2	45	7.5	45.67
15	1	60	5	48.18
16	3	60	10	46.75
17	3	30	5	44.26
18	2	45	7.5	45.23
19	2	45	7.5	46.87
20	2	45	7.5	45.43
21	3	60	10	45.75
22	3	60	5	45.76
23	1	60	10	49.85
24	3	60	10	45.8
25	3	30	10	41.18
26	3	30	10	41.37
27	1	30	10	48.08
28	1	60	10	49.87
29	1	30	10	46.88
30	1	60	5	47.78

Analysis of variance (ANOVA) of CS-WRME thin films

The ANOVA analysis for antioxidant activity shows that the p-value (<0.0001) is below the 0.001 threshold, and the model's F-value of 66.82 both demonstrate the model's relevance (Table 3). This ANOVA analysis indicates a mere 0.01% probability that an F-value of this magnitude could arise from random noise. In this scenario, the model terms A, B, C, AC, BC, and ABC are significant. The model demonstrates a strong fit to the data, as indicated by the coefficient of determination (R^2). The R^2 (0.9551), which exceeds 0.09, indicates a strong fit between the experimental and predicted values. This finding is directly attributable to the high concentration of antioxidant-rich phenolic compounds in the WRME, with its highest level yielding the most pronounced antioxidant effect. The relationship between antioxidant activity and the three parameters is represented by the codified linear regressions in the following equation.

$$\text{Antioxidant activity (\%)} = 45.83 - 1.50A + 1.42B + 0.3104C - 0.2213AB - 1.00AC \\ + 0.4512BC + 0.7738ABC \tag{4}$$

Table 3 *ANOVA results for CS-WMRE thin film formulation.*

Source	Sum of Squares	Mean Square	F-value	P-value	
Model	149.42	21.35	66.82	< 0.0001	significant
A-Chitosan	53.79	53.79	168.37	< 0.0001	
B-Glycerol	48.71	48.71	152.46	< 0.0001	
C-WMRE	2.31	2.31	7.24	0.0134	
AB	1.17	1.17	3.68	0.0682	
AC	24.18	24.18	75.69	< 0.0001	
BC	4.89	4.89	15.30	0.0007	
ABC	14.37	14.37	44.98	< 0.0001	
Residual	7.03	0.3195			
Lack of Fit	0.1394	0.1394	0.4249	0.5216	not significant
Pure Error	6.89	0.3280			
Cor Total	156.45				
R^2	0.9551				
Adjusted R^2	0.9170				

Normal plot of each parameter

As shown in Fig. 1, glycerol (B) and WMRE (C) are the most critical parameters for antioxidant activity, with highly significant effects. They also demonstrated important synergistic effects through BC and ABC interactions. In contrast, chitosan (A) and its interactions with the other factors (AB and AC) had little impact on activity. This data suggests that although chitosan serves as a structural backbone, its influence on antioxidant activity is limited compared to WMRE, likely because WMRE's intrinsic phytochemical composition plays a significant role in radical scavenging.

Functional groups and WVP

A broad band was observed at 3254.65–3257.02 cm^{-1} corresponding to the O–H stretching vibration, suggesting the presence of hydrogen bonding. A peak also observed at 2931.19–2933.61 cm^{-1} is associated with C–H stretching, whereas the band at 1405.86–1407.85 cm^{-1} is linked to O–H bending of phenolic groups, thereby confirming the existence of phenolic compounds in the thin film. Phenolic groups enhance the antioxidant activity of the film, while hydrogen bonding interactions are expected to influence the film's matrix integrity [13,14].

W ater vapour permeability (WVP) is a key factor in assessing the moisture-barrier performance of thin films. As shown in Table 4, WVP decreased with increasing amount of WMRE, indicating enhanced resistance to moisture transport. Similar results have been reported in films containing plant-based extracts, where such additives disrupt polymer chains and create longer pathways for water diffusion [15]. The incorporation of WMRE may also enhance hydrogen bonding within the film, limiting molecular movement and further reducing permeability [16]. These data demonstrate WMRE's potential to improve barrier properties while upcycling agricultural waste.

Separation Technology - ICoST 2025
Materials Research Proceedings 59 (2026) 64-71

Materials Research Forum LLC
https://doi.org/10.21741/9781644903957-9

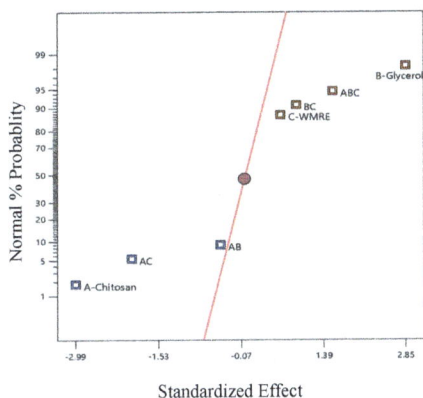

Figure 1 Normal probability plot of antioxidant activity of CS-WMRE thin film formulation.

Table 4 WVP of CS–WMRE thin films at different formulations.

Thin film sample	WVP (g/mm kPa.m^2.h)
Low range (CH:1%, GLY:30%, WMRE: 5%)	1.4003
Mid range (CH:2%, GLY:45%, WMRE: 7.5%)	2.5080
High range (CH:3%, GLY:60%, WMRE: 10%)	2.6502

Notes: CH=chitosan; GLY=glycerol

Conclusion

This study demonstrates the potential of CS–WMRE thin films as sustainable materials for active biopolymer packaging. Incorporating watermelon rind extract (WMRE) enhanced film functionality while promoting waste valorization from agricultural residues. Extraction with 80% methanol using hot plate stirring produced the highest Total Phenolic Content (TPC) (5.68 ± 0.47 mg GAE/g) and antioxidant activity ($16.49 \pm 0.37\%$). The addition of glycerol and chitosan improved structural integrity and barrier properties. Screening via a Two-level factorial design identified significant parameters affecting antioxidant activity, while FTIR confirmed the incorporation of phenolic compounds. The water vapour permeability test indicated enhanced barrier performance with the addition of WMRE. Overall, the results suggest that CS–WMRE thin films exhibit favourable functional and environmental characteristics, underscoring their potential as packaging materials that require moisture-barrier properties.

Acknowledgment

The authors express their gratitude to the Faculty of Chemical Engineering & Technology for the facilities and support provided, and to the Ministry of Higher Education for financing this work through a research grant (FRGS/1/2015/ST04/UNIMAP/03/1).

References

[1] Tabassum, Z., Girdhar, M., Kumar, A., Malik, T., & Mohan, A. (2023). ZnO Nanoparticles-Reinforced Chitosan-Xanthan Gum Blend Novel Film with Enhanced Properties and Degradability for Application in Food Packaging. ACS Omega, 8(34), 31318–31332. https://doi.org/10.1021/acsomega.3c03763

[2] Reshmy, R., Philip, E., Vaisakh, P. H., Raj, S., Paul, S. A., Madhavan, A., Sindhu, R., Binod, P., Sirohi, R., Pugazhendhi, A., & Pandey, A. (2021). Development of an eco-friendly biodegradable plastic from jack fruit peel cellulose with different plasticisers and Boswellia serrata as filler. Science of the Total Environment, 767. https://doi.org/10.1016/j.scitotenv.2020.144285

[3] Wang, F., Xie, C., Ye, R., Tang, H., Jiang, L., & Liu, Y. (2023). Development of active packaging with chitosan, guar gum and watermelon rind extract: Characterisation, application and performance improvement mechanism. International Journal of Biological Macromolecules, 227, 711–725. https://doi.org/10.1016/j.ijbiomac.2022.12.210.

[4] Rachtanapun, P., Klunklin, W., Jantrawut, P., Jantnasakulwong, K., Phimolsiripol, Y., Seesuriyachan, P., Leksawasdi, N., Chaiyaso, T., Ruksiriwanich, W., Phongthai, S., Sommano, S. R., Punyodom, W., Reungsang, A., & Ngo, T. M. P. (2021). Characterisation of chitosan film incorporated with curcumin extract. Polymers, 13(6). https://doi.org/10.3390/polym13060963

[5] Nxumalo, K. A., Fawole, O. A., & Aremu, A. O. (2024). Development of Chitosan-Based Active Films with Medicinal Plant Extracts for Potential Food Packaging Applications. Processes, 12(1). https://doi.org/10.3390/pr12010023

[6] Kahya, N., Kestir, S. M., Öztürk, S., Yolaç, A., Torlak, E., Kalaycıoğlu, Z., Akın-Evingür, G., & Erim, F. B. (2022). Antioxidant and antimicrobial chitosan films enriched with aqueous sage and rosemary extracts as food coating materials: Characterisation of the films and detection of rosmarinic acid release. International Journal of Biological Macromolecules, 217, 470–480. https://doi.org/10.1016/j.ijbiomac.2022.07.073

[7] Abazari, M., Badeleh, S. M., Barkhordari, S., Mohammed, S. S., Ahmed, M. M., Mohammed Ameen, M. S., & Abdollahi, H. (2025). Preparation and characterisation of polyvinyl alcohol thin films: Rheological, physicochemical and functional properties. Heliyon, 11(12). https://doi.org/10.1016/j.heliyon.2025.e43604

[8] Laporan Interim Buah-buahan. (2025). https://open.dosm.gov.my

[9] Zia, S., Khan, M. R., Aadil, R. M., & Medina-Meza, I. G. (2024). Bioactive Recovery from Watermelon Rind Waste Using Ultrasound-Assisted Extraction. ACS Food Science and Technology, 4(3), 687–699. https://doi.org/10.1021/acsfoodscitech.3c00601

[10] Zhang, X., Liu, J., Yong, H., Qin, Y., Liu, J., & Jin, C. (2020). Development of antioxidant and antimicrobial packaging films based on chitosan and mangosteen (Garcinia mangostana L.) rind powder. In International Journal of Biological Macromolecules, 145, 1129–1139.

[11] Susmitha, A., Sasikumar, K., Rajan, D., Padmakumar M, A., & Madhavan Nampoothiri, K. (2021). Development and characterisation of corn starch-gelatin based edible films incorporated with mango and pineapple for active packaging. Food Bioscience, 41, 100977.

[12] Iamareerat, B., Singh, M., & Bilal, M. (2018). Reinforced cassava starch based edible film incorporated with essential oil and sodium bentonite nanoclay as food packaging material. Journal of Food Science and Technology, 55(5), 1953-1959.

[13] Singh, A. K., Kim, J. Y., & Lee, Y. S. (2022). Phenolic Compounds in Active Packaging and Edible Films/Coatings: Natural Bioactive Molecules and Novel Packaging Ingredients. In Molecules (Vol. 27, Issue 21). MDPI. https://doi.org/10.3390/molecules27217513

[14] Yuan, G., Lv, H., Yang, B., Chen, X., & Sun, H. (2015). Physical properties, antioxidant and antimicrobial activity of chitosan films containing carvacrol and pomegranate peel extract. Molecules, 20(6), 11034–11045. https://doi.org/10.3390/molecules200611034

[15] Han, H. S., & Song, K. B. (2020). Antioxidant activities of mandarin (Citrus unshiu) peel pectin films containing sage (Salvia officinalis) leaf extract. International Journal of Food Science & Technology, 55(9), 3173-3181.

[16] Lei, Y., Wu, H., Jiao, C., Jiang, Y., Liu, R., Xiao, D., Lu, J., Zhang, Z., Shen, G., & Li, S. (2019). Food Hydrocolloids Investigation of the structural and physical properties, antioxidant and antimicrobial activity of pectin-konjac glucomannan composite edible films incorporated with tea polyphenol. Food Hydrocolloids, 94, 128–135.

Separation Technology - ICoST 2025
Materials Research Proceedings 59 (2026) 72-79

Materials Research Forum LLC
https://doi.org/10.21741/9781644903957-10

Parametric Evaluation of Subcritical Water Extraction of Oleoresin from *Syzygium Aromaticum* via Factorial Design Approach

Noor Amirah ABDUL HALIM[1,2,a*], Sharifah Zati Hanani SYED ZUBER[2,3,b], Rozaini ABDULLAH[2,3,c], Ain Nur Najwa ABDULLAH[2,d]

[1]Centre of Excellent for Biomass Utilization, Universiti Malaysia Perlis (UniMAP), Kompleks Pusat Pengajian Jejawi 3, 02600, Arau, Perlis, Malaysia

[2]Faculty of Chemical Engineering &Technology, Universiti Malaysia Perlis, 02600 Arau, Perlis, Malaysia

[3]Centre of Excellent for Frontier Material Research, Universiti Malaysia Perlis (UniMAP), Kompleks Pusat Pengajian Jejawi 3, 02600, Arau, Perlis, Malaysia

[a]amirahhalim@unimap.edu.my, [b]sharifahzati@unimap,edu.my , [c]rozainiabdullah@unimap.edu.my

Keywords: Subcritical Water Extraction, *Syzigium Aromaticum*, Oleoresin, Factorial Design

Abstract. The growing demand for natural products underscores the need for efficient and sustainable extraction of bioactives. Oleoresin from *Syzygium Aromaticum* (*S.aromaticum*) is highly valued for its functional and therapeutic properties, with promising applications in the food and nutraceuticals. This study employed subcritical water extraction (SWE) and full factorial design (FFD) to assess four variables; particle size (0.2–15 mm), sample-to-solvent ratio (0.05–0.25 g mL^{-1}), extraction temperature (120–180 °C), and extraction time (10–50 min) on the oleoresin yield. The highest yield (53.9%) was obtained at 0.2 mm, 0.05 g mL^{-1}, 120 °C, and 50 min. Statistical analysis revealed that particle size and sample-to-solvent ratio as the most significant factors, with a notable interaction between them. Main effects and interactions of all factors were evaluated to provide mechanistic insights into the SWE process. The developed regression model showed excellent predictive performance (R^2 = 0.97), supporting the robustness of the factorial design for modelling *S.aromaticum* oleoresin extraction through SWE.

Introduction

Syzygium aromaticum (*S.aromaticum*) or locally known as clove is a widely used culinary and medicinal spice. Its bioactivity is primarily attributed to phenylpropanoids (e.g. eugenol, eugenyl acetate and β-caryophyllene, making clove extracts attractive for food, nutraceutical and pharmaceutical application [1,2]. Oleoresin extracted from *S.aromaticum* is a viscous concentrate that delivers the full clove flavour and exhibits significant antioxidant and antimicrobial activity, largely attributable to eugenol, its major constituent. It contains both volatile essential oils (aroma/flavour) and non-volatile resinous compounds (pigments, pungency, bioactives), which are highly valuable for its flavour strength, stability and formulation versatility [3]. However, due to its hydrophobicity, this oleoresin requires nonpolar organic solvents (e.g., hexane, ethyl acetate) for extraction, which are typically harmful in nature.

Conventional solvent extraction can be efficient, but raises concerns over solvent residues, safety and environmental burden, motivating green extraction strategies that reduce solvent use and energy while maintaining product quality [4]. Subcritical water extraction (SWE) emerges as promising alternative to conventional solvent extraction. Operating at 100–374 °C under pressure, SWE exploits the temperature-driven decrease in water's dielectric constant and viscosity to 'tune' solvent strength and accelerate mass transfer. This enables solubilisation from polar to moderately

Separation Technology - ICoST 2025
Materials Research Proceedings 59 (2026) 72-79

Materials Research Forum LLC
https://doi.org/10.21741/9781644903957-10

non-polar analytes while eliminating the need for harmful organic solvents, thereby reducing solvent handling, cost, and environmental burden and enhancing product recovery [5,6]. These features make SWE promising for *S.aromaticum* oleoresin, where simultaneous recovery of aroma compounds and phenolics is desirable.

However, SWE performance depends strongly on operating factors such as temperature, pressure, extraction time, liquid-to-solid ratio, particle size and potential thermal degradation, necessitates statistically efficient experimentation. Factorial design allows concurrent evaluation of main effects and interactions with minimal runs, offering robust insight for parameter screening and process optimisation compared with one-factor-at-a-time trials. Accordingly, this study applies factorial design to identify critical factors and interactions governing SWE of *S.aromaticum* oleoresin, supporting greener, organic-solvent-free processing for food and health products.

Materials and Methods
Fresh *S.aromaticum* buds were procured from a local supplier. Approximately 1.0 kg of material were washed and rinsed to remove visible particulates, followed by air-dried to remove surface moisture, then oven-dried at 40 °C for 24 h to constant mass to preserve thermo-labile constituents. Dried samples were categorised as whole buds, coarse grind, and fine powder, with average particle sizes of approximately of 15 mm, 1mm, and 0.2 mm, respectively. The prepared material was stored in an airtight container at 4 °C until extraction.

Extraction of oleoresin via subcritical water extraction (SWE)
The SWE unit comprised an insulated, temperature-controlled oil bath. Silicone oil served as the heat-transfer medium. Extractions were performed in a cylindrical 50 mL stainless-steel extractor cell used to load and mix the sample and solvent. Auxiliary components included an ice bath for rapid post-heating cooling of the extraction cell [7]. *S.aromaticum* sample was loaded in the extraction cell and filled with ultrapure water at the specified solid–liquid ratio. A fixed solvent volume of 20 mL was used, with sample mass adjusted accordingly. The silicone-oil bath was preheated to 120–180 °C (above water's the boiling point and below its critical point) which the cell pressure conformed to saturation-steam properties around 2 to 10 bar. Once reached the set temperature, the cell was immersed at predetermined extraction time, then was removed and rapidly quenched in an ice bath. Subsequently, the extract was transferred to centrifuge tubes and centrifuged at 4,000 rpm for 15 min to separate the oleoresin-rich supernatant from solid matrices. The liquid fraction was collected, transferred to a clean bottle and the solvent was removed by evaporation [7]. The oleoresin yield was determined gravimetrically using Eq. 1.

$$\text{Oleoresin Yield}\ \left(\%\frac{w}{w}\right) = \frac{\text{Weight of oleoresin (g)}}{\text{Weigh of } S.aromaticum \text{ (g)}} \times 100 \tag{1}$$

Parametric studies by factorial design
A two-level full factorial design (FFD) was employed to quantify the effects of four process factors tabulated in Table 1 on *S.aromaticum* oleoresin yield under SWE. Low (−1) and high (+1) levels of each factor were selected from preliminary trials to span a practical operating window. Pressure, solvent volume, and all other conditions were held constant to isolate the effects of A–D. All the analyses were generated in Design-Expert® v13 (Stat-Ease, Minneapolis, MN, USA).

Separation Technology - ICoST 2025
Materials Research Proceedings 59 (2026) 72-79

Materials Research Forum LLC
https://doi.org/10.21741/9781644903957-10

Table 1 *List of SWE factors and the corresponding treatment levels.*

Notation	Factor	Unit	Low Level (-1)	Centre Points (0)	High Level (+1)
A	Average Particle Size	mm	0.2	1.0	15.0
B	Sample-to-Solvent Ratio	g/mL	0.05	0.15	0.25
C	Extraction Temperature	°C	120	150	180
D	Extraction Time	min	10	30	50

Result and Discussion

A randomised 19-run factorial including three replicated centre points tabulated in Table 2 were executed as a single block to assess how the studied factors (A-D) can affect the experimental response, Y (oleoresin yield).

Table 2 *Design matrix for 2^4 FFD and the experimental responses (oleoresin yield).*

	Factor 1	Factor 2	Factor 3	Factor 4	Response
Std	A: Average Particle Size	B: Sample-to-Solvent Ratio	C: Extraction Temperature	D: Extraction Time	Y: Oleoresin Yield
	mm	g/mL	°C	min	%
1	0.2	0.05	120	10	44.19
2	15	0.05	120	10	4.40
3	0.2	0.25	120	10	22.99
4	15	0.25	120	10	0.92
5	0.2	0.05	180	10	46.40
6	15	0.05	180	10	4.40
7	0.2	0.25	180	10	20.26
8	15	0.25	180	10	2.28
9	0.2	0.05	120	50	53.90
10	15	0.05	120	50	14.00
11	0.2	0.25	120	50	29.46
12	15	0.25	120	50	5.00
13	0.2	0.05	180	50	39.90
14	15	0.05	180	50	27.00
15	0.2	0.25	180	50	16.06
16	15	0.25	180	50	17.16
17	1.0	0.15	150	30	37.96
18	1.0	0.15	150	30	40.83
19	1.0	0.15	150	30	39.40

Oleoresin yield (%Y) varied widely across the factorial runs ranging from 0.92% to 53.9%. The maximum yield (53.9%) occurred at 120 °C, 50 min, 0.2 mm particle size and a sample-to-solvent ratio of 0.05 g mL^{-1}. Conversely, the minimum yield (0.92%) was obtained at 120 °C, 10 min, 15 mm particles size and a ratio of 0.25 g mL^{-1}.

Evaluation of main effects and interaction effects

The Pareto chart ranks standardised effect magnitudes (main factors and interactions) from largest to smallest. Effects exceeding the Bonferroni limit are unequivocally significant, those between

Separation Technology - ICoST 2025   Materials Research Forum LLC
Materials Research Proceedings 59 (2026) 72-79   https://doi.org/10.21741/9781644903957-10

the Bonferroni limit and the t-value line are significant at $\alpha = 0.05$ without Bonferroni adjustment and those below both cut-offs are treated as non-significant (noise). From Fig. 1, the effects on oleoresin yield, in descending importance is A > B > AB > D > AC > ACD > AD. CD and C are non-significant effects but were retained under the model-hierarchy principle as components of significant higher-order interactions (e.g., ACD).

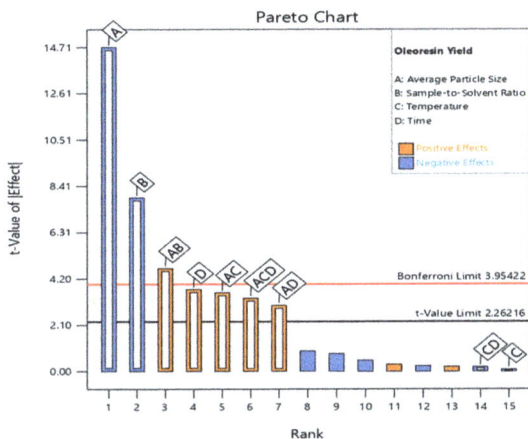

Figure 1 Pareto chart of effects.

The chart shows significant negative main effects for A and B: larger particles and higher sample loads each reduce oleoresin yield. Factor A (particle size) was highly significant, mainly due to wide size variation tested (whole buds, coarse grind, fine powder). For *S.aromaticum*, smaller particles generally increase yield by reducing intraparticle diffusion resistance [1]. The next most significant main factor, B, reveals that a high sample-to-solvent ratio (less solvent) lowers yield by reducing the concentration driving force and promoting early solvent saturation/equilibrium [8].The AB interaction is negatively significant, implying that, combining large particles with limited solvent, depresses yield more than the sum of individual effects. This synergistic is consistent with compounded mass-transfer limitations (reduced surface area, longer diffusion paths) [1,8], together with early solvent saturation and poorer wetting under low solvent availability [8].

The statistical significance was evaluated at $\alpha = 0.05$; terms with $p > 0.05$ were treated as non-significant. The ANOVA in Table 3 indicates that the factorial model is highly significant (F = 38.59 p < 0.0001) and fits well ($R^2 = 0.97$; adj-$R^2 = 0.95$; pred-$R^2 = 0.92$). The small adj-R^2–pred-R^2 difference (< 0.2) and Adeq. Precision of 20.16 (> 4) indicate a strong signal. Besides, lack-of-fit is not significant (F = 8.76, p = 0.1062), supporting the adequacy of linear-by-interaction model.

Effect estimates align with the Pareto ranking; the largest negative main effects are the average particle size (A) and sample-to-solvent ratio (B), with a strong negative AB interaction. Extraction time (D) is positive and significant, as are AC, AD, and ACD, indicating that higher temperature and longer time are most beneficial when particles are smaller.

Separation Technology - ICoST 2025

Materials Research Proceedings 59 (2026) 72-79

Materials Research Forum LLC

https://doi.org/10.21741/9781644903957-10

Table 3 *Analysis of variance (ANOVA).*

Source	Sum of Squares	Mean Square	F-value	p-value	
Model	5032.68	559.19	38.59	< 0.0001	significant
A-Average Particle Size	3137.72	3137.72	216.52	< 0.0001	
B-Sample-to-Solvent Ratio	900.90	900.90	62.17	< 0.0001	
C-Temperature	0.1225	0.1225	0.0085	0.9288	
D-Time	200.51	200.51	13.84	0.0048	
AB	316.66	316.66	21.85	0.0012	
AC	185.23	185.23	12.78	0.0060	
AD	130.42	130.42	9.00	0.0150	
ACD	160.53	160.53	11.08	0.0088	
Lack of Fit	126.31	18.04	8.76	0.1062	not significant

Std. Dev.	3.81	R^2		0.97
Mean	24.55	Adjusted R^2		0.95
C.V. %	15.50	Predicted R^2		0.92
		Adequate Precision		20.16

Two-factor interaction (2FI) plots

Fig. 2(a–c) shows two-factor interaction (2FI) plots (AB, AC, AD) with significant effects ($P < 0.05$). Non-parallel and non-crossing lines indicate an ordinal interaction, crossing lines denote a crossover (qualitative) interaction and parallel lines imply no interaction. The slope of a line shows how strongly the response changes with that factor at a fixed level of the other factor, and the vertical gap between lines represents the size of the other factor's main effect.

Fig. 2(a) shows non-parallel lines for the two sample-to-solvent (B) levels (0.05 and 0.25) g/mL at fixed temperature (C, 150°C) and time (D, 30 mins) indicating a significant AB interaction and implying that the effect of particle size (A) depends on the sample-to-solvent ratio. Across B levels, larger particle size yields less oleoresin as comminution increases external interfacial area and shortens diffusion paths, reducing internal mass-transfer resistance and accelerating solute release [8]. Yield is higher at low B (0.05 g mL^{-1}; more solvent), and lower at high B (0.25 g mL^{-1}; less solvent). More solvent sustains the concentration driving force, improves wetting, and delays solvent saturation. This trend is widely reported in solid–liquid extraction up to practical and economic limits [9,10]. The slope for A is steeply negative at low B but only mildly negative at high B, showing particle size matters most when sufficient solvent is available. Mechanistically, more solvent removes external-phase limitations, so the benefit of smaller particles (greater area, shorter diffusion length) is fully expressed. When solvent is limited, early saturation and increased slurry viscosity suppress mass transfer, so further size reduction has a smaller incremental effect [11].

Fig. 2(b) shows two nearly crossing lines for the two temperature levels (C = 120 and 180) °C at fixed B (0.15 g mL^{-1}) and D (30 min). The non-parallel trend indicates an AC interaction, implying that the effect of particle size depends on temperature. At both temperatures, yield decreases as particles size increase, with the highest yield from fine particles (~0.2 mm). The temperature effect is modest and size-dependent. This aligns with solid–liquid extraction fundamentals; size reduction increases interfacial area and shortens diffusion paths, lowering internal mass-transfer resistance [12]. In this study, temperature effects are small and context-dependent. Generally, raising the temperature in SWE lowers water's viscosity and dielectric constant (reducing polarity), which will improve the diffusivity (mass transfer) and increase the

Separation Technology - ICoST 2025
Materials Research Forum LLC
Materials Research Proceedings 59 (2026) 72-79
https://doi.org/10.21741/9781644903957-10

solubility of nonpolar compounds such as oleoresins, respectively [13]. However, at the higher setting (180 °C), partial degradation of thermolabile phenolics may offset gains [14].

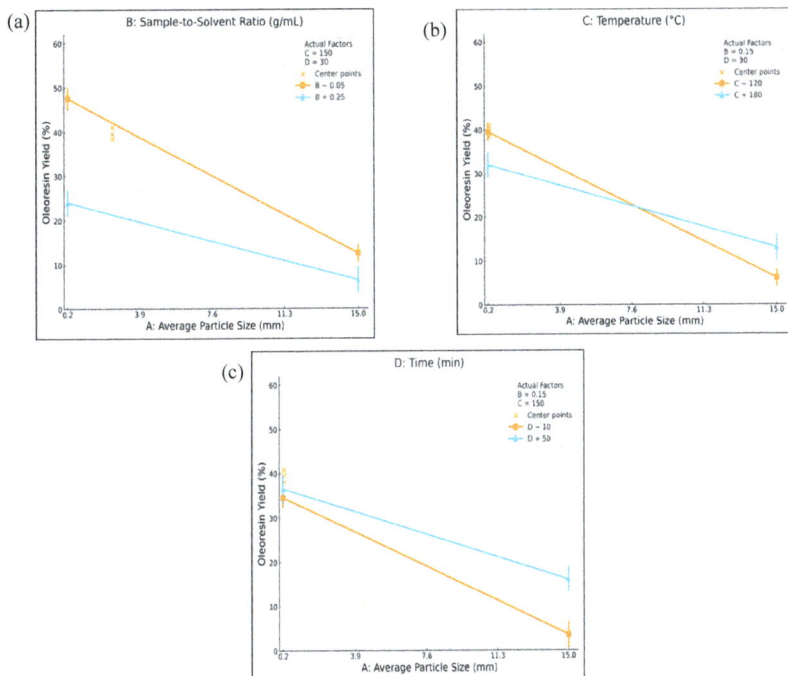

Figure 2 *Interaction plots of (a) AB, (b) AC and (c) AD.*

At small particle size, the 120 °C line lies slightly above 180 °C, suggesting that when diffusion paths are reduced, pushing temperature higher offers limited benefit and may begin to incur thermal or hydrolytic losses. At large particle size, the 180 °C point sits above 120 °C, consistent with temperature partly compensating stronger internal-diffusion limits of coarse particles barrier by improving solvent transport properties. Hence, the non-parallel (crossover) lines reflect two regimes; - (i) diffusion-controlled for coarse particles where higher temperature aid, and (ii) near-equilibrium/chemistry-limited for fine particles where extra heating yields diminishing returns [13].

Figure 2(c) shows two non-parallel lines for the time (D = 10 vs 50 min) at fixed B (0.15 g/mL) and C (150 °C.) Oleoresin yields decrease as particles size (A) increase, and the gap between lines widens at large A. With B and C held constant, extending time from 10 to 50 min yield only a modest gain because extraction is rapid initially but then slows as equilibrium approached and intraparticle diffusion becomes rate-limiting [15]. The time effect is minor for small particles but pronounced for large particles. With small A, short diffusion paths allow most solute to be removed within 10 min, so extending to 50 min has little impact. With large A, slower internal diffusion means extra time partially compensates, increasing yield and producing the diverging pattern. This

Separation Technology - ICoST 2025   Materials Research Forum LLC
Materials Research Proceedings 59 (2026) 72-79   https://doi.org/10.21741/9781644903957-10

'fast initial stage and slower diffusion-controlled stage' behaviour is well documented in plant-matrix extraction kinetics, explaining the observed AD interaction [15,16].

Conclusion

SWE is an efficient, organic-solvent-free route to recover *S.aromaticum* oleoresin, delivering yields exceeding 50% within the studied range. FFD showed that particle size (A) and sample-to-solvent ratio (B) dominate the response, with a strong negative AB interaction that makes the coarse-particle/high-solids region least favourable. Extraction time (D) increases yield by overcoming intraparticle diffusion limits, whereas temperature (C) has a modest, context-dependent effect across 120–180 °C. Practically, fine particles, ample solvent and sufficient time are recommended, with temperature tuned to balance mass-transfer gains against potential compositional changes. Model fit and diagnostics were significant ($R^2 = 0.97$) with non-significant lack of fit. Future work should incorporate multi-response objectives (e.g. phenolic profile, eugenol retention, antioxidant capacity), optimisation, as well as assessing scale-up variables to guide industrial implementation.

Acknowledgement

This research was supported by Ministry of Higher Education (MOHE) through Fundamental Research Grant Scheme (FRGS/1/2023/TK05/UNIMAP/02/10) and Universiti Malaysia Perlis (UniMAP) through research vote (29000).

References

[1] J.N. Haro-González, G.A. Castillo-Herrera, M. Martínez-Velázquez, H. Espinosa-Andrews, Clove Essential Oil (*Syzygium aromaticum L. Myrtaceae*): Extraction, Chemical Composition, Food Applications, and Essential Bioactivity for Human Health, Molecules 26 (2021) 6387. https://doi.org/10.3390/molecules26216387

[2] D.F. Cortés-Rojas, C.R.F. De Souza, W.P. Oliveira, Clove (*Syzygium aromaticum*): a precious spice, Asian Pacific Journal of Tropical Biomedicine 4 (2014) 90–96. https://doi.org/10.1016/S2221-1691(14)60215-X

[3] F.R. Procopio, M.C. Ferraz, B.N. Paulino, P.J. Do Amaral Sobral, M.D. Hubinger, Spice oleoresins as value-added ingredient for food industry: Recent advances and perspectives., Trends in Food Science & Technology 122 (2022) 123–139. https://doi.org/10.1016/j.tifs.2022.02.010

[4] F. Chemat, M. Abert Vian, H.K. Ravi, B. Khadhraoui, S. Hilali, S. Perino, A.-S. Fabiano Tixier, Review of Alternative Solvents for Green Extraction of Food and Natural Products: Panorama, Principles, Applications and Prospects, Molecules 24 (2019) 3007. https://doi.org/10.3390/molecules24163007

[5] Y. Cheng, F. Xue, S. Yu, S. Du, Y. Yang, Subcritical Water Extraction of Natural Products, Molecules 26 (2021) 4004. https://doi.org/10.3390/molecules26134004

[6] B. Díaz-Reinoso, S. Rivas, J. Rivas, H. Domínguez, Subcritical water extraction of essential oils and plant oils, Sustainable Chemistry and Pharmacy 36 (2023) 101332. https://doi.org/10.1016/j.scp.2023.101332

[7] N.A.A. Halim, Z.Z. Abidin, S.I. Siajam, C.G. Hean, M.R. Harun, Optimization studies and compositional analysis of subcritical water extraction of essential oil from *Citrus hystrix* DC. leaves, The Journal of Supercritical Fluids 178 (2021) 105384. https://doi.org/10.1016/j.supflu.2021.105384

[8] Q.-W. Zhang, L.-G. Lin, W.-C. Ye, Techniques for extraction and isolation of natural products: a comprehensive review, Chin Med 13 (2018) 20. https://doi.org/10.1186/s13020-018-0177-x

[9] O.R. Alara, N.H. Abdurahman, C.I. Ukaegbu, Extraction of phenolic compounds: A review, Current Research in Food Science 4 (2021) 200–214. https://doi.org/10.1016/j.crfs.2021.03.011

[10] N. Jiménez-Moreno, F. Volpe, J.A. Moler, I. Esparza, C. Ancín-Azpilicueta, Impact of Extraction Conditions on the Phenolic Composition and Antioxidant Capacity of Grape Stem Extracts, Antioxidants 8 (2019) 597. https://doi.org/10.3390/antiox8120597

[11] A.S. Da Silva, R.P. Espinheira, R.S.S. Teixeira, M.F. De Souza, V. Ferreira-Leitão, E.P.S. Bon, Constraints and advances in high-solids enzymatic hydrolysis of lignocellulosic biomass: a critical review, Biotechnol Biofuels 13 (2020) 58. https://doi.org/10.1186/s13068-020-01697-w

[12] N.R. Putra, D.N. Rizkiyah, A.H.A. Aziz, H. Mamat, W.M.S.W. Jusoh, Z. Idham, M.A.C. Yunus, I. Irianto, Influence of particle size in supercritical carbon dioxide extraction of roselle (*Hibiscus sabdariffa*) on bioactive compound recovery, extraction rate, diffusivity, and solubility, Sci Rep 13 (2023) 10871. https://doi.org/10.1038/s41598-023-32181-8

[13] L. Barp, A.M. Višnjevec, S. Moret, Pressurized Liquid Extraction: A Powerful Tool to Implement Extraction and Purification of Food Contaminants, Foods 12 (2023) 2017. https://doi.org/10.3390/foods12102017

[14] X. Yu, P. Zhu, Q. Zhong, M. Li, H. Ma, Subcritical water extraction of antioxidant phenolic compounds from XiLan olive fruit dreg, J Food Sci Technol 52 (2015) 5012–5020. https://doi.org/10.1007/s13197-014-1551-z

[15] I. Psarrou, A. Oreopoulou, D. Tsimogiannis, V. Oreopoulou, Extraction Kinetics of Phenolic Antioxidants from the Hydro Distillation Residues of Rosemary and Effect of Pretreatment and Extraction Parameters, Molecules 25 (2020) 4520. https://doi.org/10.3390/molecules25194520

[16] H. Xu, Q. Fei, S. Manickam, D. Li, H. Xiao, Y. Han, P.L. Show, G. Zhang, Y. Tao, Mechanistic study of the solid-liquid extraction of phenolics from walnut pellicle fibers enhanced by ultrasound, microwave and mechanical agitation forces, Chemosphere 309 (2022) 136451. https://doi.org/10.1016/j.chemosphere.2022.136451

Separation Technology - ICoST 2025
Materials Research Proceedings 59 (2026) 80-86

Materials Research Forum LLC
https://doi.org/10.21741/9781644903957-11

Kinetic Modelling of Supercritical CO_2 Extraction from *Swietenia macrophylla* Seeds Extracts Using the Single Sphere Model

Mohammad Lokman HILMI[1,a], Liza Md SALLEH[2,b*], Hasmida Mohd NASIR[2,c], Noor Azwani Mohd RASIDEK[2,d]

[1]Faculty of Chemical and Energy Engineering, Universiti Teknologi Malaysia, 81310 Johor Bahru, Johor, Malaysia

[2]Centre of Lipids Engineering & Applied Research (CLEAR), Ibnu Sina Institute for Scientific and Industrial Research, Universiti Teknologi Malaysia, 81310 Johor Bahru, Johor, Malaysia

[a]m.lokmanhilmi@gmail.com, [b]r-liza@utm.my, [c]hasmidanasir@gmail.com, [d]noor.azwani@utm.my

Keywords: Supercritical Carbon Dioxide Extraction, Swietenia Macrophylla, Single Sphere Model, Diffusion, Mass Transfer

Abstract. *Swietenia macrophylla*, commonly known as "Tunjuk Langit," has been reported to exhibit antidiabetic properties, attributed to its high phytosterol content, with β-sitosterol comprising approximately 66% of the total sterols in the extract. In this study, supercritical carbon dioxide (SC-CO_2) was used to extract oil from *S.macrophylla*, and the mass transfer behaviour was modelled using the single sphere model (SSM). Key parameters namely the diffusion coefficient (D_e) and external mass transfer coefficient (k_f) were estimated based on experimental data. The D_e values ranged from 1.0104×10^{-13} to 17.912×10^{-13} m²/s, while k_f varied between 0.7115×10^{-6} and 3.9919×10^{-6} m/s. Biot numbers consistently exceeded 500 across all parameters, confirming that internal diffusion governs the SC-CO_2 extraction process of *S.macrophylla*. The results highlight the importance of considering both internal diffusion and external film resistance when modelling extraction kinetics to achieve accurate process predictions.

Introduction

Plants are a rich source of diverse bioactive compounds such as phytochemicals, flavors, perfumes, and pigments that play a crucial role in the production of nutraceuticals. In fact, the global market for plant extracts is projected to expand at a compound annual growth rate (CAGR) of 11.2% from 2024 to 2031 [1]. A clear example is *S.macrophylla*, commonly known as tropical mahogany, is a versatile species of the Meliaceae family that thrives in well-drained soils while tolerating a broad range of environmental conditions; native to Southeast Asia and now widely cultivated across tropical regions including Africa and the Pacific Islands [2]. This species is highly valued not only for its premium timber but also for its significant medicinal properties. Notably, the presence of β-sitosterol, a phytosterol acclaimed for its cholesterol-lowering [3], anti-inflammatory [4], and antioxidant effects [5], further underscores the *S.macrophylla* medicinal potential. These multifaceted benefits not only validate the longstanding traditional use of *S. macrophylla* in diverse therapeutic practices but also underscore their potential as a natural resource for developing innovative, health-promoting interventions.

Supercritical carbon dioxide (SC-CO_2) extraction has been widely explored for its effectiveness in isolating bioactive compounds from natural plant matrices [6,7]. As a non-toxic, non-flammable, and chemically inert solvent, CO_2 provides a clean extraction medium that can be easily removed from the final product through depressurization, minimizing residual solvent contamination [8]. Its classification as generally recognized as safe (GRAS) by both the European Food Safety Authority (EFSA) and the U.S. Food and Drug Administration (FDA) further underscores its suitability for applications in food, pharmaceutical, and nutraceutical industries

Separation Technology - ICoST 2025   Materials Research Forum LLC
Materials Research Proceedings 59 (2026) 80-86   https://doi.org/10.21741/9781644903957-11

[9,10]. One of the key advantages of SC-CO$_2$ extraction lies in its relatively mild critical pressure (7.38 MPa) and temperature (31.1 °C), which create a thermally stable environment conducive to preserving heat-sensitive bioactive compounds. Furthermore, the mass transfer properties of supercritical fluids are significantly more favorable than those of conventional liquid solvents, as solute diffusivities in SC-CO$_2$ are typically an order of magnitude higher, and its viscosity is considerably lower [11]. As a result, SC-CO$_2$ extraction not only enhances solute transport within the plant matrix but also accelerates the overall process, often achieving complete extraction within a fraction of the time required for traditional liquid solvent methods. These advantages collectively position SC-CO$_2$ extraction as a highly efficient, sustainable, and scalable technique for obtaining high-purity bioactive compounds from plant materials. Moreover, selecting a model that accurately reflects the actual solute distribution is essential for reliably determining key parameters, such as diffusion and mass transfer coefficients. Among the various approaches, the single sphere model (SSM) has emerged as the preferred method due to its simplicity and effectiveness in representing the distribution of extractable oil within the particle structure, whether on the surface, within pores, or uniformly dispersed in cells. The aim of this work is to determine diffusion and mass transfer coefficient of the extraction process of the galls by fitting the SSM on the experimental data.

Single Sphere Model Fitting

The single sphere model (SSM), introduced by Crank in 1975, examines the relationship between particle size and the diffusion coefficient. In this model, a high diffusion coefficient signifies an efficient extraction process. The model is developed based on the following assumptions [12,13]:

(a) The solid matrix is represented as a homogeneous sphere, with particles of uniform size and an even distribution of extractable material at the beginning of extraction.
(b) The mass transfer resistance of the fluid is zero.
(c) The extraction stages are uniform across all particles in the bed.
(d) The solute to be extracted moves through the matrix in a manner similar to diffusion.
(e) The extraction process is primarily governed by internal diffusion control, with intraparticle diffusion being the key factor.

The rate of solute that transport radially across an internal surface within spherical and isotropic particles takes the form of Fick's law,

$$J = -4\pi r^2 De\left(\frac{\partial C}{\partial r}\right) \tag{1}$$

where,
J = mass flux (kg/m^2 s)
C = solute concentration inside the particle at radius r (kg of solute/m^3 of particle)
D_e = diffusivity or coefficient of diffusion (m^2/s)
The following initial and boundary conditions are applied based on the stated assumptions:

$$C = C_0, \qquad 0 < r < R, t = 0 \tag{2}$$

$$\frac{\partial C}{\partial r} = 0, r = 0 \quad all\ t \tag{3}$$

$$q = 0, r = R \quad all\ t \tag{4}$$

If the solute concentration in the bulk solvent phase is low and the solute film resistance is negligible, then the equation derived by [14] to solve the diffusion coefficient is,

$$Y' = \frac{M_t}{M_\infty} = 1 - \frac{6}{\pi^2}\sum_{n-1}^{\infty}\frac{1}{n^2}exp\frac{D_e t n^2 \pi^2}{R^2} \tag{5}$$

where,
M_t = total amount of solute diffused at time t (g)
M_∞ = total amount of solute (g)
n = number of iterations

Separation Technology - ICoST 2025   Materials Research Forum LLC
Materials Research Proceedings 59 (2026) 80-86   https://doi.org/10.21741/9781644903957-11

R = radius of sphere (m)
t = time (s)

A precise diffusion equation may not always provide an accurate representation of the extraction phenomena occurring in natural materials, as these processes are often complex and not well-defined. Consequently, a linear driving force approach proves to be a more practical and effective simplification method for describing the extraction of natural materials. In this case, it is assumed that both internal and external mass transfer processes can be reasonably modeled by the linear driving force approximation, which is derived based on a parabolic concentration profile within the particle [15]. For spherical particles, the overall mass transfer coefficient, k_p is defined in Eq. 7, and is determined using the Biot (Bi) number.

$$B_i = \frac{k_f r}{D_e} \tag{6}$$

$$k_p = \frac{k_f}{1 + \frac{B_i}{5}} \tag{7}$$

The external mass transfer coefficient (k_f) can be determined through the Sherwood number (Sh).

$$Sh = \frac{k_f d}{D_e} \tag{8}$$

Typically, the Sherwood (Sh) number is expressed as a function of the Reynolds number, $Re = \frac{\rho v d}{\mu}$ and the Schmidt (Sc) number, where $Sc = \frac{\mu}{\rho D_e}$. The correlation applicable over the range of Re from 3 to 3000 and Sc from 0.5 to 10,000 is provided by [16].

$$Sh = 2 + 1.1 Re^{0.6} Sc^{0.33} \tag{9}$$

For conditions involving supercritical velocities ranging from 4.4×10^{-3} to 3.1×10^{-2} cm/s , and Re values from 2 to 40 and Sc from 2 to 40, the correlation is given by [6].

$$Sh = 0.38 Re^{0.83} Sc^{0.33} \tag{10}$$

To compute the extraction yield, various parameters, including physical and mass transfer properties, must be determined based on the relevant equations. The physical properties of the seeds, such as particle density (ρ_s) and bulk density (ρ_b), were measured experimentally. The porosity of the ground-dried raw material was then calculated using the relation, $\varepsilon = 1 - \left(\frac{\rho_b}{\rho_s}\right)$.

Results and Discussion

The accuracy of the SSM fitting was assessed using the average absolute relative deviation (AARD%). Based on Table 1, all parameter showed AARD values ranged from approximately 4.7% to 14.19%, which is within the acceptable range for extraction modelling reported by [17]. Highest AARD (14.19%) was recorded for the run at 50 °C, 20 MPa, and 0.5 mm particle size, indicating that under these specific conditions, the model assumptions did not fully capture the actual extraction dynamics. On the other hand, the lowest AARD value of 4.72% was observed at 60 °C, 25 MPa, and 0.5 mm. This contrasts with a study on extraction of sinensetin from *Orthosiphon stamineus* leaves, which reported good fits for small particles even at low pressures [10].

An increase in temperature typically led to higher D_e values as seen in Table 1, supporting better internal mass transfer. For instance, at 20 MPa and 1.5 mm, D_e increased from 9.04×10^{-13} m²/s at 40 °C to 11.36×10^{-13} m²/s at 60 °C, matching the known effect of temperature reducing CO_2 viscosity and enhancing diffusivity [18,19]. However, this trend is not uniform and it can be observed at 0.5 mm and 20 MPa, D_e was relatively low (1.21×10^{-13} m²/s) at 60 °C, even though it reached 13.37×10^{-13} m²/s at 50 °C. In general, a higher diffusivity coefficient reflects more high mass transfer between SC-CO_2 and solutes. When D_e increases, it helps solutes migrate more quickly from the solid matrix into the solvent, accelerating extraction. Conversely, lower D_e values suggest that the solvent struggles to penetrate and diffuse through the particle pores, pointing to stronger resistance within the solid phase itself [20]. This suggests that despite temperature benefits, matrix structure or compactness of raw material can limit diffusion at small particle sizes, consistent with findings [21].

The data also reveal how pressure effect mass transfer. Theoretically, higher pressure should increase CO_2 density and raise D_e, but the effect depends on particle size and temperature. For instance, at 25 MPa, 40 °C and 1.5 mm, D_e reached 17.91×10^{-13} m²/s, the highest measured yet at 30 MPa, D_e dropped slightly to 15.35×10^{-13} m²/s. For smaller particles, pressure had a clearer effect whereby at (20 MPa, 40 °C, 1 mm), D_e increased from 2.86×10^{-13} m²/s to 4.67×10^{-13} m²/s at 30 MPa. This shows that increasing pressure often improves diffusion, but compaction or reduced porosity due to compress from high pressure may slightly limit D_e [22].

The extraction kinetics of *S.macrophylla* seeds using SC-CO_2 reveal how temperature, pressure, and particle size interact to influence both internal and external mass transfer. Based on the data, the effective diffusivity (D_e) ranged broadly, from 1.01×10^{-13} to 17.91×10^{-13} m²/s, while the external mass transfer coefficient (k_f) varied between 0.71×10^{-6} and 3.99×10^{-6} m/s. Bi numbers mostly exceeded 500, indicating internal diffusion dominates overall control, yet variations in k_f and D_e reveal significant external film resistance under certain conditions.

Table 1 Calculated mass transfer parameters using single sphere model.

P (MPa)	T (°C)	d (mm)	$D_e \times 10^{-13}$ (m²s⁻¹)	Re	$Sc \times 10^6$	$Sh \times 10^3$	$K_f \times 10^{-6}$ (ms⁻¹)	$K_p \times 10^{-8}$ (ms⁻¹)	$Bi \times 10^3$	AARD (%)
20	40	0.5	1.4281	92.4	5.4410	2.9277	0.8362	0.2846	1.4639	6.08
20	40	1.5	9.0409	216.3	0.8595	2.6360	1.5888	0.6004	1.3180	6.80
20	40	1	2.8614	157.62	2.7157	3.1979	0.9150	0.2852	1.5989	7.43
20	50	1	6.4295	175.02	1.0885	2.5111	1.6145	0.6404	1.2555	5.87
20	50	0.5	13.374	102.67	0.5233	1.4293	3.8233	2.6563	0.7147	14.19
20	50	1.5	10.978	240.26	0.6375	2.5408	1.8596	0.7290	1.2704	6.40
20	60	1.5	11.360	259.31	0.5708	2.5636	1.9416	0.7544	1.2818	6.07
20	60	0.5	1.2073	110.81	5.3706	3.2492	0.7845	0.2407	1.6246	5.84
20	60	1	2.2410	188.89	2.8934	3.6406	0.8158	0.2234	1.8203	7.22
25	40	1	6.9537	164.49	1.0708	2.4063	1.6732	0.6924	1.2031	5.23
25	40	0.5	14.626	96.50	0.5091	1.3647	3.9919	2.9039	0.6823	5.23
25	40	1.5	17.912	225.81	0.4157	2.1231	2.5354	1.1885	1.0616	5.16
25	50	1	4.9954	185.72	1.3202	2.7748	1.3861	0.4977	1.3874	5.59
25	50	0.5	1.8070	108.95	3.6496	2.8279	1.0220	0.3601	1.4140	5.48
25	50	1.5	16.390	254.95	0.4024	2.2588	2.4681	1.0878	1.1294	5.39
25	60	1	2.2410	204.39	2.6740	3.7179	0.8332	0.2235	1.8589	5.50
25	60	0.5	1.0103	119.91	5.9310	3.5210	0.7115	0.2015	1.7605	4.72
25	60	1.5	11.338	280.59	0.5285	2.6198	1.9802	0.7530	1.3099	5.98
30	40	1.5	15.353	233.94	0.4681	2.2562	2.3092	1.0190	1.1281	5.88
30	40	0.5	1.8261	99.977	3.9358	2.7542	1.0059	0.3639	1.3771	5.31
30	40	1	4.6742	170.41	1.5377	2.7727	1.2960	0.4657	1.3863	5.59

30	50	1	6.9122	193.94	0.9137	2.5193	1.7413	0.6884	1.2596	5.81
30	50	0.5	1.8728	113.78	3.3721	2.8269	1.0588	0.3732	1.4135	5.64
30	50	1.5	15.893	266.25	0.3974	2.3086	2.4460	1.0549	1.1543	5.81
30	60	0.5	1.4281	92.473	5.4410	2.9277	0.8362	0.3494	1.4639	5.88
30	60	1.5	9.0409	216.38	0.8595	2.6360	1.5888	0.9384	1.3180	5.93
30	60	1	2.8614	157.62	2.7157	3.1979	0.9150	0.6199	1.5989	5.90

Fig. 1 at 50 °C, discrepancies become more pronounced, especially at lower pressure and small particle size, leading to the highest AARD. Steeper initial slopes in the curve for 20 MPa, 50 °C and 0.5 mm reflect faster solute release and improved mass transfer, supported by D_e values reaching up to 13.37×10^{-13} m²/s however is only evident in the model data and does not appear clearly in the experimental curve. Fig. 2 shows pressure effect, where the model closely follows experimental data at the beginning and for smaller particle sizes. However, noticeable differences appear over longer extraction times, reflecting slower internal diffusion [23,24].

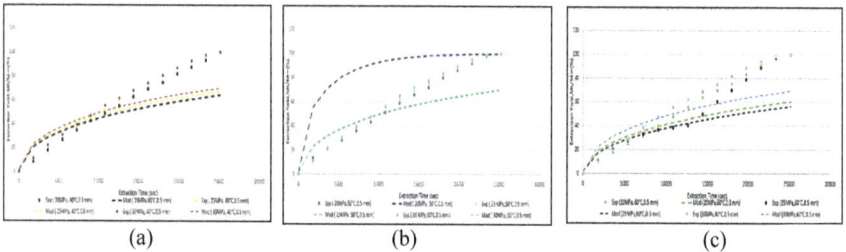

Figure 1 Single sphere model fitted the mass transfer of extraction yield at varying temperatures: (a) 40 °C, 0.5 mm, (b) 50 °C, 0.5 mm and (c) 60 °C, 0.5 mm.

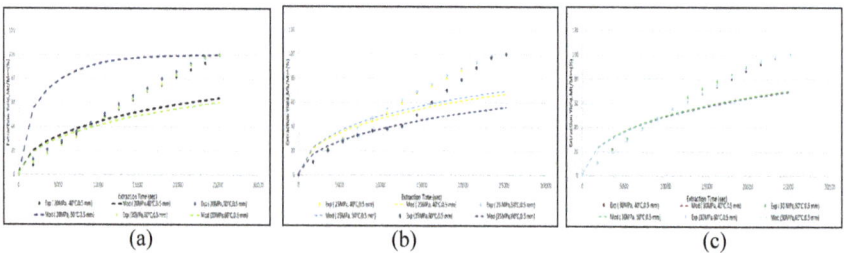

Figure 2 Single sphere model fitted the mass transfer of extraction yield at varying pressures: (a) 20 MPa, 0.5 mm, (b) 25 MPa, 0.5 mm and (c) 30 MPa, 0.5 mm.

Similar trends were reported in other studies under comparable low-pressure conditions, indicating that external film resistance can influence model predictions even when internal diffusion appears dominant [10]. At higher temperatures (60 °C), the model and experimental data exhibit closer agreement across all pressure and particle size conditions, with AARD values ranging from 4.71% to 7.20%. Asides that, this pattern supports earlier findings that higher temperature improves internal diffusion by reducing CO_2 viscosity [18,19,25], consistent with reported D_e values under high temperature and moderate pressure

Separation Technology - ICoST 2025
Materials Research Proceedings 59 (2026) 80-86

Materials Research Forum LLC
https://doi.org/10.21741/9781644903957-11

Conclusion

This study showed that the SSM effectively described the SC-CO_2 extraction of *S.macrophylla* seeds, with Biot numbers above 500 indicating internal diffusion as the dominant controlling mechanism. The D_e ranged from 1.0104×10^{-13} to 17.912×10^{-13} m²/s, while k_f varied between approximately 0.711×10^{-6} and 3.991×10^{-6} m/s. Across the tested conditions, the AARD remained mostly below 15%, confirming good agreement between model and experimental data. Both D_e and k_f generally increased with higher temperature and pressure, supporting higher k_p. However, at high pressure, D_e sometimes decreased slightly due to pore compaction effects, highlighting that optimal process conditions require balancing internal diffusion and external transfer. Overall, the results confirm that while internal diffusion mainly governs extraction, external film resistance remains significant under certain conditions, and jointly optimizing these factors is essential for improved efficiency.

Acknowledgment

The authors acknowledge Ministry of Higher Education, Malaysia and Universiti Teknologi Malaysia for funding this project by Fundamental Research Grant Scheme (R.J130000.7846.5F615), and Centre of Lipid Engineering and Applied Research (CLEAR) for the use of equipment.

References

[1] P. Poddar, "Plant Extract Market Analysis," *Coherent Market Insights*, 2024.

[2] K. H., K. M.H., and K. M., "Swietenia macrophylla King: Ecology, silviculture and productivity," *Swietenia macrophylla King Ecol. Silvic. Product.*, 2011. https://doi.org/10.17528/cifor/003395

[3] N. S. Md Norodin *et al.*, "Inhibitory effects of swietenia mahagoni seeds extract on A-glucosidase and A-amylase," *Int. J. Eng. Trans. B Appl.*, vol. 31, no. 8, pp. 1308–1317, 2018. https://doi.org/10.5829/ije.2018.31.08b.20

[4] A. Mishra, S. Das, and S. Kumari, "Potential role of herbal plants and β sitosterol as a bioactive constituent in circumventing Alzheimer's Disease," *Plant Sci. Today*, vol. 11, no. 1, pp. 454–465, 2024. https://doi.org/10.14719/pst.2420

[5] P. Zhang *et al.*, "Anti-Inflammatory and Antioxidant Properties of Squalene in Copper Sulfate-Induced Inflammation in Zebrafish (Danio rerio)," *Int. J. Mol. Sci.*, vol. 24, no. 10, 2023. https://doi.org/10.3390/ijms24108518

[6] E. Reverchon and F. Senatore, "Isolation of rosemary oil: Comparison between hydrodistillation and supercritical CO2 extraction," *Flavour Fragr. J.*, vol. 7, no. 4, pp. 227–230, 1992. https://doi.org/10.1002/ffj.2730070411

[7] A. K. Jha and N. Sit, "Extraction of bioactive compounds from plant materials using combination of various novel methods: A review," *Trends Food Sci. Technol.*, vol. 119, no. October 2021, pp. 579–591, 2022. https://doi.org/10.1016/j.tifs.2021.11.019

[8] T. Arumugham, R. K, S. W. Hasan, P. L. Show, J. Rinklebe, and F. Banat, "Supercritical carbon dioxide extraction of plant phytochemicals for biological and environmental applications – A review," *Chemosphere*, vol. 271, 2021. https://doi.org/10.1016/j.chemosphere.2020.129525

[9] "SCOGS (Select Committee on GRAS Substances)," *U.S. FOOD & DRUG ADMINISTRATION*, 1979. https://www.hfpappexternal.fda.gov/scripts/fdcc/index.cfm?set=SCOGS&sort=Sortsubstance&order=ASC&startrow=1&type=basic&search=CARBON DIOXIDE

[10] A. H. Abdul Aziz, N. R. Putra, H. Kong, and M. A. Che Yunus, "Supercritical Carbon

Dioxide Extraction of Sinensetin, Isosinensetin, and Rosmarinic Acid from Orthosiphon stamineus Leaves: Optimization and Modeling," *Arab. J. Sci. Eng.*, vol. 45, no. 9, pp. 7467–7476, 2020. https://doi.org/10.1007/s13369-020-04584-6

[11] K. I. Amajuoyi, "Behavior and elimination of pesticide residues during supercritical carbon dioxide extraction of essential oils of spice plants and analysis of pesticides in high-lipid-content plant extracts," 2001, [Online]. Available: https://mediatum.ub.tum.de/?id=603193%0Ahttps://lens.org/145-139-566-767-513

[12] K. D. Bartle, A. A. Clifford, S. B. Hawthorne, J. J. Langenfeld, D. J. Miller, and R. Robinson, "A model for dynamic extraction using a supercritical fluid," *J. Supercrit. Fluids*, vol. 3, no. 3, pp. 143–149, 1990. https://doi.org/10.1016/0896-8446(90)90039-O

[13] E. Reverchon, G. Donsi, and L. S. Osséo, "Modeling of Supercritical Fluid Extraction from Herbaceous Matrices," *Ind. Eng. Chem. Res.*, vol. 32, no. 11, pp. 2721–2726, 1993. https://doi.org/10.1021/ie00023a039

[14] J.Crank, *The Mathematics of Diffusion*, vol. 79, no. 2nd edition. 1975.

[15] M. Goto, J. M. Smith, and B. J. McCoy, "Parabolic profile approximation (linear driving-force model) for chemical reactions," *Chem. Eng. Sci.*, vol. 45, no. 2, pp. 443–448, 1990. https://doi.org/10.1016/0009-2509(90)87030-V

[16] N. Wakao and T. Funazkri, "Effect of fluid dispersion coefficients on particle-to-fluid mass transfer coefficients in packed beds. Correlation of sherwood numbers," *Chem. Eng. Sci.*, vol. 33, no. 10, pp. 1375–1384, 1978. https://doi.org/10.1016/0009-2509(78)85120-3

[17] R. Muhammad Syafiq Hazwan, "Empirical and Kinetic Modelling on Supercritical Fluid Extraction of Areca Catechu Nuts," no. August, p. PhD Thesis., 2016, [Online]. Available: PhD Thesis.

[18] S. Zarghami, F. Boukadi, and Y. Al-Wahaibi, "Diffusion of carbon dioxide in formation water as a result of CO2 enhanced oil recovery and CO2 sequestration," *J. Pet. Explor. Prod. Technol.*, vol. 7, no. 1, pp. 161–168, 2017. https://doi.org/10.1007/s13202-016-0261-7

[19] V. K. Mark Mchugh, *Supercritical Fluid Extracttion*. 1994.

[20] H. Mohd Nasir, L. Md Salleh, M. S. H. Ruslan, and M. A. Mohamed Zahari, "Single sphere model fitting of supercritical carbon dioxide extraction from Quercus infectoria galls," *Malaysian J. Fundam. Appl. Sci.*, vol. 13, no. 4, pp. 821–824, 2017. https://doi.org/10.11113/mjfas.v13n4.920

[21] L. N. Yian *et al.*, "Supercritical carbon dioxide extraction of hevea brasiliensis seeds: Influence of particle size on to oil seed recovery and its kinetic," *Malaysian J. Fundam. Appl. Sci.*, vol. 17, no. 3, pp. 253–261, 2021. https://doi.org/10.11113/MJFAS.V17N3.2073

[22] L. Wang *et al.*, "Effect of supercritical carbon dioxide on pore structure and methane adsorption of shale with different particle sizes," *J. Supercrit. Fluids*, vol. 212, no. March, p. 106343, 2024. https://doi.org/10.1016/j.supflu.2024.106343

[23] N. R. Putra *et al.*, "Influence of particle size in supercritical carbon dioxide extraction of roselle (Hibiscus sabdariffa) on bioactive compound recovery, extraction rate, diffusivity, and solubility," *Sci. Rep.*, vol. 13, no. 1, pp. 1–18, 2023. https://doi.org/10.1038/s41598-023-32181-8

[24] E. S. Prasedya *et al.*, "Effect of particle size on phytochemical composition and antioxidant properties of Sargassum cristaefolium ethanol extract," *Sci. Rep.*, vol. 11, no. 1, pp. 1–9, 2021. https://doi.org/10.1038/s41598-021-95769-y

[25] M. M. Esquível, M. G. Bernardo-Gil, and M. B. King, "Esquível_1999_The-Journal-of-Supercritical-Fluids.pdf," *J. Supercrit. Fluids*, vol. 16, pp. 43–58, 1999

Separation Technology - ICoST 2025
Materials Research Proceedings 59 (2026) 87-97

Materials Research Forum LLC
https://doi.org/10.21741/9781644903957-12

Optimization of Amygdalin Extraction from *Prunus Armeniaca* Kernels for Antioxidant and Anti-Inflammatory Potentials

Sayed Ibrahim WAFA[1,a], Lee Suan CHUA[1,b] *, Roshafima RASIT ALI[2,c]

[1]Faculty of Chemical & Energy Engineering, Universiti Teknologi Malaysia, 81310 UTM Johor Bharu, Johor, Malaysia

[2]Malaysia-Japan International Institute of Technology (MJIIT), Universiti Teknologi Malaysia 54100 UTM Kuala Lumpur, Malaysia

[a]sayedibrahimwafa@graduate.utm.my, [b]chualeesuan@utm.my, [c]roshafima@utm.my

Keywords: Amygdalin, Prunus Armeniaca, Anti-Oxidant, Anti-Inflammatory, Ultrasonic Assisted Extraction

Abstract. Amygdalin is one of the compounds contributing to the major pharmacological properties of *Prunus armeniaca* kernels. Clinical trials have proved the anticancer activity of amygdalin, and its pharmacological activities such as antioxidant, anti-inflammatory, anti-tumor, anti-bacterial, anti-fibrotic and many more. This study was to investigate the effect of extraction parameters such as temperature, solvent ratio and particles size on ground *P. armeniaca* kernels powder to obtain the high extraction yield of amygdalin. Ultrasonic assisted extraction (UAE) was utilized to recover amygdalin from the kernels. The operating variables such as temperatures (27-60°C), ethanol concentration (10-90%) and particle size (20-40 mesh) were varied using the technique of one-factor-at-a-time (OFAT) for optimization. The results found that the extraction temperature of 60°C in 70% ethanol using mesh 40 (400 μm) of particle size exhibited the highest concentration of amygdalin 326.92±3.11 mg/g of extract powder. The lowest and the highest of amygdalin content of samples were also compared for their anti-oxidant potential by correlating their radical scavenging ability (DPPH and ABTS) and reducing power (FRAP). The kernels extract with the highest amygdalin content showed higher anti-oxidant capacity in both DPPH and FRAP assays with the effective concentration at 137.92±21.87 mg GAE/100g and 101.19±0.14 mg GAE/100g, respectively. However, there was no significant difference for the ABTS results between the lowest and highest amygdalin content of *P. armeniaca* kernels extract. Albumin assay showed the significant anti-inflammatory action (32.48±609.70 g DCFE/100g). As a conclusion, the optimization of UAE produced higher content of amygdalin and enhanced better anti-oxidant and anti-inflammatory properties of *P. armeniaca* kernels.

Introduction

Prunus armeniaca is commonly known as apricot from the genus of prunus in the Rosaceae family. *P. armeniaca* has been cultivated in China domestically over 3,000 years ago. It has been introduced to Europe through Greece and Italy by the Romans and also brought into North America by English travelers. Hence, it is extensively cultivated and widely found throughout Middle Asia, Caucasus, Iran, Pakistan and China nowadays. *P. armeniaca* is a species surviving in the cool temperate climate region. It survive in winter hardy with temperatures down to -30°C, but it is sensitive to unstable and fluctuations of winter temperature [1].

   P. armeniaca seeds are rich in proteins, carbohydrates, phenolic compounds, flavonoids or polyphenols, carotenoids, unsaturated fatty acids, phytosterols and many more [2]. The bitterness of *P. armeniaca* kernels is the sign of the existence of amygdalin [3]. Amygdalin is an active substance present in the seeds of about 800 plants and mostly exist in plant of the Rosaceae or Prunus seeds. Other than prunus seeds, amygdalin presents in the seeds of grapes, olives and

Separation Technology - ICoST 2025
Materials Research Forum LLC
Materials Research Proceedings 59 (2026) 87-97
https://doi.org/10.21741/9781644903957-12

buckwheat [4]. Amygdalin is one of the major compounds in *P. armeniaca* seeds. It has been reviewed in about 100 over papers for its *in vivo* and *in vitro* studies. These studies included its pharmacological activities such as antioxidant, anti-inflammatory, anti-tumor, anti-bacterial, anti-fibrotic, immunoregulatory, immunomodulatory and anti-apoptotic properties to improve the neurodegeneration and myocardial hypertrophy, as well as reducing blood glucose [5]. Scientists have intensively carried out studies on the pharmacological properties of amygdalin. Clinical trials have proved the anticancer properties of amygdalin without any harmful effects on the human body and healthy cells. Amygdalin has been proven for its activities inhabiting abnormal cells and metabolic diseases [6].

Many extraction techniques have been applied to extract amygdalin from *P. armeniaca* kernels. Most techniques showed the importance of extraction variables such as temperature, solvent and particle size of *P. armeniaca* kernel in affecting the extraction efficiency. The influence of temperature is important to optimize the efficiency of extraction method. Elevated temperature is possible to increase the solubility rate to a higher extent by changing the extractant polarity [7]. However, too high temperature might cause some inevitable consequence like thermal degradation [8]. Previous studies reported that amygdalin was not soluble in non-polar solvents, moderately soluble in water, and highly soluble in ethanol [9]. Ethanol could be the choice of solvent in term of its effectiveness and less toxicity. Particle size affects the extraction yield by influencing the mass transfer of amygdalin during extraction. Greater surface area would consequently increase the mass transfer and enhance the penetration into the sample matrix of kernels. However, optimization is strongly required as extremely coarse and fine particles caused a decline in isoflavone recovery from soybean extract [10].

In this study, ultrasonic assisted extraction (UAE) was utilized to recover amygdalin from *P. armeniaca* kernels. It possesses many advantages including as a cost-effective green extraction technique, besides contributing to eco-friendliness, cost effectiveness, safety, rapidity, versatility and simplicity. UAE has less consumption of energy and time, and thus reducing the usage of expensive organic solvents [8]. The optimum parameters such as temperature, solvent ratio and particles size were optimized to recover the highest yield of amygdalin along with its high antioxidant and anti-inflammatory potentials. The finding of this study would benefit the pharmaceutical industries in order to obtain the highest amygdalin content of *P. armeniaca* extract. The amygdalin extract can be efficiently utilized in treating various types of metabolic diseases, as an additive to cosmetics and body care products.

Methodology

The *P. armeniaca* kernels were purchased from Gilgit–Baltistan, northern Pakistan. A kilogram of the kernels was ground and sieved into the particle sizes of 20-60 mesh. The ground particles were packed in vacuum and stored in a chiller at 5°C Gasparini et al. [11]. Standard amygdalin (99.8% purity) was purchased from Merck (Malaysia). Reagents such as 2,2-diphenyl-1-picrylhydrazyl (DPPH) and 2,2'-azinobis-(3-ethylbenzothiazoline-6-sulfonic acid) (ABTS) were sourced from Sigma Aldrich. Sodium acetate anhydrous was bought from LabChem. Phosphate buffer saline (PBS) was purchased from R&M Chemicals Evergreen Malaysia. The other chemicals such as potassium persulfate, hydrochloric acid (37%, 12M), 2,4,6-Tri(2-pyridyl)-s-triazine (TPTZ) and Iron (III) chloride were purchased from Merck (Malaysia). Ethanol (absolute 99.8%) and acetic acid glacial were purchased from ChemAR SYSTERM. HPLC grade methanol was bought from Fisher Scientific (Malaysia).

QSonica MSX-Q500200 sonicator was used to extract amygdalin from ground *P. armeniaca* kernels (10 g). OFAT method was used to optimize the extraction process. The extraction variables such as temperature (27-60°C), ethanol concentration (10 to 90%) and particle size (20-60 mesh) were varied at the fixed extraction time, 15 min. After extraction, samples were transferred into the centrifugal tubes and proceeded for centrifugation at 10,000 rpm using a centrifuge (Kubota

Separation Technology - ICoST 2025   Materials Research Forum LLC
Materials Research Proceedings 59 (2026) 87-97   https://doi.org/10.21741/9781644903957-12

Refrigerated Centrifuge model 5922) for 10 min. The harvested supernatant was concentrated under reduced pressure using a vacuum rotary evaporator (Heidolph brand, Germany) at 40°C with 170 rpm and 90 Mbar of vacuum pressure. The concentrated extract was frozen for -20°C and proceeded for freeze drying using a Labogene freeze drier from Denmark for 24 h at temperature -44°C to -51°C. The weight of the dried extract was recorded. *P. armeniaca* extracts were kept in amber glass bottles and stored in a refrigerator at -4°C until further analysis.

High performance liquid chromatograph (Perkin Elmer A10 Altus,) was used to separate and quantify amygdalin in samples. This analysis was performed according to the method described by Ozturk et al. [12]. *P. armeniaca* extracts and standard amygdalin were reconstituted in 50% ethanol separately. They were diluted to various concentrations from 2500 to 10,000 mg/L and 10 to 2,000 mg/L, respectively. The separation of amygdalin was executed by passing through the *P. armeniaca* kernel extract solution through a column (4.6 × 150 mm) at 0.7 mL/min. The mobile phase was methanol and distilled water at a ratio of 20:80. The total run time was 15 min. The column temperature was set at 25°C. Amygdalin was detected by a photodiode array detector (PDA) at 215 nm. The detection of amygdalin was based on the retention time as detected for standard amygdalin. The sample solution was filtered with 0.22 μm nylon filter before injection. The injection volume was 10 μL.

The antioxidant activity of *P. armeniaca* kernels extracts were determined using free radical scavenging activity assay. DPPH (2,2-diphenyl-2-picrylhydrazyl) was used to generate free radicals in the assay. This study was performed by referring to the method described by Qin et al. [13] with some modifications. The presence of anti-oxidative compounds in *P. armeniaca* kernel extracts would decolorize the DPPH reagent. DPPH solution (0.1 mM) was prepared in 50 mL methanol (99.9%). The extracts were reconstituted in methanol and prepared at the different concentrations ranged from 200-1000 mg/L by serial dilution. An aliquot of 0.75 mL DPPH solution was added into 2.25 mL sample in a tube. The solution was mixed well and incubated for 30 min in the dark condition. The absorbance of sample was measured using an ultraviolet-visible spectrophotometer (SHIMADZU UV-2600, Kyoto, Japan) at 517 nm. The experiments were performed in triplicate. The antioxidant activity was calculated in the percentage of radical scavenging activity as expressed in Eq. (1), where $A_{control}$ is the absorbance of DPPH solution without sample and A_{sample} is the absorbance of DPPH solution in the presence of sample solution.

$$\text{DPPH Radical Scavenging Activity} = \frac{Acontrol - Asample}{Acontrol} \times 100 \tag{1}$$

The concentration of *P. armeniaca* kernel extract required to scavenge 50% of DPPH free radicals was expressed as IC50. Gallic acid was used as the standard chemical for the assay. The results are expressed in Gallic Acid Equivalent (mg GAE/100g).

ABTS assay was used to determine the reducing ability potential in *P. armeniaca* kernel extract. The experiment was executed based on the method performed by Qin et al. [13] with some modifications. ABTS and potassium persulfate reagents were dissolved in distilled water and prepared at the concentration of 7 mM and 2.45 mM, respectively. The solutions were mixed up thoroughly with the same volume and stored in a dark place for 16 h at room temperature. The solution was diluted with approximately 65 mL distilled water to the absorbance of 0.750±0.005 at 734 nm using a UV-Vis spectrophotometer. Sample was reconstituted in distilled water and prepared to the concentration 125 to 2000 mg/L with serial dilution. Then, an aliquot of 0.75 mL sample was added to 2.25 mL ABTS solution and incubated in the dark place at room temperature for 7 min. The mixture was measured for its absorbance at 734 nm. The experiment was performed in triplicate. The reducing effect of samples against ABTS radical cations was calculated in the percentage of radical cation reducing activity as expressed in Eq. (2), where A_{contol} is the

Separation Technology - ICoST 2025
Materials Research Proceedings 59 (2026) 87-97

Materials Research Forum LLC
https://doi.org/10.21741/9781644903957-12

absorbance of ABTS without sample and A_{sample} is the absorbance of the ABTS solution in the presence of *P. armeniaca* kernel extract.

$$\text{ABTS Radical Cation Reducing Activity} = \frac{Acontrol - Asample}{Acontrol} \times 100\% \qquad (2)$$

The concentration of *P. armeniaca* kernel extract required to reduce 50% of the radical cations was expressed as IC_{50}. Gallic acid was used as the standard chemical for the assay. The results are expressed in Gallic Acid Equivalent (mg GAE/100g).

FRAP assay is a colorimetric method to measure the antioxidant ability of sample to reduce colourless complex (Fe^{3+}-TPTZ) to ferrous blue coloured Fe^{2+}. The experiment was conducted based on the procedure described by Skroza et al. [14] with modification. FRAP reagent was prepared freshly by mixing 10 mM TPTZ solution into 40 mM hydrochloric acid (HCl, 37%), 0.3 M acetate buffer at pH 3.6 and 20 mM ferric chloride ($FeCl_3$) in the ratio of 1:10:1. The FRAP solution was wrapped with aluminium foil and incubated in the dark place at 37°C. *P. armeniaca* kernel extract was reconstituted in distilled water and prepared in the concentration from 125 - 2000 mg/L by serial dilution. An aliquot of 1.5 mL extract was added into 1.5 mL FRAP reagent solution and the mixture was left for 6 min. The absorbance of the mixture was measured using a UV-Vis spectrophotometer at 593 nm. The experiment was executed in triplicate. The reducing ability of sample was measured in the percentage of ferric reducing power according to the formula as stated in Eq. (3) where $A_{control}$ was the absorbance of the mixture of FRAP reagent with 2.5 mg/L of gallic acid, whereas A_{blank} was the absorbance of FRAP solution without sample and A_{sample} was the absorbance in the presence of sample.

$$\text{FRAP Reducing Activity} = \frac{Asample - Ablank}{Acontrol - Ablank} \times 100\% \qquad (3)$$

The antioxidative ability of sample against FRAP reagent was expressed in the required concentration of samples to reduce 50% of Fe^{3+} to Fe^{2+}. The results are expressed as IC50. The reducing ability of gallic acid was used as positive control.

Albumin is coagulable protein and it was used to evaluate the anti-inflammatory potential of samples. The assay of egg albumin denaturization was used to access the anti-inflammatory activity. This was carried out by inducing the denaturization of globular in egg albumin protein. The procedure of the assay was carried out referring to the method described by Malik et al. [15]. Egg albumin (10%) solution was homogenized in 0.2 mL distilled water and topped up to 2.8 mL phosphate buffer saline (PBS at pH 7.4). Sample was reconstituted in distilled water and prepared in the concentration of 250 to 2000 mg/L by serial dilution. An aliquot of 2 mL extract solution was added into 3 mL albumin in PBS. The mixture was incubated for 20 min at 37°C to enable the interaction between sample and albumin. Subsequently, the mixture was heated at 70°C for 5 min in water bath in order to induce the denaturation of globular in egg albumin. The heated solution was measured for its absorbance using a UV-Vis spectrophotometer at 660 nm. This experiment was proceeded in triplicate.

The anti-inflammatory potential is the ability of sample to inhibit the denaturation of protein albumin in term of biological activity and chemical structure when the albumin exposed to heat. Optical observation was used to measure the extent of albumin denaturization based on the transition of clear solution to cloudy solution. The presence of anti-inflammatory compounds would be able to remain the clarity of solution after heating. The absorbance of the mixture was measured at 660 nm. The inhibition of albumin denaturization can be calculated using the formula as stated in the Eq. (4) where $A_{control}$ is the absorbance of denatured albumin protein without sample and A_{sample} is the absorbance of albumin in the presence of sample.

$$\text{Albumin Protein Denature} = \frac{Acontrol - Asample}{Acontrol} \times 100\% \tag{4}$$

The required concentration of *P. armeniaca* kernel extract to denature 50% of egg albumin was expressed as IC_{50}. The value was furthermore compared to diclofenac which was used as positive control. The results are expressed as Diclofenac Equivalent (mg DE/100g).

Results and Discussion
In this study, UAE was chosen as the extraction method for *P. armeniaca* kernel. The method has been used widely due to its green process. The operating variables such as temperatures (27-60°C), ethanol concentration (0-100%) and particle size (20-40 mesh) of ground kernel were varied to determine their effects on the extraction yield and amygdalin content.

Pure amygdalin was prepared in six concentrations ranged from 10–2000 mg/L. The chromatogram of the standard solutions are shown with the peak area of 122.64×10^3 to 23.63×10^6 AU whereby it exhibited at 10.387-10.430 min of retention time. The calibration curve of the pure amygdalin is well presented as a regression linear with $R^2 = 0.9999$. The S15 represents sample without amygdalin and S35 exhibits the highest content of amygdalin. The extract samples were analyzed at three concentrations ranged from 2.5–10 g/L. The results showed that the extraction of S15 sample at the particular extraction conditions was unable to isolate amygdalin content from the ground apricot kernels. The extraction of S35 sample exerted a better extraction result of 9.36×10^6 to 36.09×10^6 AU of peak area at 10.421 to 10.427 min of retention time with the well fitted linear regression where $R^2 = 0.9999$.

Temperature plays a vital role in optimizing the extraction process. High temperature might degrade the thermolabile of bioactive compound and contaminate the extraction yield [16]. Amygdalin content at 60°C gave the highest amount (212.92±33.05 mg/g) in *P. Armeniaca* extract. The extraction at 27°C exerted the lowest content of amygdalin (22.63±10.07 mg/g) in the extract despite it presented the highest yield of *P. armeniaca* extract (125.67±12.50 mg/g). The extraction of *P. armeniaca* kernels with UAE at 27 to 60°C was able to increase the amount of amygdalin. The increased temperature enhanced the solubility and extractant polarity rate to a higher extent and subsequently increased the compound diffusivity to improve its mass transfer from matrices. The lowest concentration of amygdalin in the present study where extracted at 27°C in 50% ethanol was still higher than the value reported by Wang et al. [17] who harvested only 5.586 mg/g amygdalin after soaking *P. armeniaca* kernels for 24 h in 40 mL water at 30°C.

Furthermore, the extract at low extraction temperature was hygroscopic, possibly due to the presence of sugars or polysaccharides in the extract. The stickiness was also observed for the 40°C extract. UAE extraction at higher temperature 50°C and 60°C was sufficient to transmit the energy and pressure to produce cavitation and form bubbles in liquid to interact with the sample matrices, subsequently affects the increased of solubility and reduced the viscosity of solvent, and thus increasing the mass transfer of solutes. The microbubbles created during cavitation destroyed the cell wall of matrices and broken the bonding of polysaccharides into smaller sugar units such as glucoses. It was found that the extracts at 50°C and 60°C were less sticky than the extracts at lower temperatures. The optimum temperature was 60°C which was near to the boiling point (78°C) of solvent, 50% ethanol.

In the subsequent optimization process, the ethanol concentration was varied from (10%-90% ethanol), while the other parameters were kept constant at 60°C using the particle size less than 400 µm (mesh ≥ 40). The amygdalin content of *P. armeniaca* kernel extract varied among four solvent compositions. The 90% ethanol extract exhibited the highest concentration of amygdalin (314.76±14.91 mg/g). It was found that the second highest amygdalin content was from the 70% ethanol extract which contained about 6.17% slightly lower of amygdalin (295.35±12.63 mg/g) than the 90% ethanol extract. Despite the variance was statistically insignificant, 70% ethanol extract was preferable due to its extract yield (111.27±9.40 mg/g) was exerted about 43% higher than 90% ethanol extract (77.25±25.28 mg/g). The results revealed that lower ethanol concentration, as 10% and 30% ethanol in this study, was not effective to extract amygdalin from *P. armeniaca* kernel. A higher ethanol concentration was preferable for amygdalin extraction as reported by Savic et al. [18]. This study proved the importance of solvent selection to efficiently extract amygdalin from *P. armeniaca* kernels. The results revealed that 70% ethanol was the most suitable solvent composition which was then fixed for the following study. The studies of Ozturk et al. [12] reported that the application of UAE at 35 kHz for extracting amygdalin from different species of bitter almond kernels. They found that 25 citric acid could produce comparable result (36.79±1.47 mg/g) with the sample in the present study, even though with 20 kHz sonication. However, they applied longer extraction time (30 min) which was double than the present study (15 min). They also reported that acetonitrile and water were unable to extract amygdalin.

Amygdalin was not soluble in non-polar solvent, severely soluble in ethanol and moderately soluble in water [9]. Therefore, the binary solvent system of aqueous ethanol was chosen in the present study. The results revealed that higher composition of ethanol was more favorable for amygdalin extraction. The work of Savic et al. [18] used reflux method to perform amygdalin extraction in 100% ethanol at 34.4°C for 2 h. They obtained high yield of 253 mg/g amygdalin in their experiments [18]. Ramadan et al. [19] compared the amygdalin content of *P. armeniaca* kernel extract extracted from 99.9% ethanol and 70% ethanol with 6% citric acid by soaking at 55°C for 100 min. Although they used high ethanol concentration in the solvent system, the results were lower which were 57.20 mg/g and 102.20 mg/g extract, respectively than the present study [19]. UAE was more effective than the soaking method as the cavitation generated by bubbles enhancing extraction by promoting mass transfer and improving diffusion.

Amygdalin content in mesh 40 indicated the highest of amygdalin content (326.92±3.11 mg/g) whereby it was slightly higher (1.28%) than mesh 30 (322.79±12.29 mg/g) in the extract whereas mesh 30 was presented the highest extract yield (112.00±8.89 mg/g) where it was 1.49% higher than mesh 40 (110.33± 4.51 mg/g). Both meshes of particle size of 20 mesh and 25 mesh were revealed the lowest of harvested of extract yield and amygdalin content. The smaller particle size with the mesh between 30 and 40 produced higher extract yield and amygdalin content. A smaller particle size would create higher surface area which facilitates the diffusion of amygdalin during extraction. Statistical comparison indicated that amygdalin content in the kernel extract showed a significant difference between mesh 20 and 40, but insignificant for the rest. It also statistically analysed that the variance for extract yield was insignificant except for mesh 25 and mesh 40. These interpretations clearly defined that mesh 40 was more reliable and practical compared to mesh 30 due to mesh 40 has the best of standard deviation among four meshes variable in extract yield and amygdalin content. By this study proven that the different particle size 20 mesh (841 μm) to 40 mesh (400 μm) able to influence the efficiency of extraction which contributing the increased of extract yield powder and amygdalin content although there were insignificant different between the variable particle size but it contributes towards the optimization, stability and efficiency of the extraction.

The extraction of *P. armeniaca* was optimized by variable of temperature (27-60°C), solvent composition (10%-90%) of ethanol and particle size (20–40 mesh) by OFAT method. The variables were represented by 36 samples which were experimented in triplicate for each parameter of extraction. It indicates that the elevated temperature (27-60°C) exerted the increment of amygdalin content (22.63±10.07 to 212.92±33.05 mg/g), elevation of ethanol composition (10%-90%) promoted the increased of amygdalin content (0 to 314.76±14.91 mg/g) and the smaller of particle size 20–40 mesh (841-400 μm) influence the escalation of amygdalin content (289.18±20.39 to 326.92±3.11 mg/g) in *P. armeniaca* extract. Fig. 1 demonstrates amygdalin content in 36 samples contributed into 12 parameters in vary which was *P. armeniaca* extract powder. In this study two samples with the lowest and highest amygdalin content have to be chosen with the data justification to be studied for bio assay in term of anti-oxidant and anti-inflammatory effect.

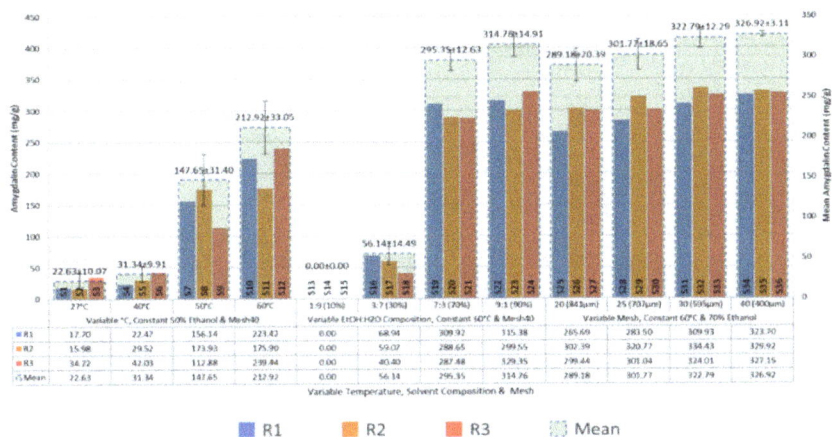

	Variable °C, Constant 50% Ethanol & Mesh40				Variable EtOH H2O Composition, Constant 60°C & Mesh40				Variable Mesh, Constant 60°C & 70% Ethanol			
	27°C	40°C	50°C	60°C	1:9 (10%)	3:7 (30%)	7:3 (70%)	9:1 (90%)	20 (841μm)	25 (707μm)	30 (50%μm)	40 (400μm)
R1	17.70	22.47	156.14	223.42	8.00	68.94	309.92	315.38	265.69	283.50	309.93	323.70
R2	15.98	29.52	173.93	175.90	0.00	59.07	288.65	299.55	302.39	320.77	334.43	329.92
R3	34.72	42.03	112.88	299.84	0.00	40.40	287.88	329.35	299.44	301.04	324.01	327.15
Mean	22.63	31.34	147.65	212.92	0.00	56.14	295.35	314.76	289.18	305.77	322.79	326.92

Variable Temperature, Solvent Composition & Mesh

■ R1   ■ R2   ■ R3   ▨ Mean

Figure 1 36 Samples of amygdalin content (mg/g) in P. armeniaca extraction powder with variable temperature, solvent composition and mesh.

As discussed earlier, extraction parameter of temperature 60°C, solvent 70% ethanol-water and mesh 40 of particle size demonstrated the highest content of amygdalin in *P. armeniaca* extract. This parameter was represented by S34, S35 and S36 whereby S35 demonstrated the highest content of amygdalin 329.92 mg/g in extract powder. Therefore, S35 has been chosen for most practical sample in term of the best amygdalin content recovery to be experimented for bio assay. In other hand, S15 has been selected as the worst amygdalin content recovery to be experimented for bio assay.

Comparison between the highest (S35) and the lowest (S15) content of amygdalin in term of extraction parameter was fortified in the bio-assay experiment in order to certify the optimization of the UAE extraction parameter. The best and the worst parameter of extraction were chosen and studied for bio-assay which involved anti-oxidant and anti-inflammatory. DPPH solution was decolorized for the samples tested from violet to yellow during the mixture of DPPH solution with the extract solution. Decolorized from violet to yellow was the physical sign towards the scavenging effect of the sample over anti-oxidant activity. Scavenging effect of S35 and S15 were analyzed and detected by UV-Vis at 517 nm wavelength exerted significant value (P<0.05) where the P value was 1.39×10^{-3}. Linear graph for different concentration for both samples demonstrated well regression of $R^2 = 0.9939$ and 0.9777. Radical scavenging of DPPH assay exhibited that S35 resulted the higher value 137.92±21.87 mgGAE/100g than S15 which is 35.89±4.76 mgGAE/100g as shown in Fig. 2 subsequently validated that S35 gave more effect towards DPPH radical scavenging antioxidant activity.

ABTS solution was decolorized from bluish green (turquoise) to lighter color upon the mixture of ABTS solution to *P. armeniaca* extract solution. Decolorized transition exhibits the physical sign towards the reducing effect of the sample over ABTS radical cation thus determined the value of anti-oxidant activity. Reducing effect was analyzed and detected by UV-Vis at 734nm wavelength with different concentration where the linear graph was plotted with well regression

Separation Technology - ICoST 2025
Materials Research Proceedings 59 (2026) 87-97

Materials Research Forum LLC
https://doi.org/10.21741/9781644903957-12

obtained for both samples $R^2 = 0.9903$ and 0.9862. Both samples were indicated insignificant (P>0.05) different for ABTS antioxidant value where P = 0.577. Fig. 2(a) showed the value of ABTS antioxidant assay where S35 resulted the slightly lower value 147.09 ± 0.30 mgGAE/100g than S15 which was 147.53 ± 1.23 mgGAE/100g. Both values have validated that S15 exhibited insignificant and slightly higher (0.44 mgGAE/100g) and better value for reducing ABTS radical cation thus determined that S35 and S15 gave an even effect towards antioxidant activity on ABTS assay.

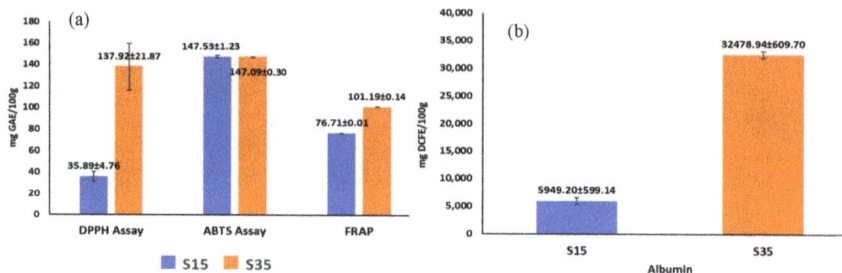

Figure 2 (a) Anti-Oxidant bio assay (mgGAE/100g) of DPPH, ABTS and FRAP for samples S15 and S35, (b) anti-inflammatory bio assay (mgDCFE/100g) of albumin protein denature for samples S15 and S35.

FRAP solution was colorized from clear to blue colour upon the mixture of Ferric solution to *P. armeniaca* extract solution. Colorized transition exhibits the physical sign towards the ferric reducing effect of the sample over anti-oxidant activity. Both samples presented that the reducing of ferric exceeded 50% of reduction but ANOVA analysis indicates a significant (P<0.05) different with P = 7.90×10^{-10}. Linear graph for different concentration at absorbance of 593 nm has exhibited well regression for FRAP assay to both samples S35 and S15 with $R^2 = 0.9995$ and 0.9980. Fig. 2(a) shows the reducing power for FRAP assay value where S15 resulted much lower value 76.71 ± 0.01 mgGAE/100g than S35 which was 101.19 ± 0.14 mgGAE/100g. Both values indicated that S35 gave much higher and better value for ferric reducing power potential therefore determined that S35 gave far more potential towards antioxidant activity.

Albumin protein denature is the method that utilized egg albumin denaturation assay in order to observe that the sample of extract yield able to inhibit egg albumin change to denatured under particular condition. Since albumin is coagulable protein, it was used to experiment the inflammatory potential effect of the particular samples. In other words, the anti-inflammatory effect is the capability of the sample to stop the protein albumin changes in term of biological activity and chemical structure. Heat was used as denaturing factor to which was applied to denature albumin by transforming its biological structure. The inhibition of Albumin protein denatured by S35 and S15 has been observed and clarified by UV-Vis absorption spectra graph at the absorbance of 660 nm. S35 has shown great inhibition towards albumin denaturization with 93.72% inhibition at 2000mg/L with well regression of $R^2 = 0.9822$ on linear graph plotting toward multiple concentration and it was exceeding 50% of inhibition. S15 has demonstrated opposite with irregular effect 6.86% of inhibition at 2000 mg/L concentration with irregular and poorly regression of $R^2 = 0.3421$ on the linear graph.

Fig. 2(b) presented the value of anti-inflammatory effect over *P. armeniaca* extract in diclofenac equivalent for albumin protease denaturation assay. This experiment clarified that a significant different (P<0.05) for the capability to inhibits the albumin protein denaturization with

$P = 7.17 \times 10^{-7}$ in between S35 and S15. The figure shows that sample S35 exerted the higher value of anti-inflammatory effect 32.48 ± 609.70 gDCFE/100g as compared with S15 which was 5.95 ± 599.14 gDCFE/100g. Both values indicates that S35 has far better value for inhibiting the albumin protease denaturation. Horozić et al. [20] was reporting that *P. armeniaca* extraction by maceration, UAE and Soxhlet extraction techniques using ultrapure water solvent obtained the highest result of DPPH 99.76 mgGAE/100g. Alam et al. [21] was experimented towards Clinacanthus nutans Lindau on the arial part using methanol, ethanol and petroleum ether reported antioxidant result on DPPH 99.42 ± 6.59 mgGAE/100g. In other experiment done by Pérez-Jiménez et al. [22] shows that the free radical scavenging activity of DPPH towards *P. armeniaca* extract around 99 mgGAE/100g. Wu et al. [23] gone through the experiment on *P. armeniaca* from different cultivars and origins also indicates the potential of antioxidant free radical scavenging of DPPH value from 99.47 ± 2.70 to 114.71 ± 0.02 mgGAE/100g. Others antioxidant assay indicated by Fan et al. [24] towards antioxidant using ABTS assay shows the significant of capability apricot kernels to reduce ABTS radical with antioxidant value 158.50 ± 0.80 mgGAE/100g as compared to its pulp and peel.

From our studies indicates that the optimized UAE extraction method with optimized parameter of temperature extraction, solvent ethanol-water composition and mesh of particle size able to affect the higher amount of amygdalin, antioxidant and anti-inflammatory. These indications were validated towards S15 sample which was presented the lowest amount of amygdalin demonstrated the lower value of DPPH and FRAP antioxidant assay and presented as well towards the lower value inhibition of albumin protein denature for anti-inflammatory potential. As compared to S35 sample which was presented the highest amount of amygdalin demonstrated the higher value of DPPH and FRAP although ABTS antioxidant assay showed the insignificant different value between S15 and S35 which were the lowest and the highest of amygdalin content respectively. Anti-inflammatory effect also was validated the significant higher value of albumin protein denature for S35 as compared to S15. In the study, a few types of apricot kernels were extracted using water as a solvent at a maintained temperature of 40°C for 3 h extraction under the optimized condition. This reviewed study presented the highest amygdalin amount obtained by Zerdali $(50.67 \pm 0.08$ mg/g) and the lowest was Kabasi apricot kernels $(1.40 \pm 0.06$ mg/g) subsequently show the significant different with FRAP assay where Ferric reducing antioxidant assay presented significant value 2.63 ± 0.39 Zerdali and 1.01 ± 0.65 FE/g for the Kabasi. It was also significant different value for ABTS antioxidant assay where 1155.5 ± 10 and 915.5 ± 15 TE/g were reported as the value to Zerdali and Kabasi where were the highest and lowest amount of amygdalin respectively. But the result of DPPH antioxidant assay was insignificant in term of different value for both Zerdali and Kabasi which were 220.1 ± 5 and 209.2 ± 4 TE/g, respectively [25].

Conclusion

Ultrasonic assisted extraction (UAE) with optimization of parameter temperature, solvent composition and particle able to increase the amount of amygdalin thus enhance the capability of antioxidant and anti-inflammatory potential. From the study proven that increased of temperature to 60°C, addition of ethanol-solvent to 70% and the thinner of particle size able to extract higher of amygdalin content and elevate the antioxidant and anti-inflammatory activity. In future may study for the other solvent composition such as citric acid, methanol and acetone nitrile with higher temperature with utilization of the other modern technology of extraction that able to reduce consumption of time and solvent cost and be able to increase higher recovery of particular substances.

Acknowledgment

The authors are thankful to Checa Lab MJIIT for providing the necessary lab facilities and FCEE Universiti Teknologi Malaysia for the guidance of this research.

References

[1] Lim, T.K., Prunus armeniaca, in Edible Medicinal And Non-Medicinal Plants. 2012. p. 442-450. https://doi.org/10.1007/978-94-007-4053-2_51

[2] Sharma, R., et al., J7.5Value addition of wild apricot fruits grown in North-West Himalayan regions. -a review. J. Food Sci. Technol.,, 2014. 51, 2917-2924. https://doi.org/10.1007/s13197-012-0766-0

[3] Zhang, Q., et al., Guizhi-Shaoyao-Zhimu decoction possesses anti-arthritic effects on type II collagen-induced arthritis in rats via suppression of inflammatory reactions, inhibition of invasion & migration and induction of apoptosis in synovial fibroblasts. Biomedicine & Pharmacotherapy, 2019. 118: p. 109367. https://doi.org/10.1016/j.biopha.2019.109367

[4] Del Cueto, J., et al., Cyanogenic glucosides and derivatives in almond and sweet cherry flower buds from dormancy to flowering. Frontiers in Plant Science, 2017. 8: p. 800. https://doi.org/10.3389/fpls.2017.00800

[5] Xio Yan He, et al., Amygdalin - A pharmacological and toxicological review. J Ethnopharmacol, 2020. 254: p. 112717. https://doi.org/10.1016/j.jep.2020.112717

[6] Makarević, J., et al., Amygdalin blocks bladder cancer cell growth in vitro by diminishing cyclin A and cdk2. PloS one, 2014. 9(8): p. e105590. https://doi.org/10.1371/journal.pone.0105590

[7] Teo CC, et al., Pressurized hot water extraction (PHWE). J Chromatogr A 2010. 1217(16): p. 2484-94. https://doi.org/10.1016/j.chroma.2009.12.050

[8] Kashif Ameer, Hafiz Muhammad Shahbaz, and J.-H. Kwon, 68.Green Extraction Methods for Polyphenols fromPlant Matrices and Their Byproducts A Review 191120.pdf>. Comprehensive Reviews in Food Scienceand Food Safety, 2017. Vol.16. https://doi.org/10.1111/1541-4337.12253

[9] Salama, R.H., et al., Experimental and Therapeutic Trials of Amygdalin. International Journal of Biochemistry and Pharmacology, 2019. 1(1): p. 21-26. https://doi.org/10.18689/ijbp-1000105

[10] Zuo, Y., et al., Extraction of soybean isoflavones from soybean meal with aqueous methanol modified supercritical carbon dioxide. Journal of Food Engineering, 2008. 89(4): p. 384-389. https://doi.org/10.1016/j.jfoodeng.2008.05.004

[11] Gasparini, A., et al., Ultrasound assisted extraction of oils from apple seeds: A comparative study with supercritical fluid and conventional solvent extraction. Innovative Food Science & Emerging Technologies, 2023. 86: p. 103370. https://doi.org/10.1016/j.ifset.2023.103370

[12] Ozturk, et al., Determination of amygdalin in fifteen different fruit kernels and extraction optimization. Rev Roum Chim, 2022. 67(10-12). https://doi.org/10.33224/rrch.2022.67.10-12.03

[13] Qin, F., et al., Phenolic composition, antioxidant and antibacterial properties, and in vitro anti-HepG2 cell activities of wild apricot (Armeniaca Sibirica L. Lam) kernel skins. Food and Chemical Toxicology, 2019. 129: p. 354-364. https://doi.org/10.1016/j.fct.2019.05.007

[14] Skroza, D., et al., Investigation of antioxidant synergisms and antagonisms among phenolic acids in the model matrices using FRAP and ORAC methods. Antioxidants, 2022. 11(9): p. 1784. https://doi.org/10.3390/antiox11091784

[15] Malik, et al., Characterization of Citrus nobilis peel methanolic extract for antioxidant, antimicrobial, and anti-inflammatory activity. Molecules, 2021. 26(14): p. 4310. https://doi.org/10.3390/molecules26144310

[16] Plaza, M. and C. Turner, Pressurized hot water extraction of bioactives. TrAC Trends in Analytical Chemistry, 2015. 71: p. 39-54. https://doi.org/10.1016/j.trac.2015.02.022

[17] Wang, et al., Determination of amygdalin in Cassia obtusifolia L. by capillary electrophoresis. 2019. https://doi.org/10.1063/1.5085552

[18] Savic, I.M., et al., Optimization of technological procedure for amygdalin isolation from plum seeds (Pruni domesticae semen). Front Plant Sci, 2015. 6: p. 276. https://doi.org/10.3389/fpls.2015.00276

[19] Ramadan, et al., The pharmacological effect of apricot seeds extracts and amygdalin in experimentally induced liver damage and hepatocellular carcinoma. Journal of Herbmed Pharmacology, 2020. 9(4): p. 400-407. https://doi.org/10.34172/jhp.2020.50

[20] Horozić, E., et al., Influence of extraction technique on nutrient content, antioxidant and antimicrobial activity of aqueous extracts of commercial apricot kernels. 2020.

[21] Alam, A., et al., Clinacanthus nutans: A review of the medicinal uses, pharmacology and phytochemistry. Asian Pac J Trop Med, 2016. 9(4): p. 402-409. https://doi.org/10.1016/j.apjtm.2016.03.011

[22] Pérez-Jiménez, et al., Updated methodology to determine antioxidant capacity in plant foods, oils and beverages: Extraction, measurement and expression of results. Food research international, 2008. 41(3): p. 274-285. https://doi.org/10.1016/j.foodres.2007.12.004

[23] Wu, et al., Antioxidant and antiproliferative activities of red pitaya. Food chemistry, 2006. 95(2): p. 319-327. https://doi.org/10.1016/j.foodchem.2005.01.002

[24] Fan, et al., Polyphenol composition and antioxidant capacity in pulp and peel of apricot fruits of various varieties and maturity stages at harvest. International Journal of Food Science & Technology, 2018. 53(2): p. 327-336. https://doi.org/10.1111/ijfs.13589

[25] Aydın, et al., Evaluation of antioxidant, antimicrobial, and bioactive properties and peptide sequence composition of Malatya apricot kernels. Journal of the Science of Food and Agriculture, 2024. https://doi.org/10.1002/jsfa.13632

Separation Technology - ICoST 2025
Materials Research Proceedings 59 (2026) 98-105

Materials Research Forum LLC
https://doi.org/10.21741/9781644903957-13

Zinc Chloride Recovery for Synthesis of Porous Adsorbent from Lipid Condensate for Dye Removal

Nurul Aishah ABDUL RAHIM[1,a], Muhammad Abbas AHMAD ZAINI[2,b*], Sariah ABANG[1,c], Norlisa MILI[1,d]

[1]Department of Chemical Engineering, Faculty of Engineering, Universiti Malaysia Sarawak, 94300 Kota Samarahan, Sarawak, Malaysia

[2]Centre of Lipids Engineering & Applied Research (CLEAR), Ibnu-Sina Institute for Scientific and Industrial Research, Universiti Teknologi Malaysia, 81310 UTM Johor Bahru, Malaysia

[a]n.aishah6470.na@gmail.com, [b]abbas@cheme.utm.my, [c]asariah@unimas.my, [d]mnorlisa@unimas.my

Keywords: Zinc Chloride Recovery, Porous Adsorbent, Lipid Condensate, Methylene Blue Adsorption, Chemical Activator, Wastewater Treatment

Abstract. Zinc chloride ($ZnCl_2$) is one of the most frequently used chemical activators for the synthesis of porous adsorbent. However, $ZnCl_2$ activator is constrained by the corrosive nature of zinc cations (Zn^{2+}) and accumulation of chemical waste, posing a significant environmental concern. To address these issues, the potential of recycling the $ZnCl_2$ recovered from washing filtrate is explored as a resource utilization strategy. Lipid condensate, a by-product from the sterilization process in palm oil mill is proposed as a precursor for activation using the recovered $ZnCl_2$ washing filtrate. Preliminary insights suggest that the absorption performance of the absorbent produced with recovered $ZnCl_2$ may exhibit a comparable result to that prepared using fresh $ZnCl_2$. Nevertheless, further reuse of the activator beyond the first cycle of recovered $ZnCl_2$ may results in a significant decline of adsorption performance. This commentary highlights a practical approach for $ZnCl_2$ recovery aimed at reducing environmental burden while maximizing the utilization efficiency of $ZnCl_2$.

Introduction

Palm oil mill effluent (POME) is recognized as one of the main contributors to water pollution in Malaysia, with a pollution load reported to be up to 100 times higher than municipal sewage [1,2]. In recent years, the palm oil industry has generated between 50 to 70 million m^3 of wastewater during its product processing [3]. Lipid condensate, a brownish effluent by-product generated from sterilization process in palm oil mill is one of the major sources of POME. It is rich in organic matter and content high amounts of chemical oxygen demand (COD), biological oxygen demand (BOD) and suspended solids [4,5]. If discharged without proper treatment, this waste can cause severe environmental issues such as eutrophication and soil pollution. One of the promising pathways of lipid condensate utilization is by converting this material into activated carbon (AC).

Its utilization as a precursor for AC synthesis through $ZnCl_2$ activation holds a promising valorization pathway for palm oil waste streams. However, during the impregnation and activation stage, a large volume of $ZnCl_2$ is typically consumed, leaving significant fraction remaining in the post-treatment effluent [6]. Frequent use of single-use $ZnCl_2$ leads to accumulation of chemical-laden wastewater, posing severe risks of $ZnCl_2$ leaching into aquatic ecosystems and impairing soil health. By developing a sustainable strategy to recover and recycle $ZnCl_2$ from lipid condensate derived AC, this helps in minimizing the environmental burden while simultaneously reducing the production cost of AC.

Separation Technology - ICoST 2025
Materials Research Proceedings 59 (2026) 98-105

Materials Research Forum LLC
https://doi.org/10.21741/9781644903957-13

To address this issue, this study proposed a strategy for $ZnCl_2$ recovery from washing filtrate in post-activation stage as solution. The approach to recovered $ZnCl_2$ can still be able to perform effective activation, yielding results comparable with one produced with fresh $ZnCl_2$. Several studies have proven the effectiveness of recycled $ZnCl_2$ in activation of AC [6,7,8]. In this study, the potential of $ZnCl_2$ recovery is reviewed for converting lipid condensate into AC for dye removal. The aim of this approach is to minimize the chemical effluent accumulated associated with $ZnCl_2$, thus contributing to better sustainable management in activated carbon production. This solution aligns with the United Nations Sustainable Development Goal 12 emphasizing Responsible Consumption and Production, promoting a circular economy within the industrial sector.

Activation Mechanism of $ZnCl_2$

AC typically exhibits an amorphous and non-graphitic micro crystalline with a turbostratic structure [9,10]. This structure formed when AC undergoes two main stages: carbonization and activation [9,10,11,12]. During carbonization process, the precursor is heated at elevated temperature (500-700°C) in an inert atmosphere (e.g., nitrogen or argon) [9,13]. This process removes the volatile component within the precursor to produce a rudimentary carbon matrix. Meanwhile, activation process has role in developing porous structure and enhancing the surface area of the AC. Activation approaches are categorized into physical activation and chemical activation. Physical activation is a method of employing oxidizing gases such as CO_2 or steam at elevated temperature [14] while chemical activation involves the use of chemical activation such as $ZnCl_2$, KOH and H_2SO_4 as dehydrating catalyst for pore formation development [9,10]. Prior studies have reported that chemical activation is among the preferred methods as it yields better results based on surface areas formation, pore size distribution and adsorption capacity at lower activation temperatures compared to physical activation [11].

Activation agents are mainly responsible for altering the thermal decomposition pathway and aid in developing micropores and mesopores in the carbon char upon impregnation. Different activators formed distinctive pore types and yield [10]. Among other activators, $ZnCl_2$ has been extensively employed in AC synthesis owing to its strong dehydrating capability and ability to produce well-developed porous carbon. Its unique characteristics as Lewis acid help in lowering the decomposition temperature of lignocellulosic and cellulosic components in precursor [11,15]. This helps in facilitating the breakdown of precursor while removing moist without melting it during heat treatment (carbonization). The strong dehydration effect plays role in removing oxygen and hydrogen in the form of water while stabilizing the carbon structure. The presence of zinc ions (Zn^{2+}) is responsible for assisting the aromatization of carbon structure while ZnO prevents the carbon structure from collapsing [12]. As the structure stabilizes, more microspores and mesopores are developed within carbon structure, resulting in high surface area.

During activation stage, $ZnCl_2$ acts as dehydrating agent to react with water released from scission of glycosidic bond and elimination of carboxyl and hydroxyl groups to form intermediates $Zn_2OCl_2.2H_2O$ [6,16]. This intermediate, resembles a cement like material is mainly in amorphous phase with minor of crystalline phase, can be found within the AC at temperature of 400-500°C [16]. Subsequently, at temperature above 400°C, the intermediate is decomposed into volatile $ZnCl_2$, ZnO and steam under pyrolysis [6,16]. Eqs. (1)-(2) show the chemical reactions involved. After activation, the unreacted $ZnCl_2$ and ZnO remain within the carbon structure which needs to be removed through washing procedure.

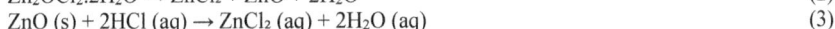

$$2ZnCl_2 + 3H_2O \rightarrow Zn_2OCl_2.2H_2O + 2HCl \qquad (1)$$
$$Zn_2OCl_2.2H_2O \rightarrow ZnCl_2 + ZnO + 2H_2O \qquad (2)$$
$$ZnO\ (s) + 2HCl\ (aq) \rightarrow ZnCl_2\ (aq) + 2H_2O\ (aq) \qquad (3)$$

Recovery of $ZnCl_2$ can be achieved by washing with either water or dilute acid (commonly HCl). When washing the with water, the unreacted $ZnCl_2$ remain within the carbon structure dissolved into the washing filtrate. While washing with acid like HCl will further converts ZnO back into ZnCl₂ as shown in Eq. (3). Thus, the washing filtrate will contain both regenerate ZnCl₂ (for acid washing) and unreacted ZnCl₂ which can be recycled for subsequent activation. This procedure also assists in clearing the blocked pore channels, exposing additional pore and helping to neutralize the AC [6].

Material and Methods

Lipid condensate was obtained from the sterilizer of palm oil mill in Johor state of Malaysia. The chemicals used, zinc chloride (ZnCl2) and methylene blue were supplied by R&M Chemicals (Essex, UK). All the chemicals were analytically graded and used directly as received.

A10 g of lipid condensate was mixed with ZnCl₂, as the activator according to predetermined ratios of 1:2 (lipid condensate: activator). The mixture was stirred thoroughly to ensure even distribution of the activation agent within the lipid condensate. The obtained samples undergo pyrolysis in a furnace at 600°C for 1.5 h, fostering porosity and the desired activation level in the carbon material. The resulting AC was washed with distilled water on a hot plate stirrer to ensure complete elimination of the activating agent. Subsequent to the washing procedure, the sample was dried in an oven overnight. The recovered ZnCl₂ from washing filtrate was reused for another activation process. A new batch of 10 g of lipid condensate was mixed with the recovered ZnCl₂ and dried in an oven overnight before undergoing activation in the furnace, following the same conditions as the initial procedure. This step was repeated for another second recovered ZnCl₂ cycle. The surface area of the lipid condensate AC was measured by BET at liquid nitrogen adsorption of 77 K, using Micromeritics ASAP 2020 analyzer.

The preparation of methylene blue (MB) solution was carried out by dissolving the dye in distilled water to produce a series of concentrations ranging from 10 mg/L to 500 mg/L. The adsorption experiment was conducted by introducing 20 mg of activated carbon into 20 mL of the MB solution. The mixture was maintained at room temperature and allowed to equilibrate for 7 days. The equilibrium adsorption capacity (q_e) was calculated using Eq. (4),

$$q_e = \frac{(C_o - C_e)V}{W} \tag{4}$$

where C_o denoted as initial concentration (mg/L), C_e as equilibrium concentration (mg/L), V as volume of dye solution (L) and W as mass of adsorbent (g).

Results and Discussion

Characteristics of AC

The yield and surface area of lipid condensate AC are presented in Table 1. Lipid condensate AC prepared with fresh ZnCl₂ produced the highest yield compared to ACs obtained using recycled ZnCl₂ (ZR1: 21.96%; ZR2: 18.64%). This trend exhibited a gradual reduction of yield in subsequent activation of AC, implying that the effective amount of ZnCl₂ available becomes lesser with each reuse. During pyrolysis process in subsequent activations, some parts of ZnCl₂ may have volatilized and evaporated while some remained trapped within the carbon channels or on the surface [17]. This reduced the recovery efficiency of the activator. Thus, limited amount of ZnCl₂ was present during activation to facilitate the release of volatile matter, which resulted in lowered yield of recovered samples.

In contrast with the surface area values, sample ZR1 generates higher surface area (205.74 m²/g) in comparison with sample Z2 (25.44 m²/g) and ZR2 (8.99 m²/g). These data suggest that the first cycle of recovered ZnCl₂ has promoted further pore development during subsequent activation. According to [6], ZnCl₂ is not consumed during activation stage as it is regarded as dehydrant. ZnCl₂ is converted into intermediates known as Zn₂OCl₂.2H₂O and further decompose into ZnO [6,16]. Thus, during washing procedure, the washing filtrate is concentrated with recoverable ZnCl₂, as large portion of unreacted ZnCl₂ dissolves in the effluent due to its high-water solubility.

Separation Technology - ICoST 2025   Materials Research Forum LLC
Materials Research Proceedings 59 (2026) 98-105   https://doi.org/10.21741/9781644903957-13

This indicate that the first cycle of recovered $ZnCl_2$ can still performed it capability in pore development. However, the surface area of second cycle of AC (ZR2) shows a drastic decline compared to other samples, likely due to partial pore collapse as low amount of $ZnCl_2$ remaining in the washing filtrate.

Table 1 Yield and surface area of lipid condensate AC samples.

Sample	This work			Palm kernel shell [8]		Lotus root [6]	
	Z2	ZR1	ZR2	F-AC[c]	R-AC[d]	F-AC[c]	R-AC[d]
Yield (%)	25.44	21.96	18.64	70.70	51.40	33.50	48.70
Surface area (m²/g)	107.62	205.74	8.99	858	345	1560	908.8

[c]AC activated by fresh activator, [d]AC activated by recovered activator

These findings highlight the importance of $ZnCl_2$ as activator in facilitating the yield and porosity development in AC synthesis. Similar findings can be observed from prior studies on the use of recovered activators in AC adsorption. A study by [8] has showed that $ZnCl_2$ can be effectively recovered and reuse through washing with water for subsequent activation. In this study, the recycled $ZnCl_2$ can still produce good surface area of 345 m²/g compared to 858 m²/g obtained using fresh $ZnCl_2$ for methylene blue removal. Although the recycled $ZnCl_2$ resulted in a lower surface area, the adsorption performance remained acceptable while significantly reduced chemical cost. Acid washing procedure, on the other hand provides higher recovery efficiency yet generates acidic wastewater that requires neutralization. A study by [6] found that activation with recycled $ZnCl_2$ through acid washing higher yield (48.7%) compared to activation from fresh $ZnCl_2$ (33.5%), with surface area of 908.08 m²/g which is comparable to the first-round activation (1560 m²/g). Another method for $ZnCl_2$ recovery includes modified vacuum pyrolysis (MVP) [7]. In his work, it reported to achieved recovery of $ZnCl_2$ over 99.99% without generating wastewater, while displaying higher adsorption capability compared to traditional pyrolysis method. However, this method is constrained by its complex vacuum system, extensive energy usage as well as tight control of temperature-pressure conditions which makes it unsuitable for low-cost operations.

Fig. 1 depicts the surface morphology of the samples obtained from scanning electron microscopy (SEM). The effectiveness of sample adsorption performance surface area corresponds to the pore size distribution and surface functionality. As observed, sample Z2 and ZR1 display clear pores, cracks and widened openings formed during pyrolysis process. This is due to the presence of $ZnCl_2$ which act as dehydrating agent and facilitates removing the volatile matters from the carbon matrix. As the result, irregular pores and non-uniform pore channels were formed within the samples. Meanwhile for sample ZR2, the sample shows a more compact surface area with fewer distinguishable pores due to partial pore collapse and limited activation in second recovery cycle. These observations correspond well with the surface area results, where sample ZR1 recorded at the highest result, followed by sample Z2 while sample ZR2 exhibits lower surface area due to limited $ZnCl_2$ availability in the washing filtrate for the second recovery cycle. Overall, with the pore structure as shown in sample Z2 and ZR1, these AC able to promote better accessibility for larger MB molecules during adsorption process.

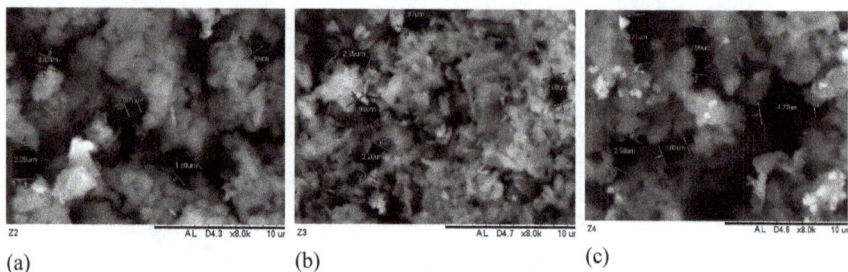

(a)   (b)   (c)

Figure 1 SEM images of (a) Z2, (b) ZR1 and (c) ZR2.

Equilibrium adsorption

Fig. 2 shows the equilibrium adsorption of MB dye by lipid condensate AC. In general, the adsorption process takes place when methylene blue molecules transfer from aqueous phase to the solid surface, where the molecules interact with active site of the AC and subsequently diffuse into the internal pores. In this work, the AC samples were tested with MB concentrations ranging from 5 to 500 mg/L. An increase in dye concentration resulted in higher adsorption capacity of AC, with dye concentration gradient acting as the driving force.

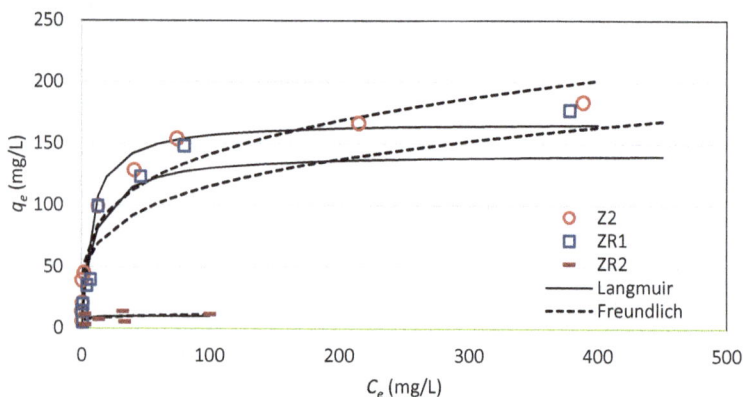

Figure 2 Equilibrium adsorption of MB dye onto AC.

Table 2 Isotherm constant for MB adsorption by lipid condensate AC.

	Z2	ZR1	ZR2
Langmuir			
q_m (mg/g)	168.669	142.911	10.24
k_L (L/mg)	0.133	0.101	0.811
R^2	0.968	0.951	0.116
Freundlich			
k_F (mg/g)(L/mg)$^{1/n}$	43.629	36.0	6.051
n	3.918	3.957	7.374
R^2	0.947	0.796	0.165

Separation Technology - ICoST 2025
Materials Research Proceedings 59 (2026) 98-105

Materials Research Forum LLC
https://doi.org/10.21741/9781644903957-13

At higher dye concentrations, the movement of dye molecules was enhanced by the stronger concentration gradient, thus overcoming the mass transfer resistance at solid phase [18,19].

The variation in adsorption performance can be observed among the AC samples even though they were prepared at the same impregnation ratio. The differences were mainly due to the recovery cycle of ZnCl₂. Sample Z2 displayed the highest maximum adsorption capacity of 183.73 mg/g, followed by ZR1 with 177.16 mg/g meanwhile ZR2 only showed low capacity of 11.74 mg/g. Sample Z2 and ZR1 performed superior results due to larger numbers of accessible pores and active site presence in the AC. While sample ZR2 poor performance is due to its underdeveloped porosity and limited surface area. These findings demonstrate the importance of pore developments in adsorption performance of AC.

The findings were further fitted to Langmuir and Freundlich models to understand the adsorption mechanism of the AC. These data were calculated through non-linear regression using *Solver* of MS Excel. Table 2 summarizes the isotherm constants of Langmuir and Freundlich models for methylene blue adsorption by lipid condensate AC. From Table 2, Langmuir model provided the best fit for adsorption data compared to Freundlich model where sample Z2 and ZR1 recorded with high coefficient of determination (R^2) of 0.968 and 0.951, respectively. This suggests that adsorption of these samples had mainly occurred through the monolayer coverage on homogeneous surface [20]. In contrast with sample ZR2, both models displayed poor agreement results, confirming that the sample is not an effective AC under the tested conditions. The Langmuir constant, k_L is a significant parameter that reflects the adsorption affinity between the adsorbent and dye molecules at low concentrations. Table 2 shows that sample Z2 ($k_L = 0.133$) and ZR1 ($k_L = 0.101$) have relatively high affinity toward methylene blue dye, indicating stronger interactions at low concentrations.

The Freundlich isotherm constants also demonstrated a favourable adsorption result where *n* values of sample Z2 and ZR1 were recorded as 3.918 and 3.957, respectively. Yet, sample ZR2 showed an unusually high *n* value (7.374) with poor correlation, suggesting unreliable fitting. The Freundlich constants, k_F values for sample Z2 (43.6) and ZR1 (36.0) were also higher in comparison to sample ZR2, validating its stronger adsorption capacity and affinity at the stated conditions.

Conclusion

This study demonstrated the recovery and reuse of ZnCl₂ for lipid condensate-based AC synthesis. The results showed that the first cycle of recovered ZnCl₂ produced AC with higher surface area (205.74 m²/g) compared to lipid condensate activated with fresh ZnCl₂ (107.62 m²/g), while maintaining comparable adsorption capacity (177.16 mg/g) and yield (21.96%). However, the subsequent activation using recovered ZnCl₂ resulted in a drastic decline in both surface area (18.64 m²/g) and adsorption capacity (11.74 mg/g), indicating the limited availability of ZnCl₂ remained in the recovered solution for effective activation. Overall, these findings highlight the potential strategy for minimizing reliance on single-use chemicals while offering a practical and scalable pathway for sustainable AC production by reducing costs, improving resource efficiency and supporting cleaner processing of palm oil industry waste.

Acknowledgement
This work was fully funded by Sarawak state government, and Land and Survey Department, Sarawak through Kursi Premier Sarawak Research Grant No. 1R037.

References
[1] N. Qurratu, ain Mohammad, N.A. Hazren Hamid, Palm Kernel Shell as Potential Adsorbent for Treatment and Decolorization of Palm Oil Mill Effluent (POME), 5 (2024) 557–562. https://doi.org/10.30880/peat.2024.05.01.059

[2] H. Kamyab, S. Chelliapan, M.F.M. Din, S. Rezania, T. Khademi, A. Kumar, Palm Oil Mill Effluent as an Environmental Pollutant, in: Palm Oil, InTech, 2018. https://doi.org/10.5772/intechopen.75811

[3] P. Nyawi, Biomass Palm oil as Renewable Energy, Malaysian Sustainable Palm Oil (2022).

[4] M.A. Bin Mohd Yusof, Y.J. Chan, C.H. Chong, C.L. Chew, Effects of operational processes and equipment in palm oil mills on characteristics of raw Palm Oil Mill Effluent (POME): A comparative study of four mills, Cleaner Waste Systems 5 (2023) 100–101. https://doi.org/10.1016/j.clwas.2023.100101

[5] N. Jusoh, M.B. Rosly, N. Othman, H.A. Rahman, N.F.M. Noah, R.N.R. Sulaiman, Selective extraction and recovery of polyphenols from palm oil mill sterilization condensate using emulsion liquid membrane process, Environmental Science and Pollution Research 27 (2020) 23246–23257. https://doi.org/10.1007/s11356-020-07972-5

[6] M. Fan, Y. Shao, Y. Wang, J. Sun, H. He, Y. Jiang, S. Zhang, Y. Wang, X. Hu, Preparation of activated carbon with recycled ZnCl2 for maximizing utilization efficiency of the activating agent and minimizing generation of liquid waste, Chemical Engineering Journal 500 (2024) 157–278. https://doi.org/10.1016/j.cej.2024.157278

[7] Q.-F. Wu, F.-S. Zhang, A clean process for activator recovery during activated carbon production from waste biomass, Fuel 94 (2012) 426–432. https://doi.org/10.1016/j.fuel.2011.08.059

[8] M.A.A. Zaini, W.M. Tan, M.J. Kamaruddin, S.H.M. Setapar, M.A.C. Yunus, Microwave-Induced Zinc Chloride Activated Palm Kernel Shell for Dye Removal, 2014.

[9] C. Xia, S.Q. Shi, Self-activation for activated carbon from biomass: theory and parameters, Green Chemistry 18 (2016) 2063–2071. https://doi.org/10.1039/C5GC02152A

[10] Y. Gao, Q. Yue, B. Gao, A. Li, Insight into activated carbon from different kinds of chemical activating agents: A review, Science of The Total Environment 746 (2020) 141094. https://doi.org/10.1016/j.scitotenv.2020.141094

[11] J. Andas, N. Safinaz Naserun, Synthesis, Parameter Optimization and Characterization of ZnCl2 Activated Carbon Derived from Waste Tamarind Seed, 2025.

[12] B. Li, C. Li, D. Li, L. Zhang, S. Zhang, Z. Cui, D. Wang, Y. Tang, X. Hu, Activation of pine needles with zinc chloride: Evolution of functionalities and structures of activated carbon versus increasing temperature, Fuel Processing Technology 252 (2023) 107987. https://doi.org/10.1016/j.fuproc.2023.107987

[13] S.G. Herawan, M.S. Hadi, Md.R. Ayob, A. Putra, Characterization of Activated Carbons from Oil-Palm Shell by CO_2 Activation with No Holding Carbonization Temperature, The Scientific World Journal 2013 (2013). https://doi.org/10.1155/2013/624865

[14] I. Yang, M. Jung, M.-S. Kim, D. Choi, J.C. Jung, Physical and chemical activation mechanisms of carbon materials based on the microdomain model, J Mater Chem A Mater 9 (2021) 9815–9825. https://doi.org/10.1039/D1TA00765C

[15] R. Farma, R.I. Julita, I. Apriyani, A. Awitdrus, E. Taer, ZnCl2-assisted synthesis of coffee bean bagasse-based activated carbon as a stable material for high-performance supercapacitors, Mater Today Proc 87 (2023) 25–31. https://doi.org/10.1016/j.matpr.2023.01.370

[16] B. Li, J. Hu, H. Xiong, Y. Xiao, Application and Properties of Microporous Carbons Activated by ZnCl $_2$: Adsorption Behavior and Activation Mechanism, ACS Omega 5 (2020) 9398–9407. https://doi.org/10.1021/acsomega.0c00461

[17] M. Abbas Ahmad Zaini, T. Wee Meng, M. Johari Kamaruddin, Microwave-Induced Zinc Chloride Activated Palm Kernel Shell for Dye Removal, 2014.

[18] P.E. Hock, M.A.A. Zaini, Zinc chloride–activated glycerine pitch distillate for methylene blue removal—isotherm, kinetics and thermodynamics, Biomass Convers Biorefin 12 (2022) 2715–2726. https://doi.org/10.1007/s13399-020-00828-5

[19] F. Amran, T. Sarawanan, Y.K. Qi, A. Azmi, A. Arsad, M.A.A. Zaini, Coconut shell carbon via phosphoric acid activation for rhodamine B, malachite green, and methylene blue adsorption–equilibrium and kinetics, Int J Phytoremediation (2024). https://doi.org/10.1080/15226514.2024.2399062

[20] M.A. Akl, A.G. Mostafa, M. Al-Awadhi, W.S. Al-Harwi, A.S. El-Zeny, Zinc chloride activated carbon derived from date pits for efficient biosorption of brilliant green: adsorption characteristics and mechanism study, Appl Water Sci 13 (2023). https://doi.org/10.1007/s13201-023-02034-w

Separation Technology - ICoST 2025
Materials Research Proceedings 59 (2026) 106-112

Materials Research Forum LLC
https://doi.org/10.21741/9781644903957-14

Simulation-Based Comparative Study of CO_2 Absorption Techniques for Biogas Purification Using Aspen HYSYS and Aspen Plus

Aishah ROSLI[1,a*], Aziatul Niza SADIKIN[1,b]

[1]Faculty of Chemical and Energy Engineering, Universiti Teknologi, Malaysia

[a]aishahrosli@utm.my, [b]aziatulniza@utm.my

Keywords: Biogas Upgrading, CO_2 Removal, Chemical Absorption, Water Scrubbing

Abstract. Upgrading biogas by removing CO_2 is essential to increase its energy value and usability as a clean fuel. This study compares three CO_2 absorption techniques using simulation: chemical absorption with monoethanolamine (MEA), diethanolamine (DEA), methyl diethanolamine (MDEA) with piperazine (PZ), and physical absorption with water. The MEA, DEA, and MDEA/PZ systems were simulated using Aspen HYSYS, while water scrubbing was simulated using Aspen Plus. The simulations used standardized biogas composition and analyzed the effect of parameters such as solvent concentration, temperature, pressure, and liquid-to-gas (L/G) ratio. MEA showed a maximum removal efficiency of 99.28% at 35 mol% and 60°C. MDEA/PZ performed well under low-pressure conditions, with improved absorption due to PZ activation. Water scrubbing showed better CO_2 removal at higher pressures and L/G ratios, but was less selective than amine solvents. The results highlight trade-offs between chemical and physical absorption: amine-based systems offer higher efficiency but may involve higher regeneration costs, while water scrubbing provides a more straightforward and cost-effective option. This study provides valuable insights for selecting efficient and sustainable biogas upgrading methods.

Introduction

The rapid depletion of fossil fuels and their associated greenhouse gas emissions have accelerated the search for cleaner and more sustainable energy alternatives. Fossil-based resources such as oil, natural gas, and coal have been extensively utilized for centuries, resulting in enormous CO_2 emissions that contribute to global warming and climate change [1]. Regions with high energy consumption, including the European Union, the USA, and China, have enacted policies to reduce reliance on fossil fuels, with targets such as increasing renewable energy use and cutting greenhouse gas emissions by 2030 [2]. In this context, biogas has emerged as a promising renewable energy source that can play an important role in reducing environmental impacts while meeting growing energy demands.

Biogas is produced through anaerobic digestion of organic matter, including agricultural residues, livestock manure, agro-industrial waste, and municipal sewage sludge [3,4]. Its composition typically consists of methane (CH_4, 35–75%) and carbon dioxide (CO_2, 15–60%), with small amounts of hydrogen sulfide (H_2S), ammonia (NH_3), water vapor, and other trace gases [5,6]. The presence of CO_2 lowers the heating value of biogas, while H_2S contributes to corrosion of engine and pipeline components. Furthermore, uncontrolled CO_2 emissions from biogas combustion can exacerbate greenhouse effects and global warming [7]. Consequently, biogas upgrading—removing CO_2 and other impurities to produce biomethane with high CH_4 content— has become a crucial step to expand the utilization of biogas in electricity generation, natural gas grid injection, transportation fuels, and fuel cells [8,9].

A variety of upgrading technologies are available, including pressure swing adsorption (PSA), membrane separation, cryogenic separation, physical absorption using water scrubbing, and chemical absorption using solvents [10–12]. Among these, water scrubbing and amine-based chemical absorption are the most widely adopted due to their high efficiency and practical

Separation Technology - ICoST 2025   Materials Research Forum LLC
Materials Research Proceedings 59 (2026) 106-112   https://doi.org/10.21741/9781644903957-14

feasibility. Water scrubbing offers simplicity, low toxicity, and the simultaneous removal of CO_2 and H_2S, though it requires significant water usage and is less selective [13,14]. Chemical absorption using alkanolamine solvents such as monoethanolamine (MEA), diethanolamine (DEA), and methyl diethanolamine (MDEA) is highly effective, providing low methane losses and high CO_2 selectivity [15–17]. Nevertheless, amine-based systems often incur high energy consumption during solvent regeneration and may lead to corrosion or solvent degradation over time [18].

Process simulation has proven to be a powerful tool to evaluate and optimize biogas upgrading techniques. Aspen HYSYS is widely used for modeling chemical absorption processes, while Aspen Plus is frequently employed for physical absorption and water scrubbing simulations [19]. Previous studies have shown that MEA achieves very high CO_2 removal efficiencies, while blends such as MDEA with piperazine (PZ) offer lower regeneration energy requirements [20]. Meanwhile, water scrubbing performance is strongly influenced by absorber pressure and liquid-to-gas (L/G) ratios [21]. Despite these advances, few studies have systematically compared amine-based and water scrubbing systems under consistent simulation frameworks, leaving a gap in understanding their relative advantages and limitations.

This study addresses this gap by conducting a simulation-based comparative analysis of chemical and physical absorption techniques for CO_2 removal from biogas. Using Aspen HYSYS, the performance of MEA, DEA, and MDEA/PZ solvents was evaluated under different operating conditions, while Aspen Plus was employed to model a water scrubbing system. Key parameters, including solvent concentration, absorber temperature, gas flow rate, absorber pressure, and liquid-to-gas ratio, were systematically investigated. The findings provide a comprehensive assessment of the strengths and weaknesses of both absorption approaches, offering practical insights for optimizing biogas upgrading strategies toward more sustainable and economically viable applications.

Process Simulation Model

This study employed Aspen HYSYS and Aspen Plus process simulators to evaluate and compare the performance of chemical and physical absorption techniques for CO_2 removal from biogas. Three simulation models were developed and validated: (i) MEA absorption in Aspen HYSYS, (ii) alternative amine solvents (DEA and MDEA with PZ) in Aspen HYSYS, and (iii) water scrubbing in Aspen Plus. The raw biogas feed was standardized across all models to enable fair comparison. Based on literature data [13], the feed composition was defined as 60% CH_4, 38.9% CO_2, 300 ppm H_2S, 0.5% N_2, and 0.5% O_2. Operating pressures and flow rates were adjusted in each case according to the process requirements.

The first simulation model focused on MEA for CO_2 capture. Aspen HYSYS v11 was used with the Acid Gas–Chemical Solvent property package. The absorber was modeled as a counter-current packed column with 10 stages. Biogas was introduced at the bottom at 4 bar, while a lean MEA stream at 45°C was fed from the top. A compressor and cooler were used to condition the biogas feed. Rich MEA exiting the absorber was regenerated in a 22-stage stripper operating at 7 bar, with a heat exchanger and reboiler facilitating solvent recovery. A recycle stream, make-up water, and solvent adjustments were incorporated to achieve convergence. Parameters studied included MEA concentration (15–35 mol%), absorber temperature (40–60°C), and gas flow rate. The performance was assessed based on CO_2 removal efficiency and CH_4 purity.

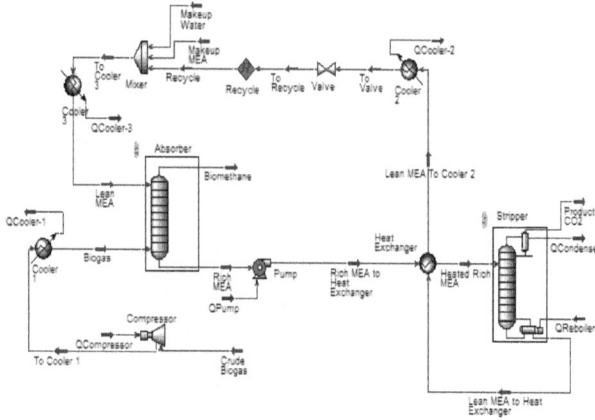

Figure 1 *Aspen HYSYS process flow diagram for CO₂ removal from biogas using MEA, showing the effects of solvent concentration, gas flow rate, and absorber temperature.*

The second Aspen HYSYS model evaluated DEA and MDEA, with the latter also tested in combination with piperazine (PZ) as an activator. The feed gas stream was set at 20°C, 18 bar, and 1000 kmol/h with 60% CH_4 and 39.9% CO_2. The lean amine stream (65 wt% H_2O, 35 wt% solvent) entered at 20°C, 18 bar, and 5000 kmol/h. The absorber was configured with 20 stages, and the rich amine was regenerated in a 20-stage distillation column at 2.5 bar with full reflux at the condenser. This model was used to compare the relative performance of DEA and MDEA/PZ in terms of CO_2 absorption capacity and methane recovery.

A high-pressure water scrubbing system was developed in Aspen Plus using the Non-Random Two-Liquid (NRTL) property package, suitable for polar mixtures at pressures up to 10 bar. Biogas was introduced at the bottom of the absorber column at 10 bar, counter-current to the water stream. Sensitivity analysis was conducted for absorber pressures ranging from 2 to 20 bar, to determine the effect on CH_4 purity and CO_2 removal efficiency. The effect of varying the L/G ratio (0.5–1.5 m³/m³) was also studied. The biogas flow rate was held constant, while water flow was adjusted. Higher water flow rates were expected to increase the mass transfer coefficient due to greater solvent availability for CO_2 absorption [21].

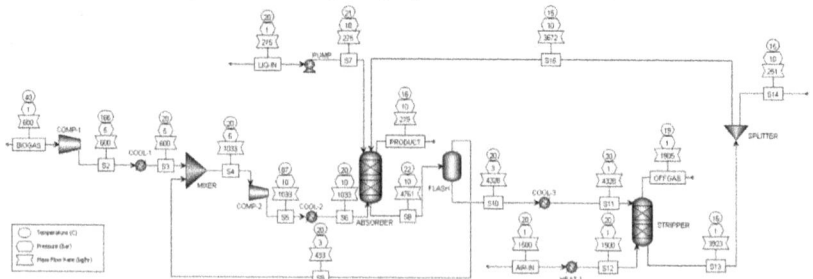

Figure 2 *Aspen Plus simulation model for CO₂ removal from biogas using water scrubbing with regeneration and recirculation of water.*

Separation Technology - ICoST 2025
Materials Research Proceedings 59 (2026) 106-112

Materials Research Forum LLC
https://doi.org/10.21741/9781644903957-14

The performance of each simulation model was evaluated in terms of CO_2 removal efficiency, methane purity in the upgraded gas, solvent or water consumption, and the qualitative assessment of energy demand. For the amine-based systems, emphasis was placed on the influence of solvent concentration, absorber temperature, and gas flow rate, while also considering the regeneration requirements associated with solvent recycling. For the water scrubbing process, attention was given to the effects of absorber pressure and liquid-to-gas ratio on CO_2 solubility and methane recovery, as well as the implications of water demand.

The outcomes of the three models were then compared to establish the relative advantages and limitations of chemical and physical absorption approaches. Amine-based systems were expected to achieve higher selectivity and CO_2 removal efficiency, albeit at higher energy costs for solvent regeneration and with potential corrosion issues. In contrast, water scrubbing was anticipated to provide a more straightforward and environmentally friendly alternative, though at the cost of greater water consumption and lower selectivity. This comparative framework allowed for a systematic assessment of the trade-offs between the two methods, offering insights into their suitability for different scales of application and operational contexts.

Results and Discussion

The simulation results from three different biogas upgrading techniques, chemical absorption using MEA, DEA, and MDEA/PZ, as well as physical absorption via water scrubbing, demonstrate distinct performance characteristics, highlighting the trade-offs between efficiency, operating conditions, and resource requirements. When comparing different amine solvents, notable differences were observed among MEA, DEA, and the MDEA/PZ blend. MEA consistently achieved high CO_2 removal efficiencies at moderate concentrations, reaching approximately 98–99% removal at 30–35 mol%. DEA, however, required a much higher concentration to achieve comparable performance; at 55 wt%, it achieved 99.86% CO_2 removal, which is still slightly below that of 35 wt% MEA. The MDEA/PZ blend showed the strongest performance among the solvents tested. A mixture containing 10 wt% PZ with 40 wt% MDEA achieved 99.69% CO_2 removal at only 10 bar, highlighting the significant promoting effect of piperazine in accelerating CO_2 absorption in tertiary amines. This demonstrates why MDEA/PZ systems are widely regarded as low-energy, high-performance absorbents.

MEA showed a strong temperature dependence, increasing from 77% at 20°C to 99% at 60°C. DEA exhibited a similar trend but with significantly lower removal at all temperatures, achieving only around 60% at 20°C and requiring temperatures above 50°C to exceed 95% removal. In contrast, the MDEA/PZ blend displayed the highest efficiency across the entire temperature range, achieving >90% removal even at 20°C and approaching 100% removal above 40°C. This highlights the strong catalytic effect of PZ in enhancing the absorption kinetics of tertiary amines, making MDEA/PZ the least temperature-dependent and most effective solvent system. Comparable results have been reported by Abu-Zahra et al. (2007), who demonstrated that increasing operating temperatures improves absorption kinetics but must be balanced with solvent regeneration requirements [15].

Fig. 4 shows the performance of MDEA blended with PZ, where adding 10 wt% PZ to 40 wt% MDEA resulted in 99.86% CO_2 removal and 99.69 mol% methane purity at only 10 bar operating pressure. This demonstrates the effectiveness of PZ as an activator, enabling higher absorption efficiency at much lower pressures than MEA (18 bar) and DEA (22 bar), thereby providing significant energy savings during operation. Such findings are consistent with Bishnoi and Rochelle (2000), who reported that piperazine significantly enhances the kinetics of CO_2 absorption in MDEA systems, thereby reducing the overall energy penalty of the process [16].

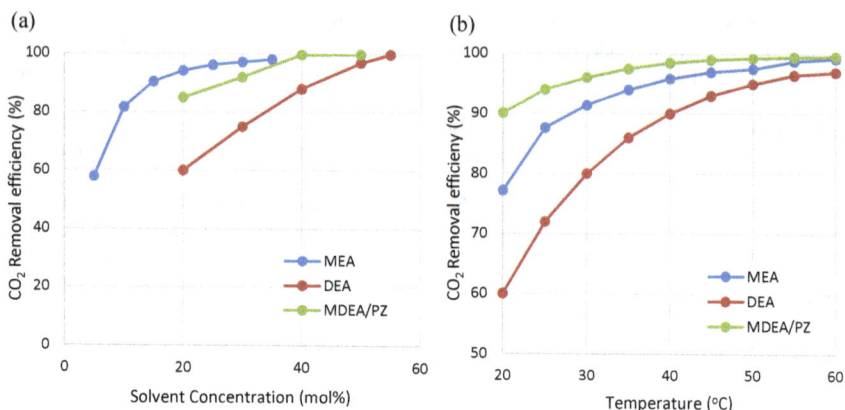

Figure 3 *(a) CO_2 removal efficiency vs. solvent concentration, (b) CO_2 removal efficiency vs. absorber temperature.*

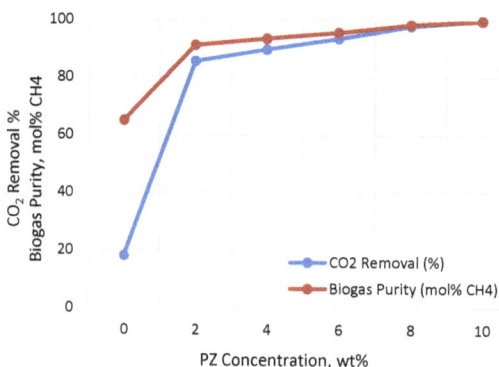

Figure 4 Effect of different concentrations of PZ in MDEA solution on CO_2

removal using the Aspen HYSYS simulator.

In contrast, water scrubbing offered a more straightforward yet equally effective option for CO_2 removal. Fig. 5 shows the impact of absorber pressure on methane purity and CO_2 removal efficiency, with performance steadily improving up to 10 bar. At this optimal point, methane purity reached 96.5 mol% while CO_2 removal was 99.96%. These results align with previous studies, which have shown that higher operating pressures increase CO_2 solubility and enhance absorption efficiency in water scrubbing systems [13]. Although increasing the liquid-to-gas (L/G) ratio further improved CO_2 absorption, it slightly decreased methane purity because CH_4 dissolved in water. This trade-off emphasizes water scrubbing's dependence on pressure and water consumption, balancing efficiency with operational costs and resource use.

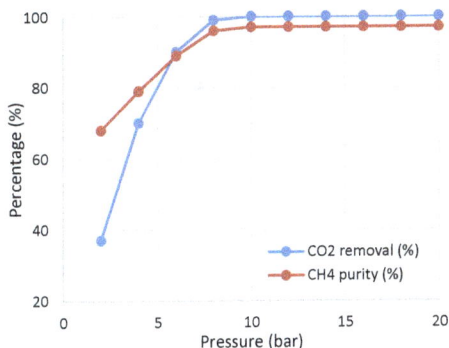

Figure 5 *Effect of pressure on CH₄ purity and CO₂ removal.*

Overall, the comparative results indicate that amine-based absorption, particularly MEA and MDEA/PZ, achieves superior methane purity and nearly complete CO_2 removal. MEA delivers high efficiency at moderate concentrations but requires more energy for regeneration, while DEA is less efficient and needs higher concentrations. The MDEA/PZ mixture stands out as the most energy-efficient chemical option because of its excellent performance at lower pressures. Water scrubbing, although less selective in terms of methane purity, achieves competitive CO_2 removal and is appealing due to its simplicity and environmental friendliness. These findings align with other comparative studies that have highlighted chemical absorption as more efficient but resource-intensive. In contrast, water scrubbing offers a cost-effective and operationally simple alternative for small- to medium-scale applications [13].

Conclusion

This study compared chemical and physical absorption techniques for biogas upgrading using Aspen HYSYS and Aspen Plus simulations. The results showed that MEA achieved a high CO_2 removal efficiency (up to 99.28%) under optimal conditions, whereas DEA required higher solvent concentrations to achieve a similar level of removal. The MDEA/PZ blend demonstrated superior performance at lower pressure, offering significant energy savings without sacrificing methane purity. Water scrubbing, although slightly less selective (methane purity of 96.5 mol%), provided nearly complete CO_2 removal (99.96%) at 10 bar and offered the advantages of simplicity and environmental friendliness. Overall, the findings suggest that amine-based absorption, particularly MDEA/PZ, is best suited for applications requiring high efficiency and lower operating pressures. At the same time, water scrubbing remains a practical and sustainable option when cost and operational simplicity are prioritized.

Acknowledgment

This work was supported/funded by the Ministry of Higher Education under Fundamental Research Grant Scheme (FRGS/1/2021/TK0/UTM/02/1), (Vot. 5F430).

References

[1] International Energy Agency (IEA). (2013). World energy outlook. Paris: OECD/IEA. https://doi.org/10.1787/weo-2013-en

[2] European Commission. (2014). 2030 framework for climate and energy policies. Brussels: European Union.

[3] Weiland, P. (2010). Biogas production: Current state and perspectives. Applied Microbiology and Biotechnology, 85(4), 849-860. https://doi.org/10.1007/s00253-009-2246-7

[4] Angelidaki, I., Treu, L., Tsapekos, P., Luo, G., Campanaro, S., Wenzel, H., & Kougias, P. G. (2018). Biogas upgrading and utilization: Current status and perspectives. Biotechnology Advances, 36(2), 452-466. https://doi.org/10.1016/j.biotechadv.2018.01.011

[5] Ryckebosch, E., Drouillon, M., & Vervaeren, H. (2011). Techniques for transformation of biogas to biomethane. Biomass and Bioenergy, 35(5), 1633-1645. https://doi.org/10.1016/j.biombioe.2011.02.033

[6] Sun, Q., Li, H., Yan, J., Liu, L., Yu, Z., & Yu, X. (2015). Review of biogas upgrading technologies and applications in China. Renewable and Sustainable Energy Reviews, 51, 521-532. https://doi.org/10.1016/j.rser.2015.06.029

[7] Abatzoglou, N., & Boivin, S. (2009). A review of biogas purification processes. Biofuels, Bioproducts and Biorefining, 3(1), 42-71. https://doi.org/10.1002/bbb.117

[8] Persson, M., Jönsson, O., & Wellinger, A. (2006). Biogas upgrading to vehicle fuel standards and grid injection. IEA Bioenergy Task 37.

[9] Bauer, F., Hulteberg, C., Persson, T., & Tamm, D. (2013). Biogas upgrading - Review of commercial technologies. SGC Rapport 2013:270.

[10] Mandal, B., & Bandyopadhyay, S. S. (2005). Absorption of carbon dioxide into aqueous blends of 2-amino-2-methyl-1-propanol and monoethanolamine. Chemical Engineering Science, 60(22), 6438-6451. https://doi.org/10.1016/j.ces.2005.02.044

[11] Khan, I. U., et al. (2017). Membrane-based biogas upgrading processes: Review of commercial applications and ongoing R&D. Separation and Purification Technology, 188, 399-425.

[12] Kapdi, S. S., Vijay, V. K., Rajesh, S. K., & Prasad, R. (2005). Biogas scrubbing, compression and storage: Perspective and prospectus in Indian context. Renewable Energy, 30(8), 1195-1202. https://doi.org/10.1016/j.renene.2004.09.012

[13] Cozma, P., Wukovits, W., & Friedl, A. (2014). Influence of process parameters on the performance of biogas upgrading by high pressure water scrubbing. Environmental Engineering and Management Journal, 13(9), 2237-2244.

[14] Rasi, S., Veijanen, A., & Rintala, J. (2007). Trace compounds of biogas from different biogas production plants. Energy, 32(8), 1375-1380. https://doi.org/10.1016/j.energy.2006.10.018

[15] Abu-Zahra, M. R. M., Schneiders, L. H. J., Niederer, J. P. M., Feron, P. H. M., & Versteeg, G. F. (2007). CO_2 capture from power plants: Part I. A parametric study of the technical performance based on monoethanolamine. International Journal of Greenhouse Gas Control, 1(1), 37-46. https://doi.org/10.1016/S1750-5836(06)00007-7

[16] Bishnoi, S., & Rochelle, G. T. (2000). Absorption of carbon dioxide into aqueous piperazine/alkanolamine solutions. Industrial & Engineering Chemistry Research, 39(11), 4441-4446.

Separation Technology - ICoST 2025
Materials Research Proceedings 59 (2026) 113-120

Materials Research Forum LLC
https://doi.org/10.21741/9781644903957-15

Supercritical–CO2 Dried Biodegradable Alginate/Zirconia Aerogels: Synthesis and Characterization

Nur Hazwani Dalili MOHAMAD[1,a], Ana Najwa MUSTAPA[1,2,b] *,
Suhaiza Hanim HANIPAH[1,c]

[1]Faculty of Chemical Engineering, Universiti Teknologi MARA, UiTM, 40450 Shah Alam, Selangor, Malaysia

[2]Integrated Separation Technology Research Group, Faculty of Chemical Engineering, Universiti Teknologi MARA, UiTM, 40450 Shah Alam, Selangor, Malaysia

[a]dalilihazwani0901@gmail.com, [b]anajwa@uitm.edu.my, [c]suhaizahanim@uitm.edu.my

Keywords: Aerogel, Alginate, Zirconia, Supercritical Drying, Thermal Insulation

Abstract. This work presents the synthesis of new biodegradable alginate-zirconia (AG-ZR) hybrid aerogels as a potential thermal insulator material. The effect of the alginate-to-zirconia ratio on their physical and chemical characteristics is investigated. Hybrid gels containing 3–5 wt% alginate and 0.05–0.2 wt% zirconia were prepared via solvent casting, followed by supercritical carbon dioxide (Sc-CO2) drying to obtain lightweight aerogels. The physical and chemical properties were characterized by Fourier-transform infrared spectroscopy (FTIR), thermogravimetric analysis (TGA), and differential scanning calorimetry (DSC) analysis. Results showed that the chemical interaction between alginate and zirconia was identified by the characteristic $Zr-O$ and $-COO-$ stretching vibrations. Thermogravimetric analysis (TGA) indicated considerable thermal stability up to 350°C, while differential scanning calorimetry (DSC) suggested that degradation commenced at approximately 270°C. Mechanical testing revealed that the aerogel formulation composed of 5 wt.% alginate and 0.2 wt.% zirconia exhibited a notable tensile strength of approximately 1.8 MPa and improved flexibility, as indicated by elongation at break. The synthesized AG-ZR hybrid aerogels showed strong chemical crosslinking, excellent thermal stability, and superior mechanical strength, particularly in AG 5:ZR 0.2 formulation. The findings demonstrate the potential of the material as a lightweight, biodegradable, and thermally stable insulation solution for advanced engineering applications.

Introduction

Advanced thermal insulation technologies have become increasingly important for improving energy efficiency and reducing greenhouse gas emissions in construction, transportation, and industrial applications. Conventional insulation materials such as fiberglass, mineral wool, expanded polystyrene, and polyurethane foams are widely used for their low cost and reasonable performance; however, they exhibit low thermal stability, weak mechanical resistance, and poor environmental compatibility due to their petroleum-based origins and lack of biodegradability [1]. These limitations have accelerated the search for sustainable, high-performance alternatives.

Aerogels, lightweight and highly porous materials produced via sol–gel processes, have emerged as next-generation insulators with ultra-low thermal conductivities (as low as 0.013 W/m·K) [2], large surface areas, and porosities exceeding 90% [3]. Biopolymer-based aerogels, in particular, represent an eco-friendly approach to thermal insulation because they combine renewability, biodegradability, and extremely low densities (0.01–0.5 g/cm³) [4] with thermal conductivities often below 0.02 W/m·K [5]. Among biopolymers, alginate, a polysaccharide extracted from brown algae is notable for its ability to form crosslinked hydrogels through carboxylate groups, offering flexibility and biodegradability. Xu et al. [6] reported that alginate aerogels reinforced with nanoclay achieved high porosity, enhanced mechanical strength, and flame-retardant properties, demonstrating that inorganic nanofillers can strengthen the network structure and improve overall performance.

While early aerogel research centered on silica systems, recent studies have extended to inorganic frameworks such as zirconia (ZrO_2), which exhibits high-temperature stability, chemical resistance, and low thermal conductivity [7,8,9]. However, pure zirconia aerogels are brittle and prone to collapse under capillary forces during conventional drying [10]. Supercritical CO_2 drying addresses this limitation by removing the solvent above its critical point, eliminating surface tension, and preserving the delicate pore structure [11,12]. Hybridizing zirconia with flexible polymers or biopolymers can further improve toughness and introduce biodegradability, creating composites with synergistic properties suitable for sustainable insulation [13,14].

In this study, alginate–zirconia (AG/ZR) hybrid aerogels were synthesized using a solvent-casting approach followed by supercritical CO_2 drying. Alginate provided the biodegradable polymeric framework, while zirconia precursors contributed to enhanced thermal stability and mechanical strength. The influence of alginate and zirconia ratios on the aerogels' physical and chemical properties was examined through Fourier-transform infrared spectroscopy (FTIR), thermogravimetric analysis (TGA), differential scanning calorimetry (DSC), and mechanical testing. This work aims to develop biodegradable, high-performance aerogels with strong thermal and mechanical resilience for advanced insulation applications.

Materials and Methods

All reagents used are analytical grade and employed without further purification. Zirconium (IV) propoxide solution (tetrapropyl zirconate, TPZ), N, N-dimethylformamide (DMF), and alginic acid sodium salt (from brown algae) were acquired from Sigma-Aldrich. Diethylamine as a catalyst and calcium chloride ($CaCl_2$) were obtained from Chemiz (Malaysia). Absolute ethanol was purchased from Systerm. Carbon dioxide (CO_2, 99.9%) used for the supercritical drying process was obtained from a certified gas supplier.

Preparation of alginate/zirconia aerogel film

Fig. 1 shows the presented work to synthesize the alginate/zirconia hybrid aerogel [15,16]. Alginate solutions (AG) with concentrations of 3 and 5 wt.% are prepared in 50 mL of distilled water and stirred until fully dissolved.

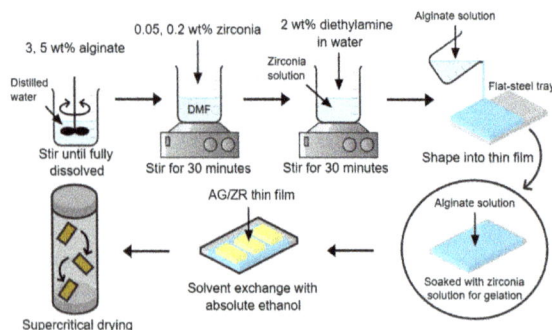

Figure 1 Schematic diagram of the preparation of alginate/zirconia aerogel film.

A 0.05 wt.% and 0.2 wt.% concentrations of zirconium (IV) propoxide solution (tetrapropyl zirconate, TPZ) are mixed into 80 mL of N, N-dimethylformamide (DMF) with constant stirring for 30 min. Then, a stoichiometric amount of distilled water containing 2 wt.% of diethylamine

Separation Technology - ICoST 2025
Materials Research Proceedings 59 (2026) 113-120

Materials Research Forum LLC
https://doi.org/10.21741/9781644903957-15

(DEA) is added to the zirconia solutions and continuously stirred for another 30 min. The prepared alginate solution is then shaped into a thin film by using a flat-steel tray and is soaked with the zirconia solution for 30 minutes to allow the complete gelation of the hydrogel. After being cut to a 6 cm × 4 cm size, the alginate/zirconia thin film was exchanged twice using pure ethanol. Finally, the samples were dried under supercritical CO_2 at 110 ± 5 bar at 40°C for a 4-5 h drying time. The formulation of the materials used for the synthesis is presented in Table 1.

Table 1 Formulation of the materials for the synthesis of alginate/zirconia aerogel film.

Alginate, AG (wt.%)	Zirconia, ZR (wt.%)	Diethylamine in water (wt.%)	Indication
3	0	-	Blank AG
3	0.05	2	AG 3: ZR 0.05
	0.2		AG 3: ZR 0.2
5	0.05	2	AG 5: ZR 0.05
	0.2		AG 5: ZR 0.2

Characterization

Fourier transform infrared spectroscopy (FTIR) was used to identify functional groups of the biopolymer films using an OPUS Optik GmBH-equipped apparatus, scanned from 400 to 4000 cm^{-1}. Thermogravimetric analysis (TGA) evaluated thermal stability between 20 and 600°C at a 10 °C/min heating rate using a Mettler Toledo SAE system. Differential scanning calorimetry (DSC) (Model Mettler Toledo) was performed at 10 °C/min from 20 to 500°C to determine thermal transitions [15]. Mechanical properties were measured at room temperature using a 5500 Instron machine following ASTM D822 standards, with a crosshead speed of 50 mm/min. Tensile strength was determined at the peak of the stress–strain curve, and three specimens were tested for each formulation [17].

Results and Discussion

FTIR spectroscopy provides insights into the chemical structure, functional groups, and intermolecular interactions within materials by analyzing the absorption of infrared radiation. The FTIR graph in Fig. 2 shows the spectra for blank AG and various AG-ZR formulations across a broad wavenumber range (4000-400 cm^{-1}).

All spectra, including the blank AG, display characteristic absorption bands typical of polysaccharides. A broad and intense band around 3500-3000 cm^{-1} is due to the stretching vibrations of O-H groups, originating from both the hydroxyl groups within the alginate structure and adsorbed water molecules [10,18]. C-H stretching vibrations are observed at bands around 2900 cm^{-1}, indicating the integrity of the carbohydrate backbone. The region between 1700 cm^{-1} and 1400 cm^{-1} is important as it is dominated by the asymmetric (around 1600 cm^{-1}) and symmetric (around 1400 cm^{-1}) stretching vibrations of the carboxylate (COO^-) groups [19]. These carboxylate groups are critical for alginate's ion-binding properties and its ability to form hydrogels [19]. The region, around 1000-1100 cm^{-1}, contains complex overlapping bands characteristic of the polysaccharide structure, including C-O stretching and C-O-C glycosidic linkages [20,21].

On the other hand, the effect of ZR concentration on these bands is subtle, yet distinguishable. As ZR concentration increases, there is a slight broadening or minor shifts in the carboxylate bands (1600 cm^{-1} and 1400 cm^{-1}). The intensity of the broad O-H band around 3500-3000 cm^{-1} exhibits slight variations, indicating possible alteration in the hydrogen bonding network or changes in the water content. The effective integration of zirconia would be indicated by the appearance of a peak or a displacement in existing peaks associated with Zr−O vibrations, generally observed within the range of $500 - 800$ cm^{-1} [20,22]. The spectrum for blank AG (black line) in the 800–500 cm^{-1}

region shows a relatively flat baseline with no distinct or prominent absorption peaks. In comparison to the blank AG, all the AG-ZR formulations show distinct absorption peaks in the 500–800 cm^{-1} region. Specifically, two characteristic peaks are observed; a prominent peak appears consistently around 730-740 cm^{-1} and another distinct peak is present around 610-620 cm^{-1}. These observed peaks are consistent with the characteristic absorption bands for Zr–O–Zr stretching vibrations, which have been reported at 738 cm^{-1} and 613 cm^{-1} [23]. Their presence in all zirconium-containing samples and absence in the blank AG confirms the incorporation and chemical interaction of zirconium within the alginate structure.

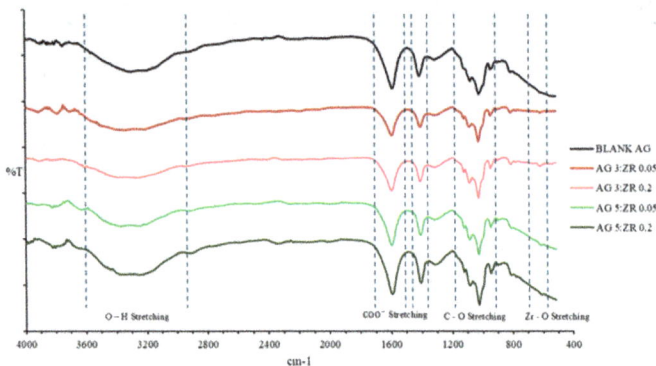

Figure 2 FTIR spectra (4000–400 cm^{-1}) of blank AG and AG–ZR films.

The TGA curves in Fig. 3(a) reveal a typical three-stage thermal degradation pattern in AG–ZR aerogels. The initial weight loss below 120°C corresponds to evaporation of adsorbed water, while the major decomposition phase above 200°C arises from alginate chain depolymerization and oxidation of organic components [24,25]. All AG–ZR samples exhibit reduced weight loss below 350°C compared to blank alginate, confirming improved thermal resistance through zirconia reinforcement.

The presence of ZrO$_2$ enhances the thermal stability of the hybrid structure by acting as an inorganic backbone. AG 3:ZR 0.05 retains the highest residue (~45%) at 600°C, while AG 5:ZR 0.2 leaves the lowest (~15%), demonstrating the role of zirconia in maintaining inorganic stability even at elevated temperatures. These findings suggest that zirconia incorporation effectively delays degradation and preserves mass at higher temperatures. Overall, increasing ZR content results in a more thermally stable network due to improved crosslinking and strong polymer–ceramic interactions. The composite aerogels resist structural collapse under heat, emphasizing the role of zirconium as a stabilizing agent that enhances the performance of alginate-based systems for high-temperature applications.

The DSC thermograms in Fig. 3(b) display broad endothermic peaks between 50–120°C for all AG–ZR formulations, corresponding to moisture evaporation [24]. Blank alginate shows endothermic decomposition beginning around 200°C, with major thermal events near 300–400°C, consistent with dehydration and depolymerization of the biopolymer matrix [25]. These transitions represent the characteristic thermal behaviour of alginate before zirconia incorporation. In contrast, the AG–ZR samples exhibit additional exothermic peaks within 200–300 °C, indicating zirconium-induced crosslinking and rearrangement reactions [9,26]. Increasing ZR concentration reduces the intensity of endothermic events while broadening exothermic transitions, implying that zirconium modifies the degradation pathway by introducing new chemical interactions and reinforcing bonds within the matrix.

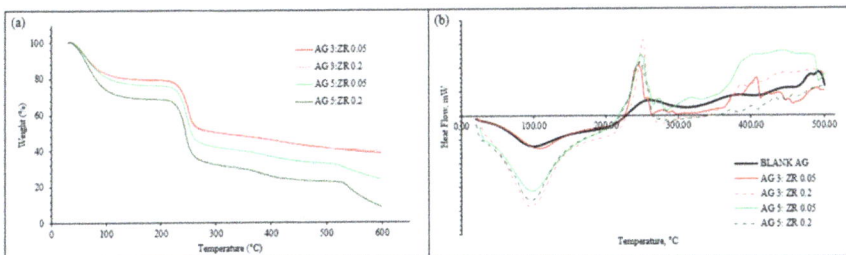

Figure 3 *(a) TGA (solid lines) measured at 10 °C·min⁻¹ under N_2, (b) DSC heat flow for AG-ZR samples.*

The appearance of these exothermic peaks demonstrates that Zr^{4+} ions promote crosslinking and charring reactions that enhance thermal resilience [9,27]. This transition from purely endothermic decomposition in blank AG to the more complex thermal behaviour observed in the AG–ZR samples suggests that zirconia actively alters the thermal degradation mechanism [28]. The appearance of exothermic peaks indicates additional reactions such as crosslinking, bond rearrangement, and char formation, all of which consume or redirect thermal energy. As a result, the composite is able to absorb and redistribute heat more effectively, improving its overall energy-dissipation capacity [29]. This thermally stabilizing effect strengthens the structural integrity of the alginate–zirconia network at high temperatures, highlighting its suitability for insulation applications.

The tensile stress–strain curves in Fig. 4 show that alginate and zirconia concentrations significantly affect the aerogels' mechanical behavior. The blank AG sample exhibits the highest tensile stress (2.8–3.0 MPa) due to strong intermolecular hydrogen bonding and the cohesive integrity of the alginate matrix [30], with its steep curve and limited elongation indicating a brittle, densely packed structure. In contrast, the AG–ZR formulations display lower tensile strength but greater flexibility, reflecting zirconia-induced modifications to the alginate network through polymer–ceramic crosslinking. The AG 3 samples show lower stiffness yet improved ductility, whereas the AG 5 samples demonstrate higher strength and rigidity, consistent with a denser, more effectively crosslinked network [31]. Among all compositions, AG 5:ZR 0.2 provides the best balance of strength and flexibility, with a tensile strength of about 1.8 MPa and improved deformability.

The improved mechanical strength with higher zirconia content results from coordination between Zr^{4+} ions and alginate carboxylate groups, forming a reinforced network that limits chain mobility and increases stiffness [26]. Higher alginate concentration further enhances performance through increased chain entanglement and cohesive interactions, promoting efficient load transfer and greater toughness. This polymer–ceramic synergy enhances both elasticity and resistance to deformation. Consequently, AG 5-based aerogels consistently outperform AG 3 at all ZR levels, demonstrating the stronger reinforcement capacity of the denser alginate matrix [6,26]. The AG 5:ZR 0.2 composition achieves the most favorable combination of tensile strength and elongation at break due to effective polymer–filler crosslinking. Overall, optimizing alginate–zirconia ratios

Separation Technology - ICoST 2025
Materials Research Proceedings 59 (2026) 113-120

Materials Research Forum LLC
https://doi.org/10.21741/9781644903957-15

enable the development of lightweight, mechanically robust aerogels suitable for advanced structural and thermal insulation applications.

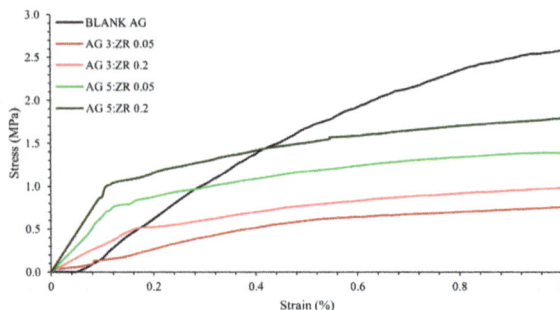

Figure 4 *Tensile stress–strain curves for AG–ZR films (n=3).*

Conclusions

The AG/ZR aerogels demonstrated a unique combination of characteristics, making them interesting candidates for advanced thermal insulation applications. FTIR analysis verified the efficient crosslinking between zirconia and alginate, indicating the creation of a homogeneous composite structure. The AG/ZR aerogels possess good thermal stability, with low weight loss up to 350°C and a decomposition starting around 270°C. Mechanical testing revealed that although the blank AG exhibited the highest tensile strength, the AG 5:ZR 0.2 formulation provided comparable strength and flexibility, indicating effective reinforcement within the hybrid network. These properties demonstrate the synergistic effect of polymer-filler interactions and the effectiveness of supercritical drying in retaining aerogel microstructures. The synthesized AG-ZR aerogels are promising candidates for sustainable thermal insulation applications, particularly where lightweight, biodegradable, and thermally stable materials are required.

Acknowledgements

The authors thank the Faculty of Chemical Engineering, Universiti Teknologi MARA Shah Alam, for providing the laboratory facilities and equipment necessary to undertake this research. Financial support for this research was provided by KPT-MYBRAIN 2.0, which is gratefully acknowledged.

References

[1] B. Abu-Jdayil, A.-H. Mourad, W. Hittini, M. Hassan, and S. Hameedi, "Traditional, state-of-the-art and renewable thermal building insulation materials: An overview," *Constr Build Mater*, vol. 214, pp. 709–735, Jul. 2019. https://doi.org/10.1016/j.conbuildmat.2019.04.102

[2] H. Maleki, "Recent advances in aerogels for environmental remediation applications: A review," *Chemical Engineering Journal*, vol. 300, pp. 98–118, Sep. 2016. https://doi.org/10.1016/j.cej.2016.04.098

[3] P. C. Thapliyal and K. Singh, "Aerogels as Promising Thermal Insulating Materials: An Overview," *J Mater*, vol. 2014, pp. 1–10, Apr. 2014. https://doi.org/10.1155/2014/127049

[4] S. Zhao, W. J. Malfait, N. Guerrero-Alburquerque, M. M. Koebel, and G. Nyström, "Biopolymer Aerogels and Foams: Chemistry, Properties, and Applications," *Angewandte Chemie International Edition*, vol. 57, no. 26, pp. 7580–7608, Jun. 2018. https://doi.org/10.1002/anie.201709014

[5] M. A. B. Meador *et al.*, "Cross-linking Amine-Modified Silica Aerogels with Epoxies: Mechanically Strong Lightweight Porous Materials," *Chemistry of Materials*, vol. 17, no. 5, pp. 1085–1098, Mar. 2005. https://doi.org/10.1021/cm048063u

[6] B.-T. Xu *et al.*, "Nanoclay-reinforced alginate aerogels: preparation and properties," *RSC Adv*, vol. 14, no. 2, pp. 954–962, Jan. 2024. https://doi.org/10.1039/D3RA07132D

[7] Y. Han, Y. Wu, H. Zhang, S. Huang, S. Wu, and Z. Liang, "A three-dimensional network modifier (dimethyldiethoxysilane) makes ZrO2-SiO2 aerogel with excellent thermal insulation performance and high-temperature stability," *Colloids Surf A Physicochem Eng Asp*, vol. 671, p. 131716, Aug. 2023. https://doi.org/10.1016/j.colsurfa.2023.131716

[8] J. He *et al.*, "Large-scale and ultra-low thermal conductivity of ZrO2 fibrofelt/ZrO2-SiO2 aerogels composites for thermal insulation," *Ceram Int*, vol. 44, no. 8, pp. 8742–8748, Jun. 2018. https://doi.org/10.1016/j.ceramint.2018.01.089

[9] R. C. Walker, A. E. Potochniak, A. P. Hyer, and J. K. Ferri, "Zirconia aerogels for thermal management: Review of synthesis, processing, and properties information architecture," *Adv Colloid Interface Sci*, vol. 295, p. 102464, Sep. 2021. https://doi.org/10.1016/j.cis.2021.102464

[10] A. K. Chitoria, A. Mir, and M. A. Shah, "A review of ZrO2 nanoparticles applications and recent advancements," *Ceram Int*, vol. 49, no. 20, pp. 32343–32358, Oct. 2023. https://doi.org/10.1016/j.ceramint.2023.06.296

[11] A. C. Pierre and G. M. Pajonk, "Chemistry of Aerogels and Their Applications," *Chem Rev*, vol. 102, no. 11, pp. 4243–4266, Nov. 2002. https://doi.org/10.1021/cr0101306

[12] S. Shafi, T. Rasheed, R. Naz, S. Majeed, and M. Bilal, "Supercritical CO2 drying of pure silica aerogels: effect of drying time on textural properties of nanoporous silica aerogels," *J Solgel Sci Technol*, vol. 98, no. 3, pp. 478–486, Jun. 2021. https://doi.org/10.1007/s10971-021-05530-0

[13] H.-B. Zhao, M. Chen, and H.-B. Chen, "Thermally Insulating and Flame-Retardant Polyaniline/Pectin Aerogels," *ACS Sustain Chem Eng*, vol. 5, no. 8, pp. 7012–7019, Aug. 2017. https://doi.org/10.1021/acssuschemeng.7b01247

[14] F. Zou *et al.*, "Maximizing sound absorption, thermal insulation, and mechanical strength of anisotropic pectin cryogels," *Chemical Engineering Journal*, vol. 462, p. 142236, Apr. 2023. https://doi.org/10.1016/j.cej.2023.142236

[15] H. Sofiah Roslan *et al.*, "Characteristics of hybrid alginate/soy protein isolate wound dressing aerogels dried by supercritical carbon dioxide," *Mater Today Proc*, May 2023. https://doi.org/10.1016/j.matpr.2023.05.047

[16] M. I. Sarwar, S. Zulfiqar, and Z. Ahmad, "Properties of polyamide-zirconia nanocomposites prepared from sol-gel technique," *Polym Compos*, vol. 30, no. 1, pp. 95–100, Jan. 2009. https://doi.org/10.1002/pc.20538

[17] B. Du *et al.*, "Ultrafast Polymerization of a Self-Adhesive and Strain Sensitive Hydrogel-Based Flexible Sensor for Human Motion Monitoring and Handwriting Recognition," *Polymers 2024, Vol. 16, Page 1595*, vol. 16, no. 11, p. 1595, Jun. 2024. https://doi.org/10.3390/POLYM16111595

[18] Y. Zhou *et al.*, "Preparation and stability characterization of soybean protein isolate/sodium alginate complexes-based nanoemulsions using high-pressure homogenization," *LWT*, vol. 154, p. 112607, Jan. 2022. https://doi.org/10.1016/j.lwt.2021.112607

[19] S. H. Rashedy, M. S. M. Abd El Hafez, M. A. Dar, J. Cotas, and L. Pereira, "Evaluation and Characterization of Alginate Extracted from Brown Seaweed Collected in the Red Sea," *Applied Sciences*, vol. 11, no. 14, p. 6290, Jul. 2021. https://doi.org/10.3390/app11146290

[20] A. M. Hezma, W. A. Shaltout, H. A. Kabary, G. S. El-Bahy, and A. B. Abdelrazzak, "Fabrication, Characterization and Adsorption Investigation of Nano Zinc Oxide–Sodium Alginate Beads for Effective Removal of Chromium (VI) from Aqueous Solution," *J Inorg Organomet Polym Mater*, vol. 33, no. 5, pp. 1400–1408, May 2023. https://doi.org/10.1007/s10904-023-02573-4

[21] S. Bhatia *et al.*, "Fabrication, Characterization, and Antioxidant Potential of Sodium Alginate/Acacia Gum Hydrogel-Based Films Loaded with Cinnamon Essential Oil," *Gels*, vol. 9, no. 4, p. 337, Apr. 2023. https://doi.org/10.3390/gels9040337

[22] S. R. Al-Mhyawi *et al.*, "Zirconium oxide with graphene oxide anchoring for improved heavy metal ions adsorption: Isotherm and kinetic study," *Journal of Materials Research and Technology*, vol. 22, pp. 3058–3074, Jan. 2023. https://doi.org/10.1016/j.jmrt.2022.11.121

[23] S. K. Papageorgiou, E. P. Kouvelos, E. P. Favvas, A. A. Sapalidis, G. E. Romanos, and F. K. Katsaros, "Metal–carboxylate interactions in metal–alginate complexes studied with FTIR spectroscopy," *Carbohydr Res*, vol. 345, no. 4, pp. 469–473, Feb. 2010. https://doi.org/10.1016/j.carres.2009.12.010

[24] L. Marangoni Júnior, R. G. da Silva, C. A. R. Anjos, R. P. Vieira, and R. M. V. Alves, "Effect of low concentrations of SiO₂ nanoparticles on the physical and chemical properties of sodium alginate-based films," *Carbohydr Polym*, vol. 269, p. 118286, Oct. 2021. https://doi.org/10.1016/j.carbpol.2021.118286

[25] R. Sabater i Serra, J. Molina-Mateo, C. Torregrosa-Cabanilles, A. Andrio-Balado, J. M. Meseguer Dueñas, and Á. Serrano-Aroca, "Bio-Nanocomposite Hydrogel Based on Zinc Alginate/Graphene Oxide: Morphology, Structural Conformation, Thermal Behavior/Degradation, and Dielectric Properties," *Polymers (Basel)*, vol. 12, no. 3, p. 702, Mar. 2020. https://doi.org/10.3390/polym12030702

[26] H. Suo *et al.*, "Preparation of ZrC@Al2O3@Carbon composite aerogel with excellent high temperature thermal insulation performance," *SN Appl Sci*, vol. 1, no. 5, p. 461, May 2019. https://doi.org/10.1007/s42452-019-0478-4

[27] L. Zhou, L. Wu, T. Wu, D. Chen, X. Yang, and G. Sui, "A 'ceramer' aerogel with unique bicontinuous inorganic–organic structure enabling super-resilience, hydrophobicity, and thermal insulation," *Mater Today Nano*, vol. 22, Jun. 2023. https://doi.org/10.1016/j.mtnano.2023.100306

[28] T. N. Rao, I. Hussain, J. E. Lee, A. Kumar, and B. H. Koo, "Enhanced Thermal Properties of Zirconia Nanoparticles and Chitosan-Based Intumescent Flame Retardant Coatings," *Applied Sciences 2019, Vol. 9, Page 3464*, vol. 9, no. 17, p. 3464, Aug. 2019. https://doi.org/10.3390/APP9173464

[29] E. K. Tonini *et al.*, "Metal-Organic Frameworks as Promising Textile Flame Retardants: Importance and Application Methods," 2024. https://doi.org/10.3390/app14178079.

[30] D. Ji *et al.*, "Superstrong, superstiff, and conductive alginate hydrogels," *Nature Communications 2022 13:1*, vol. 13, no. 1, pp. 1–10, May 2022. https://doi.org/10.1038/s41467-022-30691-z

[31] A. K. Rana, V. K. Gupta, P. Hart, and V. K. Thakur, "Cellulose-alginate hydrogels and their nanocomposites
for water remediation and biomedical applications," *Environ Res*, vol. 243, p. 117889, Feb. 2024. https://doi.org/10.1016/j.envres.2023.117889

Separation Technology - ICoST 2025
Materials Research Proceedings 59 (2026) 121-129

Materials Research Forum LLC
https://doi.org/10.21741/9781644903957-16

Influence of Feed Temperature on Progressive Freeze Concentration of Magnesium Sulphate Solutions

Hafizuddin SUTIMIN[1,a], Aishah ROSLI[1,2,b*], Mazura JUSOH[1,2,c]

[1]Faculty of Chemical and Energy Engineering, Universiti Teknologi Malaysia (UTM), Skudai, Johor, Malaysia

[2]Centre of Lipids Engineering and Applied Research (CLEAR), Universiti Teknologi Malaysia, 81310 UTM Johor Bahru, Johor, Malaysia

[a]hafizuddinsutimin@graduate.utm.my, [b]aishahrosli@utm.my, [c]mazurajusoh@utm.my

Keywords: Progressive Freeze Concentration, Magnesium Sulphate, Feed Temperature, Freezing Dynamics, Energy Consumption, Solute Recovery, Vertical Finned Crystallizer

Abstract. Progressive freeze concentration (PFC) offers low-temperature separation for water treatment, yet the role of feed temperature remains unclear. Magnesium sulphate solutions were processed in a vertical finned crystallizer (50 min cycle, 2100 mL min^{-1} circulation, $-10°C$ coolant) using either room-temperature feed (RF, 27.4±1.8°C) or pre-chilled feed (CF, 9.1±0.8°C), with triplicate runs per condition. Outcomes were standardized into two composites: (i) Separation–kinetics metric (solute recovery, ion recovery efficiency, reversed partition coefficient, freezing and solution-retention rates) and (ii) energy metric (reversed SEC per litre, reversed process energy, energy efficiency). RF showed higher separation: solute recovery (0.643 vs. 0.585), ion recovery efficiency (79.5% vs 71.9%), and freezing rate (12.1 vs. 11.8 g min^{-1}). Partition coefficients were similar (0.129 vs. 0.122). Energy results diverged: RF required more process energy (314 vs. 235 kJ) but achieved lower SEC per litre (1.82 vs. 2.00 kWh L^{-1}) and higher efficiency (14.2% vs. 10.1%). Composite scores favoured RF for separation (0.278 vs. -0.278), while energy outcomes balanced near zero. Multivariate testing found no significant overall effect (Pillai's Trace = 0.578, F (2,3) = 2.057, p = 0.274), though the separation–kinetics effect size was moderate-to-large (partial η^2 = 0.406). The results suggest that warmer feeds can improve recovery and kinetics while lowering per-litre energy demand, even though they increase absolute cooling duty. Larger studies are needed to confirm these trends and to explore adaptive flow control and expanded operating windows for scalability.

Introduction

Water separation and concentration processes are central to chemical engineering, environmental management, and sustainable water use. Thermal evaporation is effective but requires substantial heat input, which can degrade heat-sensitive solutes and cause scaling or corrosion [1]. Membrane technologies such as reverse osmosis (RO) bypass thermal degradation but face issues of fouling, high-pressure energy requirements, and limited capacity for hypersaline feeds [2]. These motivates alternative separation approaches with lower energy input and selective solute retention, aligned with SDG 6 (clean water and sanitation) and SDG 12 (responsible production and consumption).

Freeze concentration (FC) is a low-temperature route that crystallizes water as ice while leaving most solutes in the liquid phase. Furthermore, Freezing requires far less latent heat than boiling, and with heat recovery can be competitive in energy use [1]. Applications of FC range from concentrating food juices and pharmaceuticals to treating industrial brines and wastewater streams, demonstrating both versatility and potential scalability [3].

Separation Technology - ICoST 2025   Materials Research Forum LLC
Materials Research Proceedings 59 (2026) 121-129   https://doi.org/10.21741/9781644903957-16

Among FC techniques, progressive freeze concentration (PFC) is particularly attractive for engineering applications [1]. Unlike suspension freeze concentration (SFC), which produces dispersed ice crystals requiring downstream separation, PFC relies on ice layers growing progressively on a cooled surface or along a freezing front. This approach simplifies equipment design and has achieved high solute recovery in diverse liquids, including dairy, juices, and saline brines [3]. Key variables include cooling rate, agitation, and solute concentration [1,4].

Despite progress, the effect of initial feed temperature has not been well studied. Warmer feeds increase cooling load and may deepen supercooling, while pre-chilled feeds may promote controlled ice growth [4]. Earlier work showed that gradual cooling reduces supercooling and inclusion [5]. By analogy, the starting bulk feed temperature is expected to similarly influence ice nucleation and solute exclusion, yet this has not been systematically studied.

Therefore, this study addresses this gap by examining how initial feed temperature influences progressive freeze concentration of magnesium sulphate solutions. Magnesium sulphate was selected as a model solute relevant to industrial wastewater and brine streams, ensuring practical relevance for water treatment and resource recovery [6,7,8]. The study compares room-temperature feeds (RF) and pre-chilled feeds (CF) under controlled conditions in a vertical finned crystallizer which the crystallizer has extended area for ice growth rather than tubular vessel [9]. The objectives are to evaluate effects on two prespecified composite metrics, i.e., separation–kinetics (solute recovery, ion recovery efficiency, reversed partition coefficient, freezing rate, remaining solution rate) and energy (reversed specific energy consumption per litre of concentrate, reversed process energy, energy efficiency). It is hypothesized that feed temperature influences solute exclusion and freezing behaviour in distinct ways [1]. Clarifying this relationship will provides guidance on feed preparation strategies and informs the optimization of PFC for energy-efficient, sustainable water treatment.

Materials and Methods

Magnesium sulphate heptahydrate ($MgSO_4 \cdot 7H_2O$, analytical grade) and laboratory-prepared distilled water were used. Concentrations are reported in mg L^{-1}. A 1 L stock was prepared and diluted to 10 L at $[SO_4^{2-}] = 250$ mg L^{-1}. The batch was split into room-temperature feed (RF) and pre-chilled feed (CF). RF remained at ambient laboratory temperature while CF was held in a chiller overnight. Feed temperatures were verified by spot checks using a Type-K thermocouple and logger. Volumes were inferred from mass assuming $\rho \approx 1.00$ g mL^{-1} [10].

A vertical finned crystallizer (VFC) adapted from prior VFC studies was used as shown in Fig. 1 [9]. The jacketed stainless-steel body has outer diameter 21 cm, body height 30 cm, and separating-plate height 25 cm. Feed enters at the lower port and exits at the upper port: coolant flows counter-current through the external jacket. Two conditions were tested which is RF (27.36±1.76°C) and CF (9.08±0.76°C), n = 3 per condition. Fixed set-points were 50 min cycle duration, 2100 mL min^{-1} circulation, and −10°C coolant. Ice was removed, weighed, melted, and analyzed [11]. Sulphate was measured by $BaSO_4$ turbidity at 420 nm (Shimadzu UVmini-1240) [11]. Samples were thawed to room temperature. Calibration used 20–80 mg L^{-1} standards as in Eq. (1).

$$A = 0.013841C - 0.18829, \quad (R^2 = 0.9972) \tag{1}$$

Response variables were computed from masses, concentrations, run time, and metered energy. Eqs. (2) – (7) were used to determine the separation kinetics metrics. The following Eqs. (8) – (14) were used to determine the energy metrics.

Separation Technology - ICoST 2025
Materials Research Proceedings 59 (2026) 121-129

Materials Research Forum LLC
https://doi.org/10.21741/9781644903957-16

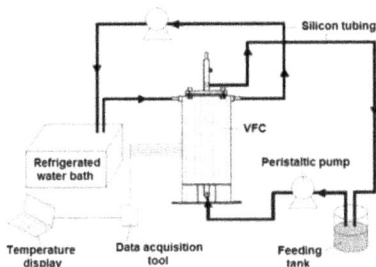

Figure 1 *Schematic of the VFC set-up and run timeline (single 50-min cycle) [9].*

$$Y_s = \frac{C_c m_c}{C_0 m_0} \tag{2}$$

$$\eta_{ion} = \frac{C_c m_c}{C_0 m_0} \times 100 \,, [\%] \tag{3}$$

$$k_{eff} = \frac{C_i}{C_c} \in [0,1] \tag{4}$$

$$K_{rev} = 1 - k_{eff} \tag{5}$$

$$R_{freeze} = \frac{m_i}{t} \,, [g\ min^{-1}] \tag{6}$$

$$R_{remain} = \frac{m_c}{t} \,, [g\ min^{-1}] \tag{7}$$

$$SEC = \frac{E_{meas}}{V_c} \,, [kWh\ L^{-1}] \tag{8}$$

$$SEC_{rev} = \frac{1}{SEC} \,, [L\ kWh^{-1}] \tag{9}$$

$$Q_{cool} = \frac{m_0 c_p (T_{start} - 0\ °C)}{1000} \,, [kJ] \tag{10}$$

$$Q_{freeze} = \frac{m_i L_f}{1000} \,, [kJ] \tag{11}$$

$$E^*_{proc} = \frac{Q_{cool} + Q_{freeze}}{3600} \,, [kJ] \tag{12}$$

$$E^*_{proc,rev} = \frac{1}{E^*_{proc}} \,, [kJ^{-1}] \tag{13}$$

$$EE = \frac{E^*_{proc}}{E_{meas}} \times 100 \,, [\%] \tag{14}$$

Where, m_0 is initial feed mass (g), m_i is ice mass (g), m_c is concentrate mass (g), V_c is concentrate volume ($\approx m_c/1000$) (L), C_0, C_i, C_c is sulphate concentration in feed, ice-melt, concentrate (mg L^{-1}) respectively, t is cycle time (min); T_{start} is initial feed temperature (°C), E_{meas} is metered chiller energy (kWh), $c_p = 4.186$ J g^{-1} and $L_f = 333.7$ J g^{-1} [12]. All response metrics from were converted to run z-scores (standardised across all six runs) for each condition and metrics. These z-scores arranged in orientation so that larger values indicate better performance by using K_{rev}, SEC_{rev}, and $E^*_{proc,rev}$. Eqs. (15) and (16) were used for separation-kinetics and energy metrics, respectively.

$$Z_{SKMetric} = z(Y_s) + z(\eta_{ion}) + z(K_{rev}) + z(R_{freeze}) + z(R_{remain}) \tag{15}$$

$$Z_{EnergyMetric} = z(SEC_{rev}) + z(E^*_{proc,rev}) + z(EE) \tag{16}$$

Separation Technology - ICoST 2025
Materials Research Proceedings 59 (2026) 121-129

Materials Research Forum LLC
https://doi.org/10.21741/9781644903957-16

Composite metrics were analyzed using GLM–MANOVA in SPSS (Pillai's Trace; Box's M for covariance equality; Levene tests for variance). Partial η^2 was reported due to n = 3 per group [13].

Results and Discussion

Fig. 2 shows RF outperformed CF on all separation-kinetics metrics across n = 3 per condition. For instance, solute recovery (SR) was higher in the RF condition (0.643±0.0145) compared with CF (0.585±0.0045). A similar pattern was observed for ion recovery efficiency (79.5±2.67 % for RF vs. 71.9±1.80 % for CF). Partition coefficients were comparable between RF (0.129±0.041) and CF (0.122±0.011). Freezing rate also tended to be higher in RF (12.09±0.25 g min⁻¹) relative to CF (11.78±0.19 g min⁻¹). Remaining-solution rates followed the same direction, with RF averaging 6.77±0.42 g min⁻¹ and CF 6.50±0.41 g min⁻¹.

Two established principles frame these findings. First, interface dynamics strongly influence progressive freeze concentration. Rapid growth and deep supercooling promote solute inclusion, whereas controlled or ramped cooling can reduce inclusion by two to threefold [4]. Second, coupled modelling and experiments have formalized how interfacial heat–mass transfer and velocity scales determine whether solutes are excluded or entrained at the ice front [14]. Although a warmer feed temperature in RF might be expected to increase supercooling and solute entrapment, the data showed higher recoveries and marginally faster kinetics. Two mechanisms account for this. At the molecular level, ion rejection at the interface improves with temperature, as ions remain more stabilized in their hydrated state relative to the ice lattice [15]. A larger ΔT enhances convection, thinning the solute boundary layer and improving exclusion [16]. Together, these effects outweigh the modest rise in growth rate, leading to more effective sulphate exclusion into the liquid and higher k_{eff} despite increased cooling duty.

The SO_4^{2-} is generally excluded from ice, with inclusion mainly from trapped brine [17]. At 250 mgL⁻¹ SO_4^{2-}, precipitation is unlikely under sub-zero conditions, so separation is potentially governed by interfacial transport and hydrodynamics. Operationally, bulk pre-chilling does not guarantee improved outcomes in PFC. Interfacial temperature control and convective renewal are key factors [18]. A warmer start with controlled cooling and mixing may therefore be favourable.

As shown in Fig. 3, the SEC per litre of concentrate was lower for RF (1.818±0.039 kWh L⁻¹) compared with CF (2.000±0.106 kWh L⁻¹). In contrast, process energy was substantially higher in RF (314.1±6.5 kJ) relative to CF (234.7±0.3 kJ). Energy efficiency was also greater under RF (14.2±0.53%) compared with CF (10.1±0.46%). The pattern indicates a normalization effect. SEC is defined as measured energy divided by concentrate volume as in Eq. (8). Therefore, increasing in concentrate volume due higher remaining solute, will reduce the SEC even if measured energy increases. RF demanded more absolute energy because the feed started warmer, requiring greater sensible cooling to 0°C, yet it also produced more concentrate.

The energy balance of freezing processes comprises sensible cooling plus latent heat of fusion of the ice produced. Because SEC is normalized to product volume, increases in remaining solute mass, m_c ($V_c \approx mc/\rho$) mechanically drive SEC downward even when measured energy (kWh L⁻¹) increases [19]. The higher process energy, E^*_{proc} under RF is consistent with the larger sensible-cooling term due to the higher T_{start}, while the higher EE reflects the larger product volume under RF [19]. No previous PFC studies were found that directly tested how initial feed temperature affects normalized energy metrics. The RF–CF energy contrast is therefore interpreted here as an inference based on the standard SEC definition and the established freezing energy balance, rather than as a direct comparison with earlier studies [1,19,20].

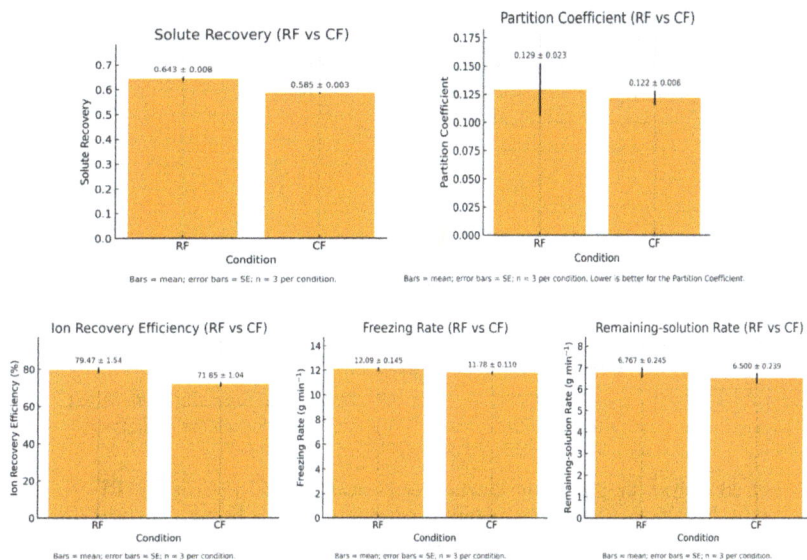

Figure 2 Separation–kinetics metrics under room-temperature feed (RF) and pre-chilled feed (CF): (a) Solute recovery, (b) ion recovery efficiency, (c) partition coefficient, (d) freezing rate, and (e) remaining-solution rate. Bars show group means with SE error bars (n = 3 per condition). Numeric labels report means ± SE. For the partition coefficient, lower values indicate better separation. No sign reversal is applied here.

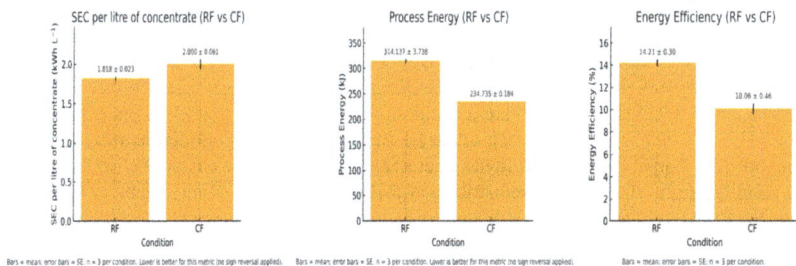

Figure 3 Energy metrics under room-temperature feed (RF) and pre-chilled feed (CF): (a) SEC per litre of concentrate, (b) process energy, and (c) energy efficiency. Bars are means with SE error bars (n = 3 per condition). Numeric labels show mean ± SE.

As shown in Fig. 4, assumption checks indicated acceptable covariance equality (Box's M = 5.306, p = 0.489). The multivariate test showed no significant overall effect of Condition (Pillai's Trace = 0.578, F(2,3) = 2.057, p = 0.274, partial η^2 = 0.578). Univariate follow-ups indicated a non-significant difference for separation–kinetics (F(1,4) = 2.734, p = 0.174, partial η^2 = 0.406) and negligible difference for energy (F(1,4) = 0.076, p = 0.797, partial η^2 = 0.019). Under

Separation Technology - ICoST 2025
Materials Research Proceedings 59 (2026) 121-129

Materials Research Forum LLC
https://doi.org/10.21741/9781644903957-16

constrained power, effect sizes and confidence intervals should be weighed alongside p-values; this aids practical interpretation of non-significant findings with moderate η^2 values [21,22].

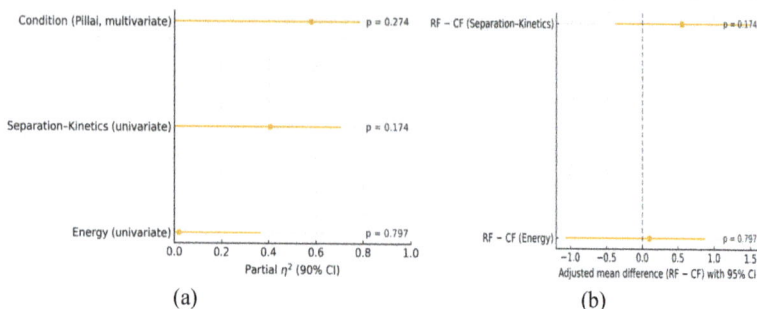

(a) (b)

Figure 4 *Multivariate analysis of feed condition (RF vs. CF), (a) Partial η^2 with 90% CI for the multivariate omnibus (Pillai) and univariate follow-ups, (b) Adjusted mean differences (RF − CF) with 95% CI from GLM pairwise contrasts.*

Across all analyses the separation–kinetics composite favoured RF (mean shift RF > CF; partial $\eta^2 = 0.406$). In contrast, the energy metric remained near neutral because the benefits of lower SEC/L and higher EE under RF were counterbalanced by higher process energy. This trade-off demonstrates how equal-weight aggregation can mask practical differences when variables move in opposite directions.

The use of z-standardized, direction-aligned composites is well established in process evaluations where multiple criteria must be integrated for comparative inference [23]. Equal-weight, mean-of-z aggregation was applied here to condense outcomes into separation-kinetics and energy metrics, following common practice once indicators are placed on a common scale and orientation [23]. Under this scheme, RF improved the separation-kinetics metric, whereas the energy metric showed little separation between conditions due to internal balancing.

The RF showed higher means for solute recovery (SR) and ion recovery efficiency (IRE) than CF (SR: 0.643±0.0145 vs. 0.585±0.0045; IRE: 79.5±2.67% vs. 71.9±1.80%). Partition coefficients were comparably low (RF 0.129±0.041; CF 0.122±0.011). RF also exhibited a higher freezing rate (12.09±0.25 vs. 11.78±0.19 g min⁻¹) and a slightly higher remaining-solution rate (6.77±0.42 vs. 6.50±0.41 g min⁻¹). Higher SR and IRE under RF indicates that a larger fraction of sulphate stayed in the unfrozen liquid, consistent with stronger solute rejection at the advancing ice front [14,24]. At a solidifying interface, solute redistribution follows the Burton–Prim–Slichter framework in which the effective distribution and boundary-layer thickness depend on interface velocity and solute diffusion in the liquid [14,25]. The RF advantage runs counter to the operational expectation that pre-chilling should deliver cleaner exclusion. Consistent with this mechanism, parametric PFC work shows that the feed temperature and the feed–coolant temperature difference are influential levers for supercooling control and salt rejection [14,26].

Conclusion

This study investigated the influence of initial feed temperature on progressive freeze concentration (PFC) of magnesium sulphate solutions using a vertical finned crystallizer. RF consistently achieved higher separation performance: solute recovery was 0.643 vs 0.585 under CF, and ion recovery efficiency was 79.5% vs. 71.9%. Freezing rates were also faster (12.1 vs 11.8 g min⁻¹), and the composite separation–kinetics score shifted positive for RF (0.278) but negative for CF (−0.278). Energy outcomes diverged: process energy was higher in RF (314 kJ vs. 235 kJ), but SEC per litre was lower (1.82 vs. 2.00 kWh L⁻¹) and efficiency higher (14.2% vs. 10.1%).

Separation Technology - ICoST 2025
Materials Research Proceedings 59 (2026) 121-129

Materials Research Forum LLC
https://doi.org/10.21741/9781644903957-16

MANOVA results did not detect a statistically significant multivariate effect ($p = 0.274$), but the effect size for separation–kinetics was moderate-to-large (partial $\eta^2 = 0.406$). Starting from room temperature improved recovery and throughput while enhancing normalized energy indices, although at the cost of higher absolute cooling demand. Future studies should expand replication, test intermediate feed temperatures, and evaluate staged cooling. Extending this work to different solute systems and scale-up conditions will help clarify whether warmer feeds can reliably deliver both separation advantages and per-litre energy efficiency in industrial PFC applications.

Acknowledgement

This work was supported/funded by the Ministry of Higher Education under Fundamental Research Grant Scheme (FRGS/1/2021/TK0/UTM/02/1), (Vot. 5F430).

References

[1] A. Najim, "A review of advances in freeze desalination and future prospects," *Nature Partner Journals Clean Water,* Review Article vol. 5, no. 1, p. 15, 2022 2022, Art no. 15. https://doi.org/10.1038/s41545-022-00158-1

[2] K. L. Foo, Yong Yeow; Lau, Woei Jye; Khan, Md Maksudur Rahman; Ahmad, Abdul Latif, "Performance of Hypersaline Brine Desalination Using Spiral Wound Membrane: A Parametric Study," *Membranes,* Research Article vol. 13, no. 2, p. 248, 2023 2023, Art no. 248. https://doi.org/10.3390/membranes13020248

[3] A. A. H. Prestes, Cristiane Vieira; Esmerino, Erick Almeida; Silva, Ramon; da Cruz, Adriano Gomes; Prudencio, Elane Schwinden, "Freeze concentration techniques as alternative methods to thermal processing in dairy manufacturing: A review," *Journal of Food Science,* Review Article vol. 87, no. 2, pp. 543-562, 2022 2022. https://doi.org/10.1111/1750-3841.16027

[4] J.-E. B. Vuist, Remko M.; Schutyser, Maarten A. I., "Solute inclusion and freezing rate during progressive freeze concentration of sucrose and maltodextrin solutions," *Drying Technology,* Research Article vol. 39, no. 10, pp. 1285–1293, 2021 2021. https://doi.org/10.1080/07373937.2020.1742151

[5] J.-E. Vuist, "Progressive freeze concentration," PhD Doctoral Thesis, Laboratory of Food Process Engineering, Wageningen University & Research, Wageningen, The Netherlands, 2021. [Online]. Available: https://doi.org/10.18174/550673

[6] D. F. Fontana, Federica; Pietrantonio, Massimiliana; Pucciarmati, Stefano; Marcoaldi, Caterina, "Magnesium recovery from seawater desalination brines: a technical review," *Environment, Development and Sustainability,* Review Article 2022 2022. https://doi.org/10.1007/s10668-022-02663-2

[7] F. R. Leon, Alejandro, "Performance Analysis of a Full-Scale Desalination Plant with Reverse Osmosis Membranes for Irrigation," *Membranes,* Research Article vol. 11, no. 10, p. 774 (article number), 2021 2021. https://doi.org/10.3390/membranes11100774

[8] I. C. Bolton & Menk, "Analyzing Alternatives for Sulfate Treatment in Municipal Wastewater," Minnesota Pollution Control Agency (MPCA), St. Paul, Minnesota, USA, Technical Report wq-rule4-15pp, 2018 2018. [Online]. Available: https://www.pca.state.mn.us/

[9] T. T. Rashid, Su Ying; Harun, Noor Hafiza; Zakaria, Zaki Yamani; Ngadi, Norzita; Mohamad, Zurina; Jusoh, Mazura, "Progressive Freeze Concentration for Leachate Treatment using Vertical Finned Crystallizer," *Chemical Engineering Transactions,* Research Article vol. 97, pp. 451–456, 2022 2022. https://doi.org/10.3303/CET2297076

[10] A. H. Harvey, "Thermodynamic Properties of Water: Tabulation from the IAPWS Formulation 1995 for the Thermodynamic Properties of Ordinary Water Substance for General and Scientific Use," National Institute of Standards and Technology (NIST), Gaithersburg, Maryland, USA, Technical Report NISTIR 5078, November 1998 1998. [Online]. Available: https://nvlpubs.nist.gov/nistpubs/Legacy/IR/nistir5078.pdf

[11] *Standard Test Method for Sulfate Ion in Water*, Standard Test Method D516-22, A. International, West Conshohocken, PA, USA, December 1, 2022 (approved) / March 2023 (published) 2023. [Online]. Available: https://doi.org/10.1520/D0516-22

[12] W. P. Wagner, A., "The IAPWS Formulation 1995 for the Thermodynamic Properties of Ordinary Water Substance for General and Scientific Use," *Journal of Physical and Chemical Reference Data,* Research Article vol. 31, no. 2, pp. 387-535, 2002 2002. https://doi.org/10.1063/1.1461829

[13] H. S.-W. Todorov, E.; Gerber, S., "Applying univariate vs. multivariate statistics to investigate therapeutic efficacy in (pre)clinical trials: A Monte Carlo simulation study on the example of a controlled preclinical neurotrauma trial," *PLoS ONE,* Research Article vol. 15, no. 3, pp. 1-20, 2020 2020. https://doi.org/10.1371/journal.pone.0230798

[14] Z. J. Zhang, M.; Vanapalli, S., "Experimental and theoretical analysis of solute redistribution during a progressive freeze concentration process," *International Communications in Heat and Mass Transfer,* Research Article vol. 152, p. 107288 (article number), 2024 2024, Art no. 107288. https://doi.org/10.1016/j.icheatmasstransfer.2024.107288

[15] S. J. Luo, Y.; Tao, R.; Li, H.; Li, C.; Wang, J.; Li, Z., "Molecular understanding of ion rejection in the freezing of aqueous solutions," *Physical Chemistry Chemical Physics,* vol. 23, pp. 13292–13299, 2021. https://doi.org/10.1039/d1cp01733k

[16] Y. W. Du, Z.; Jiang, L.; Calzavarini, E.; Sun, C., "Sea Water Freezing Modes in a Natural Convection System," *Journal of Fluid Mechanics,* 2023. https://doi.org/doi:10.1017/jfm.2023.215

[17] R. H. Wang, Y.; Yuan, X.; Chen, J.; Jiang, S.; Li, X., "Unsynchronized migrations of different salt ions and ice microstructure development during unidirectional freeze–thaw," *Desalination,* vol. 549, p. 116326, 2023 2023. https://doi.org/10.1016/j.desal.2022.116326

[18] S. Jan Eise Vuist, Remko M. Boom, "Solute inclusion during progressive freeze concentration: A state diagram approach," *Journal of Food Engineering,* vol. 320, p. 110928. https://doi.org/10.1016/j.jfoodeng.2021.110928

[19] M. H. Hendijanifard, Amir; Farhadi, Shahrokh, "Comparing energy and exergy of multiple effect freeze desalination to MEE MSF RO," *Nature Partner Journals Clean Water,* Research Article vol. 7, no. Article 95, p. 95, 2024 2024, Art no. 95. https://doi.org/10.1038/s41545-024-00395-6

[20] H. J. Zhang, I.; Hassan Ali, M. I.; Askar, K., "Freezing desalination: Heat and mass validated modeling and experimental parametric analyses," *Case Studies in Thermal Engineering,* vol. 26, p. 101189, 2021. https://doi.org/10.1016/j.csite.2021.101189

[21] R. T. S.; Carson, K., "Moving beyond P values in The Journal of Physiology: A primer on the value of effect sizes and confidence intervals," *The Journal of Physiology,* vol. 301, no. 23, pp. 5131–5133, 2023. https://doi.org/10.1113/JP285575. The Physiological Society.

Separation Technology - ICoST 2025   Materials Research Forum LLC
Materials Research Proceedings 59 (2026) 121-129   https://doi.org/10.21741/9781644903957-16

[22] A. C. M. Miola, Hélio Amante, "P-value and effect-size in clinical and experimental studies," *Journal Vascular Brasileiro (J Vasc Bras),* vol. 20, p. e20210038, 2021. https://doi.org/10.1590/1677-5449.210038

[23] N. A. J. Tarasewicz, A. M., "An ecosystem model based composite indicator, representing sustainability aspects for comparison of forest management strategies," *Ecological Indicators,* vol. 133, p. 108456, 2021. https://doi.org/10.1016/j.ecolind.2021.108456

[24] L. Phylipova, "Development of the design and determination of mode characteristics of block cryoconcentrators for pomegranate juice," *Eastern-European Journal of Enterprise Technologies,* vol. 2, no. 11 (110), pp. 6–14, 2021. https://doi.org/10.15587/1729-4061.2021.230182

[25] J. A. P. Burton, R. C.; Slichter, W. P., "The distribution of solute in crystals grown from the melt. Part I. Theoretical," *The Journal of Chemical Physics,* vol. 21, no. 11, pp. 1987–1991, 1953. https://doi.org/10.1063/1.1698728

[26] S. O. Moharramzadeh, Say Kee; Alleman, James; Cetin, Kristen S., "Parametric study of the progressive freeze concentration for desalination," *Desalination,* Research Article vol. 510, p. 115077 (article number), 2021 2021, Art no. 115077. https://doi.org/10.1016/j.desal.2021.115077

Separation Technology - ICoST 2025
Materials Research Proceedings 59 (2026) 130-137

Materials Research Forum LLC
https://doi.org/10.21741/9781644903957-17

Phenolic Compounds Recovery from Palm Oil Sterilizer Condensate using Synergistic Formulation

Norela JUSOH[1,2,a], Norasikin OTHMAN[1,2,b*], Shuhada A. IDRUS-SAIDI[1,2,c*],
Muhammad Abbas AHMAD ZAINI[1,2,d], Izzat Naim Shamsul KAHAR[1,e],
Norul Fatiha Mohamed NOAH[1,f]

[1]Faculty of Chemical and Energy Engineering, Universiti Teknologi Malaysia, 81310 Skudai, Johor Bahru, Malaysia

[2]Centre of Lipids Engineering and Applied Research (CLEAR), Ibnu Sina Institute for Scientific and Industrial Research, Universiti Teknologi Malaysia, 81310 Skudai, Johor Bahru, Malaysia

[a]norela.jusoh@gmail.com, [b]norasikin@cheme.utm.my, [c]shuhada.atika@utm.my,
[d]r-abbas@utm.my, [e]izzatnaim.sk@utm.my, [f]norulfatiha.mn@gmail.com

Keywords: Phenolic Compounds, Palm Oil Sterilizer Condensate, Extractive Extraction, Green Synergistic Formulation, Recovery

Abstract. Palm oil sterilizer condensate (POSC) has been identified as a promising source of bioactive phenolic compounds (PCs), which are rich in antioxidant properties. In this study, a green synergistic extraction approach was employed to recover PCs from POSC. The POSC was first characterized to quantify its total phenolic content (TPC). The optimization and effects of key parameters including synergistic extractant concentration, agitation speed, and feed:organic phase ratio on the extraction PCs were evaluated. Additionally, the optimal condition for recovering PCs from the loaded organic phase were investigated by varying stripping agent concentration. The result revealed that POSC contains a high concentration of TPC, approximately 2500-4200 mg GAE/L. An extraction efficiency exceeding 93% was achieved using a synergistic mixture of 0.3/0.0024 M trioctylmethylammonium chloride (TOMAC)/ D2EHPA, an agitation speed of 240 rpm, and a feed:organic of 2.5. Furthermore, More than 98% of the extracted PCs were successfully recovered using 1/0.12 M NaOH/Na$_2$CO$_3$. These findings highlight the potential of POSC as a low-cost secondary source for PCs recovery and contribute to the valorization of palm oil mill waste through sustainable resource utilization.

Introduction

Palm oil sterilization condensate (POSC) was acknowledged for its high content of phenolic compounds (PCs) that exhibit antimicrobial, antioxidant, and nutraceutical properties [1]. The efficient recovery of PCs contributes to waste valorization and support the circular economy principles in palm oil production.Various techniques have been proposed to recover PCs from agro-industrial wastewater such as adsorption [2], crystallization [3], distillation [4], forward osmosis [5], integrated systems [6], membrane filtration [7], and liquid-liquid extraction [8]. Nonetheless, many of these methods involve significant drawbacks such as long processing times, high energy requirements, elevated operational costs, and complicated procedures, making them less suitable for large-scale industrial use.

Extractive extraction is a promising technique to extract and recover PCs due to its numerous benefits, including high selectivity, low operational costs, scalability, and simplicity. It can significantly improve recovery efficiency through the careful selection and combination of diluents, extractants, and stripping agents [9]. Recently, research has been focused on utilizing synergistic formulation to improve extraction efficiency. A synergistic formulation consists of two

Separation Technology - ICoST 2025
Materials Research Proceedings 59 (2026) 130-137

Materials Research Forum LLC
https://doi.org/10.21741/9781644903957-17

or more components that lead to better extraction efficiency or selectivity compared to their individual component.

In this study, the recovery of PCs from POSC using a synergistic formulation was investigated. An initial determination of total phenolic content (TPC) was conducted to assess the potential of POSC as a secondary source for recovering PCs. The formulation consists palm/sunflower oils as diluents, TOMAC/D2EHPA as extractants, and NaOH/Na2CO3 as stripping agents. The influence of key parameters such as synergistic extractant concentration, agitation speed, and feed:organic phase ratio on optimization of PCs extraction were evaluated. Additionally, the influence of varying concentrations of NaOH and Na2CO3 in the stripping phase was examined to determine the most effective condition for recovery efficiency.

Methodology
The POSC was obtained from Kahang Palm Oil Mill, Malaysia. Di-(2-ethylhexyl) phosphoric acid (D2EHPA, 95% purity), Folin–Ciocalteu reagent, gallic acid (97.5% purity), and trioctylmethylammonium chloride (TOMAC, 99% purity) were purchased from Sigma-Aldrich. Sodium hydroxide (NaOH, 99% purity) and sodium carbonate (Na2CO3, 99% purity) were procured from Merck. Palm oil (Buruh) and sunflower oil (Sunlico) were obtained from supermarket.

Determination of TPC
The POSC was pretreated by centrifugation (Kubota 5200, Japan) at 4000 rpm for 15 min to eliminate suspended particles, followed by 30 min gravitational settling to separate oil residues. The supernatant was subjected to vacuum filtration through a cellulose acetate filter (0.45 µm) to remove fine particulates that were not fully discarded during centrifugation. The TPC was quantified using Folin-Ciocalteu calorimetric procedure [10]. The presence of PCs was also confirmed by Fourier Transform Infrared (FTIR) spectroscopy.

Synergistic extractive extraction of PCs and design of experiments
The extraction process involved mixing the diluted pretreated POSC (550 mg GAE/L feed phase) with the organic phase (synergistic mixture of TOMAC/D2EHPA in palm/sunflower oil), followed by agitation using a mechanical shaker (IKA-KS 130 basic, Germany) for 5 h [11]. After extraction, the feed and organic phases were allowed to separate by gravity for 30 min in a separating funnel. The TPCs remaining in the feed phase was determined using the Folin–Ciocalteu calorimetric procedure, and the extraction performance was determined based on the mass balance approach.

Key parameters affecting the extraction process were investigated through response surface methodology (RSM). A Box–Behnken design (BBD) was employed to generate 15 experimental runs involving three parameters: synergistic extractant concentration (X_1), agitation speed (X_2), and feed:organic phase ratio (X_3). The experimental ranges and coded levels of these parameters are displayed in Table 1. To predict the extraction efficiency under specific input conditions, a regression model for the response (Y) was developed. The accuracy, significance, and adequacy of the model were evaluated using analysis of variance (ANOVA) and Fisher's F-test.

Table 1 Experimental ranges and coded levels of these operating parameters.

Code	Operating Parameters	Unit	Range and levels		
			-1	0	+1
X_1	Synergistic extractant concentration (TOMAC/D2EHPA)	[M]	0.1/0.0008	0.25/0.002	0.4/0.0032
X_2	Agitation speed	[rpm]	160	240	320
X_3	Ratio of feed:organic	[v/v]	1	3	5

Separation Technology - ICoST 2025
Materials Research Proceedings 59 (2026) 130-137

Materials Research Forum LLC
https://doi.org/10.21741/9781644903957-17

The recovery of PCs was carried out by mixing the loaded organic phase with an equal volume of stripping agent solution at a predetermined concentration. The mixture was agitated at 320 rpm for 18 h using an IKA-KS 130 mechanical shaker (Germany). Then, it was transferred into a separating funnel and left to settle for 30 min to allow phase separation. The concentration of PCs in the stripping phase was measured. The effects of NaOH and Na_2CO_3 concentrations in the stripping solution were studied to identify the optimal condition for recovery performance. The extraction and recovery efficiency are calculated using Eqs. (1) and (2) [12],

$$\% \text{ Extraction} = \frac{TPC_{i,aq} - TPC_{f,aq}}{TPC_{i,aq}} \times 100 \tag{1}$$

$$\% \text{ Recovery} = \frac{TPC_{fs,aq}}{TPC_{i,aq} - TPC_{f,aq}} \times 100 \tag{2}$$

where $TPC_{i,aq}$, $TPC_{f,aq}$ and $TPC_{fs,aq}$ represent the TPC (mg GAE/L) in the aqueous phase prior to extraction, after extraction and after stripping, respectively.

Results and Discussion

Generally, the concentration of PCs in sterilizer condensate exhibits variability that is influenced by both the grade of FFB and the production process. The concentration of phenolic compounds in the POSC found in this study is significantly high, ranging from 2500 to 4200 mg GAE/L. A variety of phenolic compounds may be present in the sample, potentially including catechin, epicatechin, epigallocatechin, protocatechuic acid, quercetin, and p-hydroxybenzoic acid, as identified by Neo et al. [13].

FTIR analysis in the 4000 to 650 cm^{-1} range confirmed the presence of phenolic compounds, as illustrated in Fig. 1. A broad band observed arround 3304 cm^{-1} corresponds to the O–H stretching vibration, which indicates the presence of hydroxyl groups from phenolics. The strong absorption at 1637 cm^{-1} refers to C=C stretching of aromatic systems, supporting the presence of aromatic phenolics in POSC. The peak at 3264 cm^{-1} is attributed to the N–H stretching of amide groups, indicating the presence of amide-containing compounds in the POSC. A minor band near 2394 cm^{-1} corresponds to the CO_2 asymmetric stretch arising from dissolved CO_2 in the raw POSC. The weak band near 2116 cm^{-1} represent trace alkynyl (C≡C) vibration. The band at 1245 cm^{-1} likely reflects C-N stretching, and the peak at 1090 cm^{-1} is attributed to C–O stretching of carbohydrates or ethers. Features in the fingerprint region (800 to 600 cm^{-1}) indicate aromatic C–H and CH$_2$ groups from long alkyl chains, consistent with residual fatty acids.

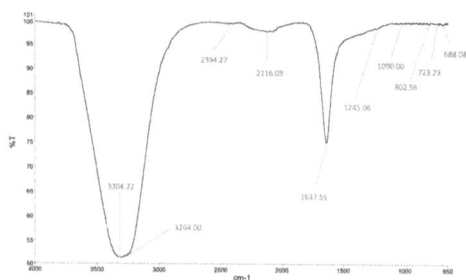

Figure 1 FTIR characterization for POSC.

Synergistic Extractive Extraction of PCs

The comparison between the actual and predicted is shown in Fig. 2. It can be seen the data points are distributed in close proximity to the 45° line, along with high coefficient of determination (R^2

Separation Technology - ICoST 2025
Materials Research Proceedings 59 (2026) 130-137

Materials Research Forum LLC
https://doi.org/10.21741/9781644903957-17

= 0.9926), indicating good agreement between the actual results and the values predicted by the BBD model. A quadratic equation model, derived through multiple regression analysis of actual experimental values is shown in Eq. 3.

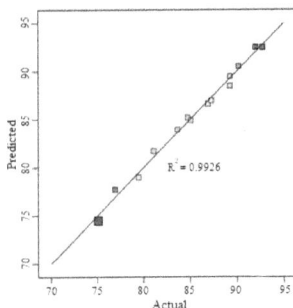

Figure 2 Predicted versus actual values for the PCs extraction performance.

$$Y = 92.38 + 3.36X_1 - 0.25X_2 - 2.02X_3 + 2.00X_1X_2 + 1.09X_1X_3 - 0.77X_2X_3 \\ - 8.08X_1^2 - 4.31X_2^2 - 0.22X_3^2 \tag{3}$$

Table 2 summarizes the ANOVA and regression analysis results for the quadratic model developed to predict the PCs extraction performance. The results show a model F-value of 74.81, which is significantly higher than the critical F-value ($F_{9,5,0.05} = 4.77$), indicating that the model is statistically significant.

Table 2 ANNOVA for quadratic model of PCs extraction.

Source	Sum of squares	df	Mean square	F-value	F-tabulated ($\alpha=0.05$)
Model	440.34	9	48.93	74.81	4.77
Residual	3.27	5	0.6540	-	-
Cor Total	443.61	14	-	-	-

The effect of individual parameters is presented in Fig. 3. It is apparent from Fig. 3(a) that the extraction efficiency increases with synergistic extractant concentration up to 0.3 M, and decreases afterwards. This result is due to higher capacity of extractant to react with PCs in the feed phase. The finding in accordance with Le Châtelier's theory, whereby an increase in extractant concentration promotes the forward progression of the reaction [14]. At higher extractant concentrations, however, the extraction efficiency decreased. This phenomenon is associated with the presence of excess extractant that initiates precipitate formation as a result of a reaction between triglycerides from the diluent and TOMAC. The presence of the precipitate hinders the complex formation between the extractant and PCs. The result is consistent with other studies by Rewatkar [15], who exerted that increasing extractant concentration generally improves extraction efficiency, but only up to a certain point, beyond which efficiency may decline due to saturation or phase imbalance.

The effect of agitation speed is demonstrated in Fig. 3(b). The results show a slight improvement in extraction efficiency is observed as agitation speed increases, up to an optimal point, after which it decreases. As shown in the figure, the extraction is less efficient at low speed (160 rpm), possibly due to incomplete phase mixing, which could limit the contact between the

Separation Technology - ICoST 2025
Materials Research Proceedings 59 (2026) 130-137

Materials Research Forum LLC
https://doi.org/10.21741/9781644903957-17

PCs and extractant. On the other hand, the excessive agitation at high speeds (320 rpm) can lead to the formation of emulsions, which may hinder phase separation and reduce extraction efficiency.

The effect of feed:organic phase ratio is presented in Fig. 3(c). At lower ratio (1:1), higher extraction was observed due to greater availability of extractant, as more organic phase was used per unit feed. For economic reasons, minimizing the organic phase volume is preferred. Hence, a higher extractant concentration is required to compensate for the larger feed phase volume. Nevertheless, increasing the feed phase volume (5:1) decreases the extraction performance due to the reduced amount of organic phase per unit feed, which leads to faster extractant saturation. This outcome match those observed in earlier study [16]. While the individual effects generate useful understanding into the role of each parameter, real-world extraction processes are driven by interdependent relationships between variables. Therefore, the following section evaluates the combined effects of these parameters to estimate optimal operating conditions.

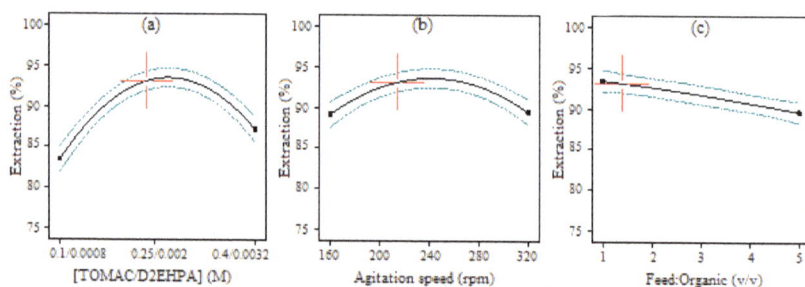

Figure 3 *Effect of individual parameters on PCs extraction, (a) synergistic extractant concentration, (b) agitation speed, and (c) feed:organic ratio.*

Interaction between parameters

Fig. 4 displays the interaction plots between main parameters. Generally, interaction is observed when the response changes according to the combinations of two parameters. The presence of non-parallel lines reflects an interaction, where the impact of one parameter changes according to the level of the other. Fig. 4(a) depicts the interaction effect between synergistic extractant concentration and agitation speed on the extraction efficiency, while keeping the feed:organic ratio constant. The black and red lines represent response at low (160 rpm) and high (320 rpm) agitation, respectively. Within the range of synergistic extractant concentration, there is a strong interaction between both parameters, owing to the intersection of the curves around 0.25 M extractant concentration, which significantly influences the mass transfer of PCs, thereby enhancing the extraction efficiency.

The interaction between synergistic extractant concentration and feed:organic phase ratio at a fixed agitation speed (240 rpm) is shown in Fig. 4(b). The interaction effect is exist as the gap between the red curves (feed:organic = 5) and black curves (feed:organic = 1) is not constant across synergistic extractant concentration. The gap is more pronounced at low synergistic extractant concentration, suggesting that extractant limitation is a stronger constraint when the volume of feed phase is high. At higher synergistic extractant concentration, the gap narrows, indicating partial compensation by the higher availability of extractant molecules.

As shown in Fig. 4(c), the extraction efficiency curve for (feed:organic = 1) and (feed:organic = 5) follow similar pattern with agitation speed. Both curves increase to a maximum near 240 rpm and then decline slightly. The relatively parallel nature of the two curves indicates a weak interaction between agitation speed and feed:organic ratio. This means that the influence of agitation speed on extraction efficiency is similar at both phase ratios, and changing feed:organic does not markedly alter the shape or position of the response to agitation.

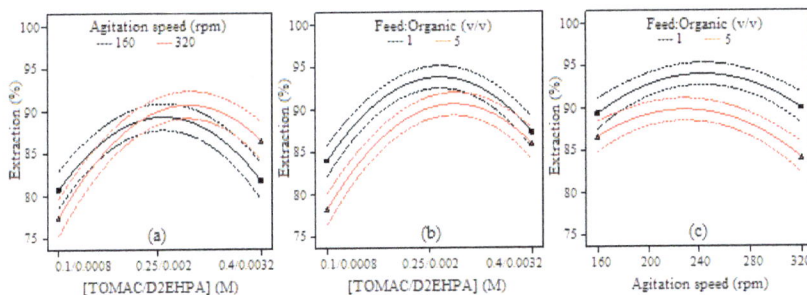

Figure 4 Interaction effect between parameters, (a) synergistic extractant concentration and agitation speed, (b) synergistic extractant concentration and feed:organic ratio and (c) agitation speed and feed:organic ratio.

The optimal conditions for PCs extraction were obtained at 0.3/0.0024 M TOMAC/D2EHPA, an agitation speed of 240 rpm, and a feed:organic ratio of 2.5. Under these conditions, the model predicted an extraction efficiency of 93.1%. Experimental verification under the same conditions yielded 93.8% extraction, resulting in deviation of less than 1%. This low deviation confirms the strong agreement between the predicted and experimental values, demonstrating that the RSM model provides a reliable representation of the system's behavior within the tested range. Therefore, the model can be applied with confidence for optimization of PCs extraction.

Recovery

The recovery of was evaluated by varying the concentrations of the $NaOH/Na_2CO_3$ mixture as shown in Fig. 5. Recovery improved from 65.9% to 98.8% as Na_2CO_3 concentration increased from 0.04 M to 0.12 M. NaOH promotes Na_2CO_3 dissociation, and the resulting higher Na^+ availability enhances stripping of PCs from the extractant complexes in the organic phase. Recovery efficiency declined to 81.4% when Na_2CO_3 concentration increased to 0.4 M. This reduction may be due to excess carbonate ions competing with PC ions for binding to Na^+ via electrostatic interactions, as also suggested by Chen et al. [17]. At concentrations above 0.16 M, a white precipitate formed at the stripping-organic interface, and at 0.4 M, a white creamy substance appeared in the organic phase, likely from soap formation. Such formation creates mass-transfer barriers that hinder the stripping reaction and may alter the chemical properties of the recovered PCs. Therefore, $1/0.12$ M $NaOH/Na_2CO_3$ was chosen as the optimal concentration of stripping agent in this study.

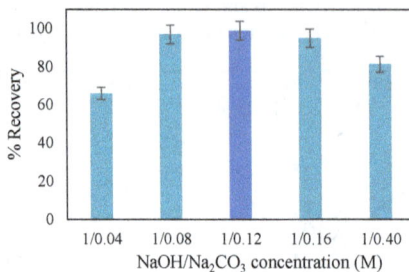

Figure 5 *Effect of Na₂CO₃ concentration on recovery efficiency at fixed NaOH concentration.*

Conclusion

This study addressed the applicability of the synergistic extractive extraction process for extracting and recovering PCs from POSC. A high concentration of PCs (2500-4200 mg GAE/L) was detected in the POSC. At the optimal conditions of 0.3/0.0024 M TOMAC/D2EHPA, an agitation speed of 240 rpm, and a feed:organic phase ratio of 2.5, 93.8% of PCs was extracted. The maximum recovery efficiency of 98.8% was obtained with 1/0.12 M NaOH/Na₂CO₃. Hence, the results of this study highlight the potential for recovering PCs from POSC, supporting circular economy principles and promoting resource valorization in vegetable oil production.

Acknowledgements

This work was supported by the Geran Kursi Premier Sarawak [SWK1.1: R.J130000.7346.1R046, SWK1.0: R.J130000.7309.1R036]; Ministry of Higher Education (MOHE); Universiti Teknologi Malaysia; and Centre of Lipids Engineering and Applied Research (CLEAR).

References

[1] P. Chantho, C. Musikavong, O. Suttinun, Removal of phenolic compounds from palm oil mill effluent by thermophilic Bacillus thermoleovorans strain A2 and their effect on anaerobic digestion, Int. Biodeterior. Biodegrad. 115 (2016) 293-301. https://doi.org/10.1016/j.ibiod.2016.09.010

[2] S. Demirci, S.S. Suner, S. Yilmaz, S. Bagdat, F. Tokay, N. Sahiner, Amine-modified halloysite nanotube embedded PEI cryogels as adsorbent nanoarchitectonics for recovery of valuable phenolic compounds from olive mill wastewater, Appl. Clay Sci. 249 (2024) 107265. https://doi.org/10.1016/j.clay.2024.107265

[3] I. Dammak, M. Neves, H. Isoda, S. Sayadi, M. Nakajima, Recovery of polyphenols from olive mill wastewater using drowning-out crystallization-based separation process, Innovative Food Sci. Emerg. Technol. 34 (2016) 326-335. https://doi.org/10.1016/j.ifset.2016.02.014

[4] P. Mastoras, S. Vakalis, M.S. Fountoulakis, G. Gatidou, P. Katsianou, G. Koulis, N.S. Thomaidis, D. Haralambopoulos, A.S. Stasinakis, Evaluation of the performance of a pilot-scale solar still for olive mill wastewater treatment, J. Cleaner Prod. 365 (2022) 132695. https://doi.org/10.1016/j.jclepro.2022.132695

[5] A.Y. Gebreyohannes, E. Curcio, T. Poerio, R. Mazzei, G. Di Profio, E. Drioli, L. Giorno, Treatment of olive mill wastewater by forward osmosis, Sep. Purif. Technol. 147 (2015) 292-302. https://doi.org/10.1016/j.seppur.2015.04.021

Separation Technology - ICoST 2025
Materials Research Proceedings 59 (2026) 130-137

Materials Research Forum LLC
https://doi.org/10.21741/9781644903957-17

[6] P.L. Pasquet, C. Bertagnolli, M. Villain-Gambier, D. Trébouet, Investigation of phenolic compounds recovery from brewery wastewater with coupled membrane and adsorption process, J. Environ. Chem. Eng. 12 (2024) 112478. https://doi.org/10.1016/j.jece.2024.112478

[7] A.V. Pablo, V. Carlotta, S.A. Carmen, C.U.B. Elena, V.V.M. Cinta, B.P. Amparo, Á.B. Silvia, Separation of phenolic compounds from canned mandarin production wastewater by ultrafiltration and nanofiltration, J. Water Process Eng. 59 (2024) 105041. https://doi.org/10.1016/j.jwpe.2024.105041

[8] B. Illoussamen, Y. Le Brech, I. Khay, M. Bakhouya, C. Paris, L. Canabady-Rochelle, F. Mutelet, Hydrophobic deep eutectic solvents as green solvents for phenolic compounds extraction from olive mill wastewater, J. Environ. Chem. Eng. 13 (2025) 116336. https://doi.org/10.1016/j.jece.2025.116336

[9] S. Tönjes, E. Uitterhaegen, K. De Winter, W. Soetaert, Reactive extraction technologies for organic acids in industrial fermentation processes – A review, Sep. Purif. Technol. 356 (2025) 129881. https://doi.org/10.1016/j.seppur.2024.129881

[10] K.H. Yim, M. Stambouli, D. Pareau, Solvent and emulsion extractions of gallic acid by tributylphosphate: mechanisms and parametric study, Solvent Extr. Ion Exch. 32 (2014) 749-762. https://doi.org/10.1080/07366299.2014.930646

[11] S.S. Suliman, N. Othman, M.A.A. Zaini, N.F.M. Noah, I.N.S. Kahar, Intensification and enhancement of phenolic compounds extraction using cooperative formulation, Chem. Eng. Process. Process Intensif. 211 (2025) 110220. https://doi.org/10.1016/j.cep.2025.110220

[12] I.N.S. Kahar, N. Othman, S.A. Idrus-Saidi, N.F.M. Noah, N.D. Nozaizeli, S.S. Suliman, Integrated emulsion liquid membrane process for enhanced silver recovery from copper-silver leached solution, Chem. Eng. Res. Des. 212 (2024) 434-444. https://doi.org/10.1016/j.cherd.2024.11.018

[13] Y.P. Neo, A. Ariffin, C.P. Tan, Y.A. tan, Determination of oil palm fruit phenolic compounds and their antioxidant activities using spectrophotometric methods, Int. J. Food Sci. Technol. 43 (2008) 1832-1837. https://doi.org/10.1111/j.1365-2621.2008.01717.x

[14] C. Homsirikamol, N. Sunsandee, U. Pancharoen, K. Nootong, Synergistic extraction of amoxicillin from aqueous solution by using binary mixtures of Aliquat 336, D2EHPA and TBP, Sep. Purif. Technol. 162 (2016) 30-36. https://doi.org/10.1016/j.seppur.2016.02.003

[15] K. Rewatkar, Reactive separation of gallic acid: Experimentation and optimization using response surface methodology and artificial neural network, Chem. Biochem. Eng. Q. 31 (2017) 33-46. https://doi.org/10.15255/cabeq.2016.931

[16] N. Jusoh, N. Othman, M.B. Rosly, Extraction and recovery of organic compounds from aqueous solution using emulsion liquid membrane process, Materials Today: Proceedings. (2021) https://doi.org/10.1016/j.matpr.2021.02.800

[17] H. Chen, J. Wang, A.E. Willaims-Jones, Q. Zhu, L. Zheng, C. Zhao, Z. Liu, W. Xu, H. Wei, L. Guo, J. Ma, Effects of carbonate species and chloride ions on calcium phosphate nucleation of biological apatite, Sci. China Mater. 66 (2023) 2872-2884. https://doi.org/10.1007/s40843-022-2424-4

Separation Technology - ICoST 2025
Materials Research Proceedings 59 (2026) 138-144

Materials Research Forum LLC
https://doi.org/10.21741/9781644903957-18

Influence of Temperature, Flow Rate and Extraction Time on Subcritical Water Extraction of Eugenol and Hydroxychavicol from Piper Betel Leaves: A Comparative Study with Soxhlet Extraction

Zuhaili IDHAM[1,a*], Mohd Sharizan MD SARIP[2,b],
Noor Azwani MOHD RASIDEK[3,c], Siti Alyani MAT[4,d], Aishah ROSLI[3,e],
Nik Zetti Amani NIK FAUDZI[5,f]

[1]Department of Deputy Vice Chancellor (Research & Innovation), Universiti Teknologi Malaysia, 81310 UTM Johor Bahru, Johor, Malaysia

[2]Faculty of Chemical Engineering & Technology, Kompleks Pusat Pengajian Jejawi 3, Universiti Malaysia Perlis (UniMAP), 02600 Arau, Perlis

[3]Centre of Lipids Engineering and Applied Research, Ibnu Sina Institute for Scientific and Industrial Research, Universiti Teknologi Malaysia, 81310 UTM Johor Bahru, Johor, Malaysia

[4]Institute of Bioproduct Development, Universiti Teknologi Malaysia, 81310 UTM Johor Bahru, Johor, Malaysia

[5]UTM Centre for Industrial and Applied Mathematics (UTM-CIAM), Ibnu Sina Institute for Scientific and Industrial Research, Universiti Teknologi Malaysia, 81310 UTM Johor Bahru, Johor, Malaysia

[a]zuhailiidham@utm.my, [b]sharizan@unimap.edu.my, [c]noor.azwani@utm.my, [d]sitialyani@utm.my, [e]aishahrosli@utm.my, [f]nikzettiamani@utm.my

Keywords: Subcritical Water Extraction (SWE), Piper Betel Leaves, Eugenol, Hydroxychavicol, Green Extraction

Abstract. Piper betel leaves are rich in phenolic compounds such as eugenol and hydroxychavicol, which possess notable antioxidant and antimicrobial properties. This study evaluated the influence of temperature (120 °C and 180 °C), flow rate (1, 3, and 5 mL/min), and extraction time (10, 20, and 30 min) on the extraction of these compounds using subcritical water extraction (SWE), and compared the results to Soxhlet extraction using ethanol at varying concentrations (25–100%). The SWE results showed that eugenol yield increased from 0.52% at 120 °C to 0.83% at 180 °C, while hydroxychavicol increased from 2.53% to 4.77%, indicating strong temperature dependence. Flow rate significantly influenced hydroxychavicol, with the highest yield (13.75%) at 3 mL/min, whereas eugenol increased steadily with higher flow rates. Extended extraction time (20–30 min) enhanced yields for both compounds, with maximum hydroxychavicol (5.17%) and eugenol (0.74%) observed at 30 min. In contrast, Soxhlet extraction required longer time (6 h) and yielded lower amounts: eugenol ranged from 0.03–0.23%, and hydroxychavicol from 0.48–2.29%, with the best result at 75% ethanol. The study demonstrates that SWE is a faster, solvent-free, and more efficient method for extracting high-value phenolics from piper betel compared to conventional extraction, highlighting its promise for sustainable phytochemical recovery.

Introduction

Piper betel L., commonly known as betel leaf, is a tropical plant widely utilized in traditional medicine across Southeast Asia. Its pharmacological properties are largely attributed to its high content of bioactive phenolic compounds, particularly eugenol and hydroxychavicol, which exhibit strong antioxidant, antimicrobial, and anti-inflammatory effects [1]. These compounds hold great promise for applications in food, cosmetics, and pharmaceutical formulations.

Separation Technology - ICoST 2025   Materials Research Forum LLC
Materials Research Proceedings 59 (2026) 138-144   https://doi.org/10.21741/9781644903957-18

However, efficient and selective extraction of these phytochemicals remains challenging due to their differing polarity, thermal stability, and susceptibility to degradation during processing [2,3]. Conventional methods such as Soxhlet extraction often involve organic solvents and prolonged heating, which can lead to bulk amount of solvent residue, lower compound stability, and potential loss of heat-labile compounds. These limitations have raised environmental and economic concerns regarding the scalability and sustainability of traditional techniques [4,5].

In recent years, subcritical water extraction (SWE) has emerged as a green and efficient alternative to conventional solvent-based techniques. SWE uses water at elevated temperatures (100–374 °C) and pressures that maintain its liquid state, reducing its dielectric constant and allowing water to act as a tuneable solvent capable of extracting a broad range of polar and semi-polar compounds without the use of hazardous chemicals [6,7]. This technique aligns well with green chemistry principles, offering improved selectivity, shorter extraction time, and minimal environmental impact.

Although SWE has been successfully applied to extract phenolic compounds from various plant matrices, limited studies have addressed its application on piper betel leaves, especially for the simultaneous extraction of both eugenol and hydroxychavicol [2,3]. Moreover, a systematic analysis of key SWE parameters such as temperature, flow rate, and extraction time and their effects on compound yield have not yet been fully established.

To fill this gap, the present study aims to evaluate the effect of temperature, flow rate, and extraction time on the SWE performance for recovering eugenol and hydroxychavicol from piper betel leaves. The study also provides a comparative assessment with Soxhlet extraction under various ethanol solvent compositions. The outcomes are expected to contribute to the development of an efficient, environmentally friendly extraction strategy for high-value bioactive compounds, promoting wider application of SWE in natural product recovery.

Methodology

Fresh piper betel leaves were collected from a local source, thoroughly washed to remove dirt, and dried under ambient conditions away from direct sunlight. Once dried, the leaves were ground into fine powder using a laboratory grinder and stored in airtight containers at 4 °C to maintain their stability. Food-grade ethanol ($\geq$99.5%) was purchased from Sigma-Aldrich (USA) for Soxhlet extraction, while standard compounds of eugenol ($\geq$99%) and hydroxychavicol (4-allylpyrocatechol; TCI, Japan) were used for quantitative analysis.

Soxhlet extraction method

A total of 20 g of powdered betel leaf was placed in a cellulose thimble and extracted using ethanol at four different concentrations: 25%, 50%, 75%, and 100% (v/v). For each run, 200 mL of solvent was added to the round-bottom flask, and Soxhlet extraction was performed continuously for 4 h using a heating mantle. After extraction, the solvent was evaporated using a MiVac vacuum concentrator. The resulting extracts were stored at 4 °C until further analysis of eugenol and hydroxychavicol content.

Subcritical water extraction

Subcritical water extraction (SWE) was conducted using a high-pressure batch system equipped with a 10 mL stainless-steel extraction vessel, a metering pump, pre-heater, and a backpressure regulator to maintain pressure. For each experiment, 5 g of powdered piper betel leaf was loaded into the vessel and extracted with deionized water at a fixed pressure of 15 MPa to maintain the subcritical condition. A single-factor experimental design (one-factor-at-a-time, OFAT) was employed to investigate the individual influence of temperature, flow rate, and extraction time. In each trial, one parameter was varied while the other two were kept constant. The SWE conditions studied were as follows:

(a) Temperature: 120 °C, 150 °C, and 180 °C (at 3 mL/min flow rate, 20 min extraction time)
(b) Flow rate: 1, 3, and 5 mL/min (at 150 °C, 20 min extraction time)
(c) Extraction time: 10, 20, and 30 min (at 150 °C, 3 mL/min flow rate)

After extraction, the aqueous extracts were collected in amber vials and stored at –20 °C to prevent degradation until analysis. Each experiment was performed in triplicate.

Determination of eugenol and hydroxychavicol

Quantification of eugenol and hydroxychavicol in the extracts was performed using high-performance liquid chromatography (HPLC) equipped with a photodiode array detector (PDA). The separation was achieved on a reverse-phase C18 column (150 mm × 4.6 mm, 5 μm particle size) using a gradient elution of 0.05% trifluoroacetic acid in water (Solvent A) and methanol (Solvent B) at a flow rate of 0.5 mL/min. The gradient profile was set as follows: 80:20 (A:B) at 0 min, linear change to 20:80 from 8 to 16 min, and back to 80:20 at 17 min. Column temperature was maintained at 30 °C, and detection was performed at 275 nm.

Extracts were redissolved in 10 mL of methanol, filtered through 0.22 μm syringe filters, and 20 μL was injected for each run. Calibration curves for quantification were established using standard solutions of eugenol (5–100 μg/mL) and hydroxychavicol (8–250 μg/mL) in methanol.

Results and Discussions

Effect of temperature on eugenol and hydroxychavicol

The influence of temperature on SWE of eugenol and hydroxychavicol from piper betel leaves was evaluated at 120 °C, 150 °C, and 180 °C, respectively with flow rate and extraction time fixed at 3 mL/min and 20 min. The results showed a consistent increase in yield for both target compounds with rising temperature (Fig. 1). Specifically, eugenol content increased from 0.52% at 120 °C to 0.74% at 150 °C, reaching the highest value of 0.83% at 180 °C. Hydroxychavicol content also followed a similar upward trend, increasing from 2.00% to 5.29%, and peaking at 7.45% at 180 °C.

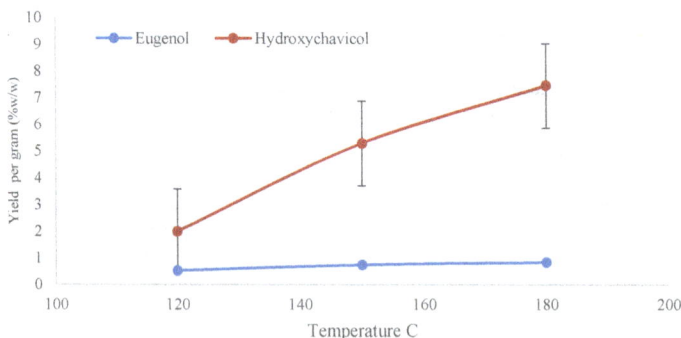

Figure 1 *Effect of temperature on the eugenol and hydroxychavicol extracted from piper betel leaves using SWE at constant flow rate (3 mL/min) and extraction time (20 min).*

This enhancement is likely attributed to the decreased dielectric constant of water at elevated temperatures, which improves the solubility of semi-polar compounds such as eugenol and polar compounds like hydroxychavicol. Additionally, higher temperatures improve mass transfer by promoting cellular disruption and diffusion of compounds from the plant matrix. These observations aligned with [8], wherein the SWE temperatures between 130 °C and 240 °C are optimal for enhancing the solubility and desorption of phenolic compounds through polarity tuning of water. Similarly, [9] reported a maximum phenolic recovery from onion peel at 180 °C using SWE, emphasizing the efficiency of elevated thermal conditions in liberating plant-based bioactives. In other work, [10] also reported significant enhancement in flavonoid yields from citrus peels with increasing SWE temperature. Thus, temperature is a key driver of extraction performance in SWE. The increased yields, especially for hydroxychavicol, confirm that higher temperatures improve recovery efficiency and offer potential for temperature-based tuning in the sustainable extraction of plant-derived phenolics.

Effect of flow rate on eugenol and hydroxychavicol
The impact of flow rate was examined at 1, 3, and 5 mL/min under constant temperature (150 °C) and extraction time (20 min), as shown in Fig. 2. The eugenol yield showed a mild increase across the flow rates, rising from 0.697% at 1 mL/min to 2.29% at 5 mL/min, with a slight decrease to 1.89% at 3 mL/min. This suggests that higher flow rates facilitate better solvent refreshment and mass transfer, leading to enhanced solubilization and transport of eugenol. In contrast, hydroxychavicol displayed a bell-shaped trend. The highest yield was obtained at 3 mL/min (13.75%), whereas lower (7.07%) and higher (12.74%) yields were observed at 1 and 5 mL/min, respectively. This pattern implies that moderate flow rates optimize the residence time of solvent within the extraction vessel, allowing sufficient contact for solute desorption, particularly for more polar and less diffusible compounds like hydroxychavicol.

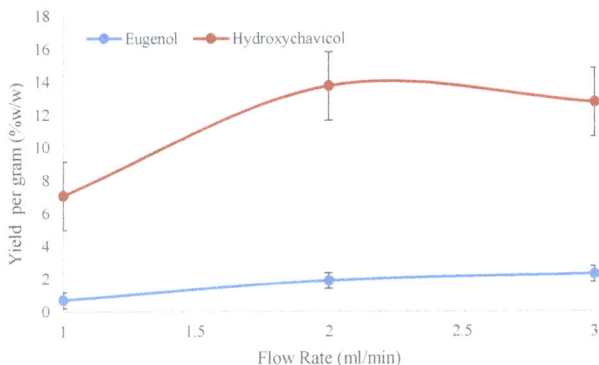

Figure 2 *Effect of flow rate on the eugenol and hydroxychavicol extracted from piper betel leaves using SWE at constant temperature (150) and extraction time (20 min).*

These findings aligned with the observations of [11], who reported that certain phenolic compounds such as scopoletin and alizarin from *Morinda citrifolia* (noni fruit) showed optimal SWE yields at intermediate flow rates (2–3 mL/min). Reference [9] also emphasized that, while higher flow rates improve extraction kinetics, they may not always translate to higher final yields due to reduced residence time and equilibrium limitations. Overall, these results suggest that while eugenol benefits from increased flow due to improved convective transport, hydroxychavicol extraction is more sensitive to residence time requiring a balanced flow rate to maximize yield without reducing solute-solvent interaction.

Effect of extraction time on eugenol and hydroxychavicol
The effect of extraction time was studied at 10, 20, and 30 min, while maintaining constant temperature (150 °C) and flow rate (3 mL/min), as shown in Fig. 3. Eugenol content increased

steadily from 0.44% at 10 min to 1.51% at 30 min. This trend indicates that extended extraction time enhances compound diffusion and facilitates the release of eugenol into the solvent phase.

Hydroxychavicol yield, on the other hand, peaked at 20 min (9.69%) and slightly decreased to 6.46% at 30 min. The initial increase reflects enhanced solute-solvent interaction and improved desorption. The subsequent decline could be due to thermal degradation or saturation effects that limit further recovery at longer durations. This behaviour is in line with the findings by [12], who noted that while longer SWE times can increase yields initially, prolonged exposure may result in compound decomposition or decreased extraction efficiency due to equilibrium constraints.

These results highlight the importance of carefully optimizing extraction time, especially for thermally sensitive compounds like hydroxychavicol. Extraction time is a key factor in ensuring efficient recovery of both eugenol and hydroxychavicol. However, optimizing it must consider both yield gains and energy or processing trade-offs, especially once yields begin to plateau.

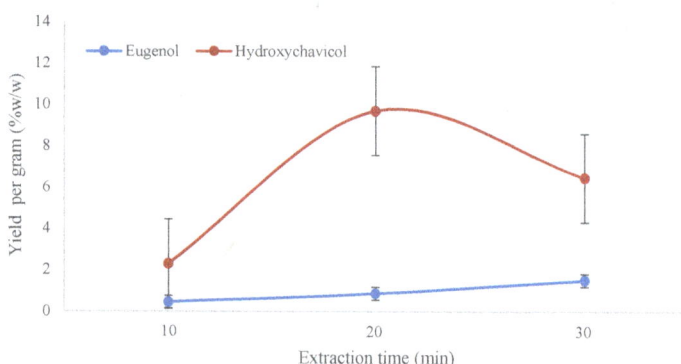

Figure 3 *Effect of extraction time on the eugenol and hydroxychavicol extracted from piper betel leaves using SWE at constant temperature (150 °C) and flow rate (3mL/min).*

Comparison of SWE and Soxhlet method

The effectiveness of SWE was compared with Soxhlet extraction method for the recovery of eugenol and hydroxychavicol from piper betel leaves, as summarized in Table 1. Based on the chemical structures of the target bioactive compounds, eugenol and hydroxychavicol, polarity plays a significant role in their solubility and extraction efficiency. Eugenol contains a methoxy group and an allyl side chain, making it relatively less polar. Meanwhile, hydroxychavicol has two hydroxyl groups, contributing to higher polarity and stronger hydrogen-bonding interactions with polar solvents. Due to this polarity difference, the extraction behaviour of the two compounds differs under various solvent environments. Polar solvents are expected to favour the extraction of hydroxychavicol, while less-polar environments support higher eugenol solubility. Therefore, the observed extraction yield trends align with the polarity-solubility relationship, suggesting that solvent polarity and extraction technique strongly influence the recovery of each phytochemical.

Soxhlet extraction was conducted using ethanol at varying concentrations (25%, 50%, 75%, and 100%). The best yields for each compound using Soxhlet were obtained with 75% ethanol, which produced 0.23% eugenol and 2.29% hydroxychavicol. In contrast, SWE demonstrated significantly superior extraction performance in a shorter time and without the use of organic solvents. Under the best SWE conditions tested (150 °C, 3 mL/min, 20 min), eugenol and hydroxychavicol yields reached up to 1.89% and 13.75%, respectively. This represents an

Separation Technology - ICoST 2025
Materials Research Proceedings 59 (2026) 138-144

Materials Research Forum LLC
https://doi.org/10.21741/9781644903957-18

approximately 8-fold increase for eugenol and a 6-fold increase for hydroxychavicol compared to the Soxhlet method.

Beyond extraction efficiency, SWE offers clear advantages in terms of environmental sustainability and process efficiency. Unlike Soxhlet, which requires long extraction times and high volumes of ethanol, SWE operates with only water as the solvent under controlled pressure and temperature. The ability to selectively recover both polar and semi-polar compounds in a single run further enhances the utility of SWE for natural product recovery.

These findings aligned with reports by [13], who highlighted SWE as an eco-friendly, high-efficiency method capable of outperforming conventional extractions for thermolabile and bioactive compounds. In addition, reference [14] emphasized SWE's potential in pharmaceutical and nutraceutical industries due to its low environmental impact and scalability. Overall, the results of this study support the adoption of SWE as a green alternative to traditional solvent extraction methods, especially for bioactive phenolics such as eugenol and hydroxychavicol from herbal sources.

Table 1 *Summary of extraction yields by SWE and Soxhlet.*

Method	Solvent / Conditions	Eugenol (%)	Hydroxychavicol (%)	Solvent used
Soxhlet	25% ethanol, 6 h	0.03	0.48	Ethanol and water
Soxhlet	50% ethanol, 6 h	0.21	1.52	Ethanol and water
Soxhlet	75% ethanol, 6 h	0.23	2.29	Ethanol and water
Soxhlet	100% ethanol, 6 h	0.23	0.85	Ethanol only
SWE	150 °C, 3 mL/min, 20 min	1.89	13.75	Water only

Conclusion

This study demonstrated the influence of key operational parameters, temperature, flow rate, and extraction time on the recovery of eugenol and hydroxychavicol from Piper betel leaves using SWE. Results indicated that increasing temperature significantly enhanced extraction yields for both target compounds, with hydroxychavicol showing greater temperature sensitivity due to its polar nature. The optimal flow rate for hydroxychavicol was found to be 3 mL/min, reflecting the importance of balancing solvent residence time and mass transfer, while extended extraction time (20–30 min) led to higher yields for both analytes. When compared to conventional Soxhlet extraction, SWE showed substantially higher extraction efficiency, shorter processing time, and eliminated the use of organic solvents. Under optimized SWE conditions (150 °C, 3 mL/min, 20 min), eugenol and hydroxychavicol yields were up to 8 and 6 times higher, respectively, than those obtained using Soxhlet with 75% ethanol. The findings support SWE as a green, rapid, and scalable alternative for the recovery of bioactive phenolics from herbal matrices. Moreover, this single-factor study provides the foundation for further optimization using response surface methodology (RSM) and for evaluating the bioactivity (antimicrobial and antibiofilm) potential of the SWE extracts in future work.

Acknowledgment

The authors gratefully acknowledge the financial support provided by the MRUN Research Officers' Grant Scheme (MROGS) under project code R.J130000.7809.5L005. The authors also wish to extend their sincere appreciation to the Centre of Lipid Engineering and Applied Research (CLEAR), Universiti Teknologi Malaysia (UTM), for access to instrumentation and research facilities that made this work possible.

References

[1] N. Singtongratana, S. Vadhanasin, J. Singkhonrat, Hydroxychavicol and eugenol profiling of betel leaves from Piper betle L. obtained by liquid-liquid extraction and supercritical fluid extraction, Agric. Nat. Resour. 47 (2013) 614-623.

[2] T.A. Musa, M.M. Sanagi, W.A. Wan Ibrahim, F. Ahmad, H.Y. Aboul-Enein, Determination of 4-allyl resorcinol and chavibetol from Piper betle leaves by subcritical water extraction combined with high-performance liquid chromatography, Food Anal. Methods. 7 (2014) 893-901. https://doi.org/10.1007/s12161-013-9697-2

[3] N.L. Rahmah, S.M.M. Kamal, A. Sulaiman, F.S. Taip, S.I. Siajam, Optimization of phenolic compounds and antioxidant extraction from Piper betle Linn. leaves using pressurized hot water, J. Appl. Sci. Eng. 26 (2022) 175-184.

[4] M. Anugrahwati, T. Purwaningsih, J.A. Manggalarini, N.B. Alnavis, D.N. Wulandari, H.D. Pranowo, Extraction of ethanolic extract of red betel leaves and its cytotoxicity test on HeLa cells, Procedia Eng. 148 (2016) 1402-1407. https://doi.org/10.1016/j.proeng.2016.06.569

[5] L. Muruganandam, A. Krishna, J. Reddy, G.S. Nirmala, Optimization studies on extraction of phytocomponents from betel leaves, Resour. Eff. Technol. 3 (2017) 385-393. https://doi.org/10.1016/j.reffit.2017.02.007

[6] D.N. Rizkiyah, N.R. Putra, Z. Idham, M.A. Che Yunus, I. Veza, I. Harny, A.H. Abdul Aziz, Optimization of red pigment anthocyanin recovery from Hibiscus sabdariffa by subcritical water extraction, Processes. 10 (2022) 2635. https://doi.org/10.3390/pr10122635

[7] Somat, H.A., Thani, N.M., Mustapha, W.A.W., Lim, S.J., Seng, N.S.S., Rahman, H.A., Razali, N.S.M., Ali, M.M. and Kamal, S.M.M., Subcritical Water Extraction of Bioactive Compounds from Plant Materials: Recent Advances, Journal of Future Foods. (2025) in press. https://doi.org/10.1016/j.jfutfo.2024.10.005

[8] Y. Cheng, X. Fumin, Y. Shuai, D. Shichao, Y. Yu, Subcritical Water Extraction of Natural Products, Molecules. 26 (2021) 13: 4004. https://doi.org/10.3390/molecules26134004

[9] Ó. Benito Román, B. Blanco, M.T. Sanz, S. Beltrán, Subcritical Water Extraction of Phenolic Compounds from Onion Skin Wastes (Allium cepa cv. Horcal): Effect of Temperature and Solvent Properties, Antioxidants. 9 (12) (2020) 202. https://doi.org/10.3390/antiox9121233

[10] D.S. Kim, S.B Lim. Kinetic study of subcritical water extraction of flavonoids from citrus unshiu peel, Separation and Purification Technology. 250 (2020) 117259. https://doi.org/10.1016/j.seppur.2020.117259

[11] R. Jamaludin, D.S. Kim, L. M. Salleh, S.B. Lim. Kinetic study of subcritical water extraction of scopoletin, alizarin, and rutin from morinda citrifolia, Foods. 10 (2021) 2260. https://doi.org/10.3390/foods10102260

[12] D.P. Xu, Y. Li, X. Meng, T. Zhou, Y. Zhou, J. Zheng, H.B. Li, Natural antioxidants in foods and medicinal plants: extraction, assessment and resources, Int J Mol Sci. 18 (1) (2017) 96. https://doi.org/10.3390/ijms18010096

[13] M. Plaza, L. M. Maria. Pressurized hot water extraction of bioactives. TrAC Trends in Analytical Chemistry. 166 (2023) 117201. https://doi.org/10.1016/j.trac.2023.117201

[14] Ž. Knez, E. Markočič, M. Leitgeb, M. Primožič, M. Knez Hrnčič, M. Škerget, Industrial applications of supercritical fluids: A review, Energy. 77 (2014) 235-243. https://doi.org/10.1016/j.energy.2014.07.044

Separation Technology - ICoST 2025
Materials Research Proceedings 59 (2026) 145-153

Materials Research Forum LLC
https://doi.org/10.21741/9781644903957-19

Subcritical Water Extraction of Protein from *Trichanthera Gigantea*: Optimization using Response Surface Methodology

Tengku Zarith Hazlin TENGKU ZAINAL ABIDIN[1,a*], Nur Hidayah ZAINAN[2,b*],
Ahmad Syahmi ZAINI[1,c], Asiah Nusaibah MASRI[1,d], Yanti Maslina MOHD JUSOH[1,e],
Nur Farzana AHMAD SARNADI[1,f], Mohd Azan MOHAMMED SAPARDI[3,g]

[1]Department of Bioprocess and Polymer Engineering, Faculty of Chemical and Energy
Engineering, Universiti Teknologi Malaysia, 81310 UTM Johor Bahru, Johor, Malaysia

[2]Centre of Lipids Engineering and Applied Research (CLEAR), Universiti Teknologi Malaysia,
81310 UTM Johor Bahru, Johor, Malaysia

[3]Department of Mechanical and Aerospace Engineering, International Islamic University
Malaysia, 53100, Gombak, Kuala Lumpur

[a]tengkuzarithhazlin@graduate.utm.my, [b]nurhidayah.zainan@utm.my,
[c]ahmadsyahmizaini@gmail.com, [d]nusaibah@utm.my, [e]yantimaslina@utm.my,
[f]nurfarzana.as@utm.my, [g]azan@iium.edu.my

Keywords: Protein, Subcritical Water Extraction, *Trichanthera Gigantea*, Optimization, Response Surface Methodology

Abstract. Rising costs of conventional poultry feed protein like soybean and fish meals highlight the need for sustainable, local protein sources. *Trichanthera gigantea* (*T. gigantea*), with around 22 % crude protein and a nutritionally rich in amino acid profile, offers potential as an alternative protein source. The direct inclusion of *T. gigantea* leaves in poultry diets is constrained by antinutritional factors, limited digestibility, and reduced shelf stability. However, extraction of its protein fraction provides a more efficient and practical avenue for nutritional utilization. This study investigates the use of subcritical water extraction (SWE) technique, an environmentally friendly and organic solvent-free technique that employs pressurized hot water to efficiently extract protein from *T. gigantea* leaves. SWE offers advantages such as enhanced extraction efficiency and reduced reliance on organic solvents, aligning with the principles of green technology. To optimize the extraction process, experiments were designed using response surface methodology (RSM) with a central composite design (CCD), evaluating the combined effects of key parameters including temperature (140–180°C), extraction time (10–40 min), and flow rate (1–3 mL/min). The highest yield of protein (29.2%) was identified at 174°C, 38 min and flow rate of 3 mL/min. ANOVA results showed that time taken was the most significant factor, followed by flow rate and temperature. The findings of this study establish a foundational approach for the scalable and efficient recovery of plant-based protein from *T. gigantea*, contributing to the diversification of sustainable poultry feed sources. This aligns with Malaysia's National Food Policy Action Plan 2021–2025, which emphasizes local feed production and food security through environmentally responsible innovation.

Introduction

Proteins are essential macromolecules that support animal growth, nutrition, and physiological functions. In poultry, adequate protein intake is particularly important for muscle development, egg production, and enzyme synthesis, making it a key determinant of performance and productivity. As poultry production expands the demand for high-quality protein sources continue to rise, driving interest in sustainability, plant-based alternatives derived from underutilized biomass.

Separation Technology - ICoST 2025
Materials Research Proceedings 59 (2026) 145-153

Materials Research Forum LLC
https://doi.org/10.21741/9781644903957-19

One promising candidate is *Trichanthera gigantea (T. gigantea)*, commonly known as Nacedero, Madre de Agua, or Ketum Ayam in Malaysia, is a fast-growing perennial shrub with relatively high crude protein content, making it a promising alternative to conventional protein sources such as soybean, corn, and fish meal. The leaves contain approximately 22% crude protein, offering considerable potential for livestock nutrition, particularly in regions where feed costs are a major constraint [1]. However, its direct use in poultry diets remains limited due to antinutritional factors, low digestibility in monogastric animals, and the short shelf life of fresh leaves. These limitations highlight the need of processing strategies to improve its nutritional quality and utilization.

Protein extraction represents one such strategy, as it not only concentrates and enhances protein availability but also minimizes the impact of antinutritional compounds. Extraction performance is influenced by both biomass composition and the extraction technique. Traditional aqueous or solvent-based methods often require long processing times and may result in protein denaturation or reduced yields [2]. Subcritical water extraction (SWE), a green extraction method, has gained attention for its ability to enhance mass transfer and cell wall disruption while operating without usage of organic solvents. In SWE, water is maintained between 100–374°C under sufficient high pressure to remain in the liquid phase, reducing its dielectric constant and improving its solvating power for polar and semi-polar compounds [3]. Compared to conventional extraction, SWE offers faster extraction, reduced solvent usage and potential enhancement of protein solubilization through improved cell wall disruption.

The effectiveness of SWE depends on parameters such as temperature, extraction time, and flow rate, where high temperatures improve solubility but may cause thermal degradation. Response surface methodology (RSM), especially central composite design (CCD), is widely used to optimize these parameters while minimizing experimental runs [4]. CCD has been successfully applied in protein extraction from various matrices, including duckweed (77.8% protein solubilization) [5], polysaccharide extraction from *Grifola frondosa* (25.1% yield) [6], and SWE of *Chlorella vulgaris* achieving up to 31.2 g protein per 100 g biomass [7].

This study aims to optimize the SWE conditions for protein extraction from *T. gigantea* leaves using RSM, evaluate the effect of the variables and compare the results with previous studies to assess SWE's effectiveness as an alternative protein extraction method. The findings provide valorisation insight into viable protein recovery from *T. gigantea* in animal feed production through the application of environmentally friendly extraction technologies.

Materials and Methods

In this study, the raw materials of *T. gigantea* leaves were obtained from a local farmer located at Kampung Temenin, Kota Tinggi, Johor. The leaves were dried using oven dryer at 40°C for 24 h to preserve the shelf-life before being grinded into 600 μm particle sizes. The samples were kept in an airtight container and kept in a chiller approximately 4°C. List of all chemicals and reagents that will be used in this study are as follow: sodium carbonate (Na_2CO_3, Chemiz®), MW; 105.99 g/mol, 99.5%), sodium hydroxide (NaOH, Merck Pty Ltd., MW; 40 g/mol, 99 %), copper sulphate ($CuSO_4.5H_2O$, Chemiz®, MW; 249.68 g/mol, 99.6%), sodium potassium tartrate ($C_4H_4KNaO_6.4H_2O$, Chemiz®, MW; 283.23 g/mol, 99 %), Folin Ciocalteu (Merck), bovine serum albumin (BSA, Sigma Aldrich).

Subcritical water extraction was conducted on a laboratory-scale extraction unit equipped with a stainless-steel extraction cell (20 mL capacity), and a high-pressure pump (maximum pressure 200 MPa). Fig. 1 illustrates the SWE apparatus schematically. Extraction was carried out using 2 ± 0.005 g of dried grind *T. gigantea* leaves loaded into a tea bag and placed in the 10 mL stainless-steel vessel of the extraction chamber with diameter 120 mm. The system allowed precise control of temperature ± 1°C and flow rate. Distilled water was used as the extraction solvent. After extraction, the collected extract containing protein in liquid formed was centrifuged to separate the

Separation Technology - ICoST 2025
Materials Research Proceedings 59 (2026) 145-153

Materials Research Forum LLC
https://doi.org/10.21741/9781644903957-19

liquid from the remaining solid and store at 4°C for analysis. The concentration of protein extracted in this study will be analysed using the Lowry method.

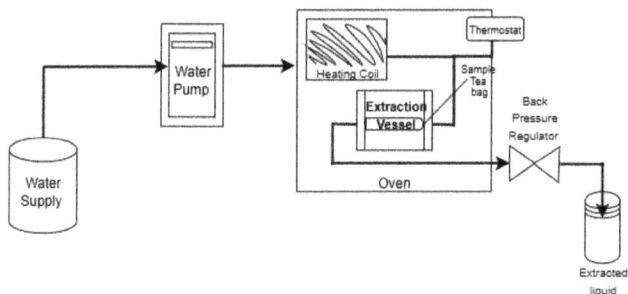

Figure 1 *Subcritical water extraction scheme diagram.*

Table 1 *SWE design of experiment.*

Run	Temperature (°C)	Time taken (min)	Flow rate (mL/min)
1	140	10	3
2	180	10	1
3	140	40	1
4	160	10	2
5	160	25	3
6	160	25	2
7	160	25	2
8	160	25	2
9	160	25	2
10	180	40	3
11	180	25	2
12	160	25	2
13	160	25	1
14	140	40	3
15	160	25	2
16	160	40	2
17	140	25	2
18	140	10	1
19	180	40	1
20	180	10	3

A face centred central composite design (FCCD) with three independent factors, temperature (140, 160, 180°C), extraction time (10, 20, 40 min) and flow rate (1.0, 2.0, 3.0 mL/min) under constant high pressure (15 MPa) was used to evaluate the effects and interactions of each parameter on protein yield. The design consisted of 20 experimental runs including six replicated at the centre point as shown in Table 1. Experimental data from the FCCD is analyzed using regression (Design Expert 13.0) and fitted to a second-order polynomial model to identify possible interactions between parameter using response function as follows,

$$Y_k = \beta_{k0} + \Sigma_{i=1}^{3}\beta_{ki} x_i + \Sigma_{i=1}^{3}\beta_{kii} x_i^2 + \Sigma_{i<j=2}^{3}\beta_{kij} x_i x_j \qquad (1)$$

where, Y is a response (%), β_0 is the intercept, and β_i, β_{ii} and β_{ij} denote the coefficients for linear, quadratic and interaction terms respectively, with x_i and x_j as the independent variables. The suitability and reliability of the quadratic model are evaluated using analysis of variance (ANOVA) [4]. After model is derived from the FCCD, a confirmation reaction is conducted to test the model validity. It is repeated for three times to evaluate the relative error. The relative error is calculated using the predicted and actual reading.

Protein obtained from SWE was determined using the Lowry method assay [5] with BSA as the standard, $y = 1.1273x + 0.0423$. Absorbance reading was taken at 660 nm using a UV-Vis spectrophotometer (Shimadzu UV-160, Japan). The total protein yield (%) is given as follows.

$$Protein\ Yield\ (\%) = \frac{Protein\ concentration\left(\frac{mg}{ml}\right) x\ solvent\ (ml)}{Biomass\ weight\ (mg)} X\ 100 \qquad (2)$$

Results and Discussion
Optimization using RSM
RSM using FCCD was to determine the optimum operating conditions for protein extraction from *T. gigantea*. The operating conditions were chosen according to the SWE system's equipment limits, while the range of independent variables results was selected based on preliminary study. Although the system allows a flow rate up to 20 mL/min, only 1–3 mL/min was used to prevent overpressure that could rupture the sample containment (tea bag) inside the extraction vessel. The quadratic regression model is fitted with the actual experimental obtained by using the central composite design method. Fig. 2 showed good correlation between the predicted and actual results of the protein yield response. The final equation in term of actual factors is as follows,

$$Y = -88.79090 + 1.09938*T - 0.152262*t - 7.93176*F + 0.005130*Tt$$
$$+ 0.056471*TF + 0.126264*tF - 0.003643*T^2 - 0.009757*t^2 + 0.216638*F^2 \qquad (3)$$

where Y is percentage protein yield, T is temperature, t is time and F is flowing rate.

The ANOVA table is shown in Table 2. The ANOVA results shows that a moderate high F-value (18.85) and a very low p-value (0.0001). The lack of fit test yielded an F-value 0.7568 with a p-value 0.6164, this implicates that the model adequately fits the experimental data [6] and unexplained variation is mainly due to random error rather than model inadequacy. Besides that, the F-value for each term namely linear (A, B, C) shows a high influence towards the protein yield, with the largest influence F-value is 78.33, with p-value < 0.00001 for the extraction time (B), followed by flow rate (C) (F = 48.76, p < 0.0001) and temperature (A) with F-value 22.69 and p-value equals to 0.0008. The most interactive (AB, AC, BC) and quadratic (A^2, B^2, C^2) were not statistically significant, excepts the interaction between time taken and flow rate (BC) (F = 5.32, p = 0.00437), suggesting a moderate higher-order effects that are still acceptable in predictive modelling [7]. Fits statistics further support model reliability, with showed high coefficient of determination ($R^2 = 0.944$), adjusted R^2 (0.894) and predicted R^2 (0.774), indicating a good agreement between observed and predicted responses and strong signal-to-noise ratio. Overall, the model is robust and suitable for optimizing SWE conditions for protein extraction from *T. gigantea*, consistent with previous RSM [8]. In this study, all of the linear factor (A = temperature, B = time taken an C = flow rate) significantly influence the response model. The influences strength is as follows: time taken > flow rate > temperature.

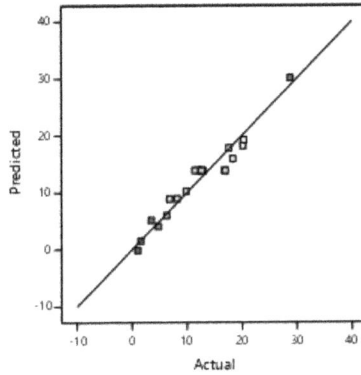

Figure 2 *The actual and predicted yield of protein extraction from T. gigantea using SWE.*

Table 2 *ANOVA for quadratic regression model for protein extraction from T. gigantea.*

Source	Sum of squares	df	Mean square	F-value	p-value	
Model	914.62	9	101.62	18.85	< 0.0001	Significant
A-Temperature	122.29	1	122.29	22.69	0.0008	
B-time taken	422.25	1	422.25	78.83	< 0.0001	
C-flow rate	262.83	1	262.83	48.76	< 0.0001	
AB	18.95	1	18.95	3.52	0.0903	
AC	10.20	1	10.20	1.89	0.1989	
BC	28.70	1	28.70	5.32	0.00437	
A^2	5.84	1	5.84	1.08	0.3225	
B^2	13.25	1	13.25	2.46	0.1479	
C^2	0.1291	0.1291	0.0239	0.8801		
Residual	53.90	10	5.39			
Lack of Fit	23.22	5	4.64	0.7568	0.6164	Not significant
Pure Error	30.68	5	6.14			
Cor Total	968.51	19				
Std. Dev.	2.32	R^2		0.9443		
Mean	12.14	Adjusted R^2		0.8943		
C.V. %	19.13	Adeq Precision		18.4220		

The effect of temperature on the protein extraction of *T. gigantea* using SWE was investigated within the range of 140°C, 160°C and 180°C, refer Fig. 3(a). An increase in temperature from 140°C to approximately 165°C to 170°C with extraction time 16 to 30 min enhanced protein recovery from around 8% to 10% to approximately around 22%. This improvement reflects greater protein solubilization and cell wall disruption under higher thermal energy and longer contact duration [9]. The interaction between temperature and time taken also detected, as the influence of temperature became more pronounced at longer extraction times. The highest yield (~22%) was obtained when both variables were simultaneously at elevated levels [10]. Overall, the dome-shaped surface suggests that optimal SWE performance lies within an intermediate range of temperature and time, beyond which excessive conditions reduce extraction efficiency.

Several studies similarly reported that protein yield increases with both extraction temperature and time taken under SWE conditions. For instance, in de-oiled rice bran, protein yield reached

219±26 mg/g at 200°C for 30 min, with extraction yield rising steadily as temperature and time taken increased [11]. Comparable findings were observed for rice bran and soybean meals, where the optimum recovery occurred after 30 min at 220°C, 210°C, and 200°C for rice bran, raw soybean, and de-oiled soybean meal, respectively, confirming the synergistic effect of elevated temperature and prolonged exposure on protein release [12].

The response surface-plot examining the combined effects of extraction time (B) and flow rate (C) on protein yield demonstrates clear linear, interaction and quadratic dependencies, refer Fig. 3(b). Protein yield ranged from approximately (~5%) at lowest flow rate (1 mL/min) and shortest time (10 min) to a maximum of roughly (~25%) at moderate flow rate (~2.5 mL/min) and longer extraction time (~35 min). The upward slope of the surface indicates a positive linear effect of flow rate, which is expected as higher flow rates enhance mass transfer that helped in improving the protein efficiency [13], while extraction time effects milder but still linear is observable. When prolonged contact allows more diffusion and solubilization of protein molecules into the solvent system [14]. In the present study, protein yield increased with both extraction time and flow rate, indicating that prolonged exposure is essential for protein unfolding and diffusion, while higher flow rates enhance solvent turnover and maintain a strong concentration gradient that facilitates solute transport [15].

The slight curvature of the surface reflects a quadratic response, where yields plateau at higher combinations of flow rate and time, suggesting diminishing returns once readily extractable proteins are removed or when diffusion becomes limiting. Similar plateauing behavior has been reported in SWE studies on *Chlorella vulgaris* and duckweed under optimized conditions [16]. Past study also reported that at highest water-to-solid ratio induced in extraction of protein content in soybean flakes for 30 min has the highest protein efficiency [17].

The response-surface plot representing the combined effects of temperature (A) and flow rate (C) on protein yield as shown in Fig. 3(c), shows a much stronger extraction response than the previous surface, with yields increasing from approximately 0% at low temperatures (140–150°C) and low flow rates (1–1.2 mL/min) to about 27–28% at the highest temperature (~180°C) and a flow rate 3 mL/min. This pattern reflects pronounced positive linear effects of both variables. Higher temperatures enhance protein release by reducing water's dielectric constant, improving its solvating power and promoting more effective cell wall disruption, allowing proteins to diffuse more readily into the solvent phase [18]. Likewise, increased flow rate improves mass transfer and solvent–sample contact, enabling faster leaching of soluble proteins from the plant matrix [10]. The curvature observed at the upper levels of both factors indicates a quadratic effect, where the yield begins to plateau as extraction approaches equilibrium or when diffusion becomes limiting [19].

The finding from this study aligns with results from Benito-Román et al. [15] where increasing flow rate from 2.5 to 6 mL/min did not significantly improve phenolic extraction under SWE, indicating that external mass transfer was not limiting under their conditions. Similarly, Jamaludin et al. (2021) [20], observed that phenolic yields from noni fruit increased substantially with flow rate from 1 to 2 mL/min, but only slightly from 2 to 3 mL/min, suggesting that beyond a certain point, further increases in flow rate produce diminishing returns due to diffusion control.

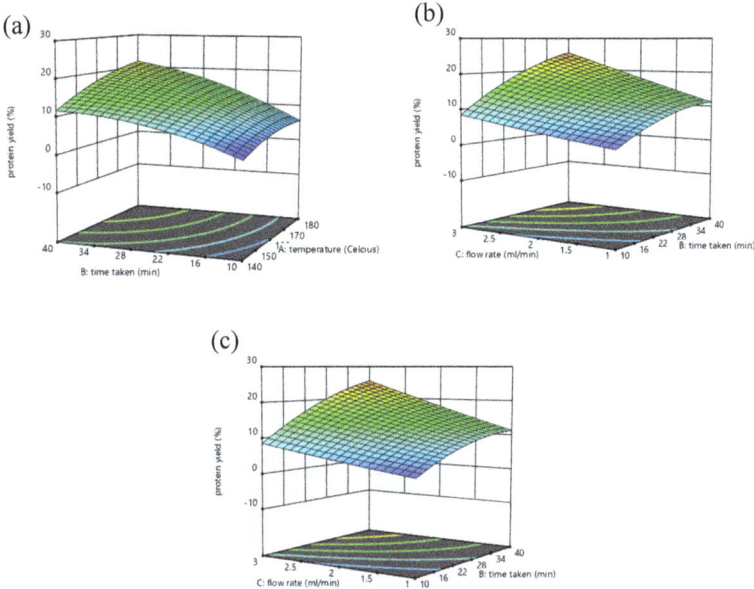

Figure 3 3D Responses of protein extraction from T. gigantea using SWE.

Table 3 Corresponding measured and predicted data for the statistical model validation.

No	Condition	Conversion [%]		Relative error [%]
		Predicted	Measured	
1	Temperature 174°C, time taken	28.3	29.0	2.51
2	38 min, flow rate 3 mL/min		29.2	3.20
3			29.1	2.63
			Average relative error [%]	2.78

Optimization process and data validation

The optimized conditions generated from the model equation, Eq. (3) were used to identify the parameter settings that maximize protein yield of *T. gigantea* by using SWE, producing 100 solution points from the numerical optimization process. The selected optimum extraction conditions capable of achieving a predicted protein yield of 28.4% were determined to be a temperature of approximately 174°C, an extraction time of 38 min, and a flow rate of 3 mL/min. The experimental verification under these optimized conditions is summarized in Table 3, which compares the actual protein yield (~ 29.2%) with the model-predicted value. A low average relative error of 2.78% was recorded, indicating strong agreement between predicted and experimental results. This result is strongly supported by literature confirming that T. gigantea leaves are high in protein biomass, typically between 17.9% and 26.5% crude protein on dry basis [21,22]. In comparison with other biomass protein extraction by using SWE, study revealed that extraction of wheat germ protein using SWE using batch reactor at 170°C, highlighted maximum protein yield of 22.93 g/100 g-WG for 30 min [23].

Conclusion

In conclusion, the subcritical water extraction method was promising in extracting protein from *T. gigantea* with maximizes protein up to ~ 29%. At the optimized conditions were found at temperature 170°C, for 38 min at flow rate 3 mL/min. Extraction time played a critical role, as prolonged treatment facilitated protein unfolding, diffusion, and solubilization, thereby enhancing yield. Similarly, increasing temperature promoted cell wall disruption and protein denaturation, which improved solubilization into the solvent, while higher flow rates enhanced solvent turnover and mass transfer, ensuring more effective washing of proteins from the matrix. These results are consistent with established trends in SWE, where elevated temperature, adequate extraction time, and moderate-to-high flow rates act synergistically to maximize protein recovery.

Acknowledgement

This work was supported by the Ministry of Higher Education under Fundamental Research Grant Scheme (FRGS/1/2023/TK05/UTM/02/9).

References

[1] S. Gilles Tran, A. Gilles Tran, and A. Dick Culbert, "Nacedero (Trichanthera gigantea) | Feedipedia," 2012. [Online]. Available: http://www.feedipedia.org/node/7270 [09/12/201614:55:28] data sheet citation

[2] D. Bhatt, H. Tomar, R. Singh, and S. Chandel, "Extraction of bioactive compounds from marine brown seaweed," *Reg Stud Mar Sci*, p. 104415, Aug. (2025). https://doi.org/10.1016/j.rsma.2025.104415

[3] M. Herrero, E. Ibáñez, A. Cifuentes, G. Reglero, and S. Santoyo, "Dunaliella salina Microalga Pressurized Liquid Extracts as Potential Antimicrobials," *J Food Prot*, vol. 69, no. 10, pp. 2471–2477, Oct. (2006). https://doi.org/10.4315/0362-028X-69.10.2471

[4] A. N. Masri, M. I. Abdul Mutalib, W. Z. N. Yahya, N. F. Aminuddin, and J. M. Leveque, "Rapid esterification of fatty acid using dicationic acidic ionic liquid catalyst via ultrasonic-assisted method," *Ultrason Sonochem*, vol. 60, p. 104732, Jan. (2020). https://doi.org/10.1016/j.ultsonch.2019.104732

[5] C.-H. Shen, "Quantification and Analysis of Proteins," in *Diagnostic Molecular Biology*, Elsevier, 2019, pp. 187–214. doi: 10.1016/b978-0-12-802823-0.00008-0

[6] K. S. Sodhi, J. Wu, A. Dalai, and S. Ghosh, "Subcritical water extraction of hydrolyzed proteins from canola meal: Optimization of recovery and physicochemical properties," *J Supercrit Fluids*, vol. 218, p. 106494, Apr. (2025). https://doi.org/10.1016/j.supflu.(2024).106494

[7] M. M. Hameed, M. K. AlOmar, W. J. Baniya, and M. A. AlSaadi, "Prediction of high-strength concrete: high-order response surface methodology modeling approach," *Eng Comput*, vol. 38, no. S2, pp. 1655–1668, Jun. (2022). https://doi.org/10.1007/s00366-021-01284-z

[8] M. Cano-Lamadrid, L. Martínez-Zamora, L. Mozafari, M. C. Bueso, M. Kessler, and F. Artés-Hernández, "Response Surface Methodology to Optimize the Extraction of Carotenoids from Horticultural By-Products—A Systematic Review," *Foods*, vol. 12, no. 24, p. 4456, Dec. (2023). https://doi.org/10.3390/foods12244456

[9] J.-K. Yan, L.-X. Wu, W.-D. Cai, G.-S. Xiao, Y. Duan, and H. Zhang, "Subcritical water extraction-based methods affect the physicochemical and functional properties of soluble dietary fibers from wheat bran," *Food Chem*, vol. 298, p. 124987, Nov. (2019). https://doi.org/10.1016/j.foodchem.2019.124987

[10] M. Álvarez-Viñas *et al.*, "Subcritical Water for the Extraction and Hydrolysis of Protein and Other Fractions in Biorefineries from Agro-food Wastes and Algae: a Review," *Food Bioproc Tech*, vol. 14, no. 3, pp. 373–387, Mar. (2021). https://doi.org/10.1007/s11947-020-02536-4

[11] R. K, C. S, and H. V, "Protein hydrolysate production from de-oiled rice bran by subcritical water hydrolysis," *Journal of Food Process Enginerring* , vol. 39, no. 6, pp. 627–635, (2016).

[12] R. K, C. S, and H. V, "Subcritical water hydrolysis for protein recovery from rice bran and soybean meal ," *Journal of Food Process Engineering* , vol. 40, no. 2, (2017).

[13] G. Náthia-Neves and E. Alonso, "Optimization of the subcritical water treatment from sunflower by-product for producing protein and sugar extracts," *Biomass Convers Biorefin*, vol. 14, no. 2, pp. 1637–1650, Jan. 2024. https://doi.org/10.1007/s13399-022-02380-w

[14] D. Solanki *et al.*, "Subcritical water hydrolysis of chia seed proteins and their functional characteristics," *Food Hydrocoll*, vol. 143, Oct. 2023. https://doi.org/10.1016/j.foodhyd.2023.108883

[15] Ó. Benito-Román, B. Blanco, M. T. Sanz, and S. Beltrán, "Subcritical Water Extraction of Phenolic Compounds from Onion Skin Wastes (Allium cepa cv. Horcal): Effect of Temperature and Solvent Properties," *Antioxidants*, vol. 9, no. 12, p. 1233, Dec. 2020. https://doi.org/10.3390/antiox9121233

[16] S. M. Zakaria, S. M. Mustapa Kamal, R. Harun, R. Omar, and S. I. Siajam, "Characterization on Phenolic Acids and Antioxidant Activity of Microalgae Chlorella sp. using Subcritical Water Extraction," *Sains Malays*, vol. 49, no. 4, pp. 765–774, Apr. 2020. https://doi.org/10.17576/jsm-2020-4904-05

[17] S. C. Ndlela, J. M. L. N. de Moura, N. K. Olson, and L. A. Johnson, "Aqueous Extraction of Oil and Protein from Soybeans with Subcritical Water," *J Am Oil Chem Soc*, vol. 89, no. 6, pp. 1145–1153, (2012). https://doi.org/10.1007/s11746-011-1993-7

[18] K. S. Sodhi, J. Wu, A. Dalai, and S. Ghosh, "Subcritical water extraction of hydrolyzed proteins from canola meal: Optimization of recovery and physicochemical properties," *J Supercrit Fluids*, vol. 218, p. 106494, Apr. (2025). https://doi.org/10.1016/j.supflu.2024.106494

[19] X. Wu *et al.*, "Optimization of subcritical water extraction of polysaccharides from Grifola frondosa using response surface methodology," *Pharmacogn Mag*, vol. 9, no. 34, p. 120, 2013. https://doi.org/10.4103/0973-1296.111262

[20] R. Jamaludin, D.-S. Kim, L. M. Salleh, and S.-B. Lim, "Kinetic Study of Subcritical Water Extraction of Scopoletin, Alizarin, and Rutin from Morinda citrifolia," *Foods*, vol. 10, no. 10, p. 2260, Sep. 2021. https://doi.org/10.3390/foods10102260

[21] M. A. S. Nasarudin *et al.*, "Alternative Plant Protein Sources Trichantera Gigantea (Ketum Ayam) for Poultry Feed: A Review," *CONSTRUCTION*, vol. 4, no. 2, pp. 238–243, Oct. 2024. https://doi.org/10.15282/construction.v4i2.10679

[22] Edwards A, Mlambo V, Lallo C, and Garcia G. W, "Yield, chemical Composition and In Vitro Ruminal Fermentation of the Leaves of Leucaena Leucocephala, Gliricidia Sepium and Trichanthera Gigantea as Influenced by Harvesting Frequency," *Journal of Animal Science Advances*, vol. 2, no. suppl.3.2, pp. 321–331, Mar. (2012).

[23] Nor Shazwani Daud, Nordin Sabli, Hiroyuki Yoshida, and Shamsul Izhar, "Wheat Germ Protein Extraction via Subcritical Water for Water Treatment Process," *Journal of Applied Science and Engineering*, vol. 28, no. 1, Apr. (2024).

Separation Technology - ICoST 2025   Materials Research Forum LLC
Materials Research Proceedings 59 (2026) 154-161   https://doi.org/10.21741/9781644903957-20

Optimization of *Mimosa Pudica Linn* Extraction at Varied Feed to Solvent Ratios and Solvent Concentrations using Response Surface Methodology

Nurul Aishah ABDUL RAHIM[1,a], Sariah ABANG[1,b*], Mohd Raziman ISMAIDI[1,c], Muhammad Abbas AHMAD ZAINI[2,d], Rubiyah BAINI[1,e], Sherena SAR-EE[1,f]

[1]Department of Chemical Engineering, Faculty of Engineering, Universiti Malaysia Sarawak, 94300 Kota Samarahan, Sarawak, Malaysia

[2]Centre of Lipids Engineering & Applied Research (CLEAR), Ibnu-Sina Institute for Scientific and Industrial Research, Universiti Teknologi Malaysia, 81310 UTM Johor Bahru, Malaysia

[a]n.aishah6470.na@gmail.com, [b]asariah@unimas.my, [c]razimanismaidi6423@gmail.com, [d]abbas@cheme.utm.my, [e]ruby@unimas.my, [f]ssherena@unimas.my

Keywords: Mimosa Pudica Linn, Response Surface Methodology (RSM), Soxhlet Extraction

Abstract. *Mimosa pudica linn*, a creeping plant with either annual or perennial flowering, is commonly known as Daun Semalu in Malaysia. This plant has gained worldwide recognition among researchers not only for its unique reaction upon touch, but also for its long-standing used in traditional medicine. In this study, Response surface methodology (RSM) was used to determine the ideal extraction conditions of the plant leaves by Soxhlet extraction. The study was conducted using ethanol as the solvent at three different concentrations (50%, 70% and 90%) with varied feed-to-solvent ratios (1:15, 1:17, 1:20). The highest extraction yield achieved is 18.61% at ratio of 1:20 with a solvent concentration of 90%. Whereby the optimum conditions predicted by RSM is at a feed-to-solvent ratio of 1:2 (0.05) and a solvent concentration of 89.59% which estimated an extraction yield of 18.72%. Further analyses using Fourier transform infrared spectroscopy (FTIR) and gas chromatography-mass spectrometry (GC-MS) were also conducted to identify the bioactive compounds present in the extracts. FTIR results indicated the presence of various functional groups such as phenolics, alcohol, and alkene groups, suggesting the presence of bioactive compounds. GC-MS analysis identified several key compounds, including benzene, phytol, and phenol, 2,4-bis(1-methyl-1-phenylethyl)-, which are known for their therapeutic properties.

Introduction

Various plant species which grow abundantly and are commonly encountered have already scientifically proven to possess medicinal properties. Among these plants is *Mimosa pudica Linn*, where the parts containing medical properties are the herbs and roots [1]. *Mimosa pudica Linn*, commonly known as 'Putri Semalu' by the locals, is a wild plant that can be found on the side of the road or in the field. The plant has round, hairy, and thorny stems with small compound leaves that fold in response to touch or physical contact [1]. *Mimosa pudica Linn* has been esteemed for its therapeutic attributes across a range of traditional medicinal practices, such as Ayurveda, Chinese, African, Korean, and American traditions [2]. As mentioned by [3], several screening studies have been carried out to investigate its antimicrobial activity, highlighting the medicinal importance of the plant's leaves, which have been used to treat diseases related to blood and bile, bilious fever, and others. *Mimosa pudica Linn* also contained antioxidant, anti-venom, anti-hepatoxic, diuretic, and wound-healing activities [3]. However, the potential of this medical plant to treat various diseases is still not widely acknowledge in Malaysia.

Separation Technology - ICoST 2025
Materials Research Proceedings 59 (2026) 154-161

Materials Research Forum LLC
https://doi.org/10.21741/9781644903957-20

Soxhlet extraction is an effective method that have been widely utilized for isolating organic compounds from both solid and liquid mixture, specifically in the research of plant analysis [4]. Soxhlet extraction often be utilized to acquire essential oils, alkaloids, tannins, and other bioactive constituents from plant leaves [5]. However, several variables such as the nature of the sample, extraction temperature and time, as well as the properties and quantity of the solvent can affect the overall extraction effectiveness and concentration yield [6].

The extraction of *Mimosa pudica Linn* had been previously studied by using several separation techniques such as maceration [7], hydrodistillation with Clevenger apparatus [8], supercritical fluid extraction (SFE) [9], and Soxhlet extraction [10]. Among other methods, Soxhlet extraction provides higher extraction efficiency, reproducibility and scalability, as well as ensuring better in both utilization of solvent and heat control during extraction process. Soxhlet extraction managed to overcome the limitations in others by producing higher yields and more consistent results, offering greater protection for thermolabile compound and having a low-cost and easily performed method. This study delves into the optimization of Soxhlet extraction of *Mimosa pudica Linn*, aiming to determine the optimum conditions based on the feed-to-solvent ratio as well as the solvent concentrations. To statistically analyse the yield of extract from this plant, response surface methodology (RSM) modelling is employed to determine the optimal extraction conditions.

Methodology

Mimosa pudica Linn leaves sample was collected in an open area within the University Malaysia Sarawak (UNIMAS) campus. The obtained leaves were then dried to remove the moisture. Afterward, the plant sample was ground into a fine powder and subsequently kept in an airtight container until further use.

A total of 10 g of dried *Mimosa pudica Linn* powder was weight and placed into a thimble made of thick filter paper. The thimble was placed in the main chamber of the Soxhlet extraction. Afterward, 170 ml of ethanol at 70% concentration was poured into the round bottom flask. The solvent was heated until it attained approximately at 80°C. This extraction process was repeated for a duration of 90 min. The extraction was conducted under the conditions specified in Table 1 provided below.

Table 1 *Factors and levels of Soxhlet extraction of Mimosa pudica Linn.*

Factors		Levels	
Feed to solvent ratio	1:15	1:17	1:20
Solvent concentration (%)	50	70	90

Once Soxhlet extraction process completed, solvent-extract mixture was transferred to a rotary evaporator. The solvent-extract mixture undergoes rotary evaporation at a temperature of 73°C and was rotated at a moderate speed to eliminate any residual solvent and water, forming a thin film of extract on the inner surface. The extract yield of *Mimosa pudica Linn* was determined by measuring the weight of the extract after solvent removal by rotary evaporator and calculated using Eq. (1).

$$\text{Percent Yield (\%)} = \frac{\text{Mass of } \textit{Mimosa pudica Linn} \text{ extract obtained}}{\text{Mass of powdered } \textit{Mimosa pudica Linn} \text{ sample}} \times 100 \tag{1}$$

RSM comprises a set of mathematical and statistical techniques employed in creating empirical models. RSM is utilized in the experiment to optimize a response influenced by multiple independent factors, where it aims to estimate the best system efficiency. A 2-factors and 1-level design was set with 13 experiment runs to analyze the effect of different operating conditions on the yield of *Mimosa pudica Linn* extract.

Results and Discussion

To optimize the extraction process of *Mimosa pudica Linn*, RSM was utilized with the assistance of Design Expert Version 13 software. The faced-centered central composite design (CCD) was selected as the experimental model due to its effectiveness in evaluating the impacts of multiple factors. Table 2 shows the extract yield of the plant after the drying process. The fresh Mimosa pudica Linn contained approximately 7% of moisture, which was removed to ensure consistent in sample mass and comparability across experimental runs.

Table 2 *Yield of Mimosa pudica Linn extraction at different feed-to-solvent ratios and different solvent concentrations.*

Run	Feed to solvent ratio	Solvent concentration (%)	Extract yield (%)		
			Actual	Predicted	Residual
1	1:17	70	14.81	11.93	2.88
2	1:15	50	0.99	4.91	-3.92
3	1:17	70	13.31	11.93	1.38
4	1:17	90	16.91	16.69	0.22
5	1:20	50	8.51	9.45	-0.94
6	1:20	70	11.71	14.20	-2.49
7	1:17	70	12.71	11.93	0.78
8	1:17	70	8.31	11.93	-3.62
9	1:17	70	12.63	11.93	0.70
10	1:17	50	12.41	7.18	5.23
11	1:20	90	18.61	18.96	-0.35
12	1:15	90	14.91	14.42	0.50
13	1:15	70	9.31	9.66	-0.35

The relation between the two manipulated variables (feed to solvent ratio and solvent concentration) and the extraction yield were analysed using a face-centered CCD with four factorial points, four axial points, and five centre points. The design models yielded a total of 13 runs with different combinations, which were utilised to construct the regression model.

The analysis of *Mimosa pudica Linn* extract was performed using multiple model types within the framework of RSM. These model summary statistics provide insights into how well the model fits the experimental data and help in assessing the model's predictive power. The study examined linear, 2FI (two-factor interaction), quadratic, and cubic models, each of which was assessed for fit and predictive power as shown in Table 3. The table shows that linear model with R^2 value of 0.6901, signifying that approximately 69.01% of the variation in extract yield can be accounted for by the model. From the table, the predicted R^2 of 0.4389, which is considerably less than the adjusted R^2, raises the possibility of problems like a big block effect, incorrect model specification, or anomalies in the data, such as outliers.

The adjusted R^2 score is reduced to 0.6282, indicating that it has been adjusted to allow for the number of predictors. The difference between the adjusted R^2 and predicted R^2 is less than 0.2 which shows a reasonable agreement between the two values. While the PRESS (predicted residual mistake sum of squares) of 135.36 signifies the model mistake in forecasting new data. A lower PRESS value indicates that the model has a better fit and is more likely to provide accurate predictions for new observations [11]. In this context, the PRESS value of 135.36 suggests that the linear model performs reasonably well in predicting new data compared to the other models evaluated.

Separation Technology - ICoST 2025
Materials Research Proceedings 59 (2026) 154-161

Materials Research Forum LLC
https://doi.org/10.21741/9781644903957-20

Table 3 *Model summary statistics.*

Source	Std. Dev	R^2	Adjusted R^2	Predicted R^2	PRESS	
Linear	2.73	0.6901	0.6282	0.4389	135.36	Suggested
2F1	2.81	0.7053	0.6070	0.1713	199.91	
Quadratic	2.64	0.7983	0.6541	-0.1635	280.67	
Cubic	2.30	0.8904	0.7370	-0.5534	374.72	Aliased

Table 4 *ANOVA for linear model.*

Source	Sum of squares (SS)	df	Mean square	F-value	p-value	
Model	166.48	2	83.24	11.14	0.0029	Significant
A-Feed to solvent ratio	30.92	1	30.92	4.14	0.0694	
B-Solvent concentration	135.57	1	135.57	18.14	0.0017	
Residual	74.75	10	7.47			
Lack of fit	51.24	6	8.54	1.45	0.3738	Not significant
Pure error	23.50	4	5.88			
Cor total	241.23	12				
R^2					0.6901	
Adeq precision					10.6953	
VIF					1.0	

Table 4 presents the results of the ANOVA linear model analysis of *Mimosa pudica Linn* Soxhlet extraction. ANOVA analysis offers crucial insights into the importance and impact of many factors on the response variable. Apart from that, as stated by [12], the F-value and p-value are the two crucial parameters for ANOVA analysis. As observed from Table 4, the model is significant, as evidenced by a p-value of 0.0029 (below than 0.05). This indicates that the combination of factors in the model explains an important part of the variation in the response variable and it also exert a statistically significant influence on the result.

When analysing the specific factor of feed-to-solvent ratio, the sum of squares (SS) is 30.92 and the mean square is also 30.92. This results in an F-value of 4.14 and a p-value of 0.0694. Although this indicates that the feed-to-solvent ratio has some effect on the response variable, it is not statistically significant at the 5% level. The marginal significance may be due to the narrow experimental range used in this experiment, which produced only minimal perturbation to the extraction system. Nonetheless, although this factor resulted in insufficient yield variation for the model to detect a significant effect, it should not be dismissed. The relatively low p-value suggests a potential pattern that this factor might still influence the response under different conditions or in combination with other factors [13].

On the other hand, the concentration of the solvent has a substantial impact on the response variable in the extraction process of *Mimosa pudica Linn*. This is evident from the SS of 135.57, the mean square of the same value (135.57), an F-value of 18.14, and a p-value of 0.0017. Therefore, the solvent concentration should be regarded as a crucial variable in the extraction process. The residuals, having SS of 74.75 and a mean square of 7.47, indicate the amount of variation in the response that is not accounted for by the model. The lack of fit, as indicated by SS of 51.24, a mean square of 8.54, an F-value of 1.45, and a p-value of 0.3738, is not significant. This suggests that the model used fits the data well, and there is no strong evidence to support the idea that a higher-order model would provide a better response [13]. The objective function as stated in Eq. (2) is obtained from the Linear model, where A represents the feed-to-solvent ratio and B represents the solvent concentration. This objective function can be a valuable tool for future optimization efforts, aiding researchers in adjusting these factors to maximize extract yield.

$$Y = 10.91935 - 267.05882\,A + 0.237667\,B \tag{2}$$

The contour plots and 3D surface diagrams in Fig. 1 shows the influence of solvent concentration and the feed-to-solvent ratio on the extraction yield. In these findings, a colour gradient was utilized where red presents the highest yield and blue indicating lowest yield. This trend indicated that the feed-to-solvent ratio does influences the extraction efficiency and the separation in Soxhlet extraction, aligning with previous studies that reported at higher feed-to-solvent ratios, the contact and mass transfer between solvent and solid material is improved, which alternately facilitated dissolution and extraction processes [14]. As clarified in Fig. 1(b), the maximum yield was obtained at a feed-to-solvent ratio of 0.05 (1:20), afterward the yield gradually declined as the ratio increased to 0.067 (1:15).

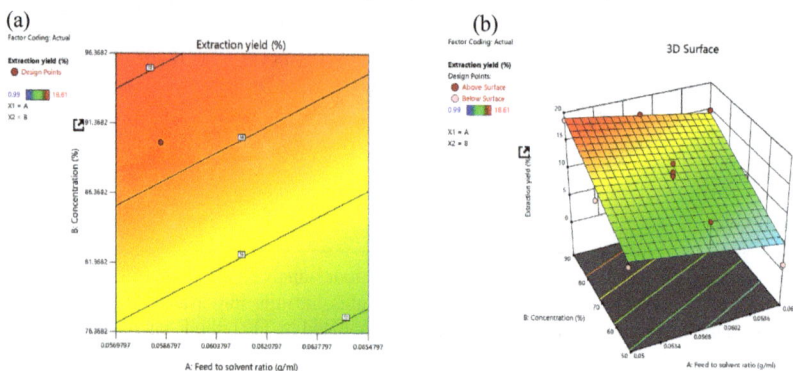

Figure 1 *(a) Contour plots of the extract yield as a function, (b) 3D diagram of the extract yield as a function.*

The extraction yield is significantly affected by the concentration of the solvent employed in the process whereby higher solvent concentrations lead to higher yields. This can be clarified by several theoretical principles. According to [15], the solubility of chemicals in a solvent is greatly influenced by the polarity of both the solvent and the compound being extracted. Higher concentrations of solvents generally result in greater purity of the solvent, which in turn enhances the solubility of the desired compounds. In addition, [16] mentioned increased solvent concentrations improve the rate at which molecules are transferred between the solid and liquid phases. The experimental results displayed in Table 2 show the yields obtained with 50% ethanol were consistently lower, regardless of the ratio of feed to solvent. While at 70% ethanol yields varied, this occurred likely due to systematic errors during the experiment such as inconsistent temperature of the heating mantle. In contrast, 90% ethanol generated greater amounts, reaching up to 18.61% at a 1:20 ratio.

Table 5 highlights FTIR analysis of *Mimosa pudica Linn* extract from Soxhlet extraction at 90% concentration of ethanol and a feed-to-solvent ratio of 1:20. Notable peaks were observed at 3240.41 cm^{-1} and 2920.23 cm^{-1} corresponding to functional group of phenolic and alkanes (C-H). These findings align with [17], who reported a close range of strong absorption band around 3476.87 cm^{-1} and 2922.86 cm^{-1} for phenolic and alkanes functional group respectively. A significant peak at 1604.77 cm^{-1} indicated the presence of primary amines (N-H), consistent with [17]'s finding of primary amines (N-H) stretching around 1654.37 cm^{-1}. Other peak at 1435.04 cm^{-1}, and 1357.89 cm^{-1} also pointed to alkanes (C-H), comparable to alkanes peaks found by [18] and [17] at stretching bond of 1474.20 cm^{-1} and 1326.97 cm^{-1} respectively. The availability of alcohol functional groups was validated by the absorption peak recorded at 1257.59 cm^{-1} (C–O

Separation Technology - ICoST 2025
Materials Research Forum LLC
Materials Research Proceedings 59 (2026) 154-161
https://doi.org/10.21741/9781644903957-20

stretching), which closely matched with the value documented by [18] at 1194.6 cm^{-1}. Overall, the *Mimosa pudica Linn* FTIR analysis was aligned with the previous research, supporting the chemical composition result of plant extracts. The spectrum identified the presence of alcohols, alkanes, phenolics, and primary amines within the extract. These groups were acknowledged for its bioactivity and potential therapeutic properties. Thus, implying the extract possessed possible medicinal benefits, consistence with its long-standing use in traditional medicine practices.

Table 5 *FTIR analysis of Mimosa pudica Linn extract.*

Absorption frequency (cm^{-1})	Possible functional groups	Stretching vibration
3240.41	Phenolic	O–H
2920.23	Alkanes	C–H
1604.77	Primary Amines	C=C
1435.04	Alkanes	C–H
1357.89	Alkanes	C–H
1257.59	Alcohols	C–O

The GC-MS analysis of the *Mimosa pudica Linn* extract revealed a complex mixture, with a total of 50 distinct peaks detected. Among these, five main peaks were identified as major compounds as summarized in Table 6. Benzene is observed at the 2nd peak, suggesting its relatively high abundance among the components. Benzene, though commonly found in various plants, is noted for its role as a precursor in the biosynthesis of more complex aromatic compounds [19]. Phytol at the 37th peak, is a diterpene alcohol that is often associated with antioxidant and anti-inflammatory properties, aligning with the medicinal uses of *Mimosa pudica Linn* [20]. Meanwhile phenol, 2,4-bis(1-methyl-1-phenylethyl at peak 50th is significant for its antiseptic and antioxidant properties, which could enhance the antimicrobial activity of the extract. The 47th peak is classified as 9-Octadecenamide, (Z)-, generally indicated to as oleamide. Oleamide was a fatty acid amide, has been informed to possess sedative and analgesic properties [21]. At the 34th peak was classified as 7,9-di-tert-butyl-1-oxaspiro (4,5) deca-6,9-diene-2,8-dione. This compound, identified for its structural stability and distinctiveness was considered to contribute to the phytochemical profile of the extract.

Table 6 *Major compounds of Mimosa pudica Linn extract identified in GC-MS.*

Peak	Area	Area [%]	Compound name
2	1606019	9.22	Benzene
34	164147	0.94	7,9-di-tert-butyl-1-oxaspiro (4,5) deca-6,9-diene-2,8-dione
37	899440	5.16	Phytol
47	26241	1.51	9-octadecenamide, (Z)-
50	510515	2.93	Phenol, 2,4-bis(1-methyl-1-phenylethyl)-

Conclusion
The optimization of *Mimosa pudica Linn* extract using the Soxhlet method was conducted by manipulating feed-to-solvent ratios and solvent concentrations. Through a series of experiments combined with RSM analysis, the objectives of this study were achieved by determining optimal conditions to obtain maximum extraction yield. The optimum conditions were determined at a feed-to-solvent ratio of 1:2 (0.05) and a solvent concentration of 89.59% which estimated an extraction yield of 18.72%. This estimated value is comparable with the maximum experimental

yield (18.61%), thus validating the reliability and accuracy of the model. FTIR and GC-MS analyses provided both qualitative and quantitative information relating to the functional groups and chemical compounds contained in the extract. The FTIR spectrum confirmed the presence of alcohols, alkanes, phenolics, and primary amines, while the GC-MS analysis classified several major compounds, including benzene, 7,9-di-tert-butyl-1-oxaspiro(4,5)deca-6,9-diene-2,8-dione, phytol, 9-octadecenamide (Z)-, and phenol, 2,4-bis(1-methyl-1-phenylethyl)-. The findings revealed the chemical composition of the extract and supported its potential applications in medicinal research.

Acknowledgement
This work was partly funded by Sarawak state government, and Land and Survey Department, Sarawak through Kursi Premier Sarawak Research Grant No. 1R037.

References
[1] S. Ramadayanti, I. Ginting, J. Naldi, S.N. Rudang, S. Ramadayanti, Sedative Test of Ethanol Extract of Putri Malu Leaves (Mimosa Pudica Linn.) In Mice (Mus Musculus) With Standardized Herbal Medicine Lelap as Comparison, Journal La Medihealtico 3 (2022) 110–115. https://doi.org/10.37899/journallamedihealtico.v3i2.579

[2] S. Kumar, S.K. Manoharan, M. Ram, K. Rao, R.K. Ranjan, M. Sathish Kumar, I. Seethalakshmi, M.R.K. Rao, Phytochemical analysis of leaves and roots of mimosa pudica collected from Kalingavaram, Tamil Nadu, Available Online Www.Jocpr.Com Journal of Chemical and Pharmaceutical Research 5 (2013) 53–55. www.jocpr.com

[3] L. Azmi, M.K. Singh, A.K. Akhtar, Pharmacological and biological overview on Mimosa pudica Linn, International Journal of Pharmacy & Life Sciences (IJPLS) 2 (2011) 1226–1234.

[4] A.R. Abubakar, M. Haque, Preparation of Medicinal Plants: Basic Extraction and Fractionation Procedures for Experimental Purposes, J Pharm Bioallied Sci 12 (2020) 1–10. https://doi.org/10.4103/jpbs.JPBS_175_19

[5] O.E. Adurosakin, E.J. Iweala, J.O. Otike, E.D. Dike, M.E. Uche, J.I. Owanta, O.C. Ugbogu, S.N. Chinedu, E.A. Ugbogu, Ethnomedicinal uses, phytochemistry, pharmacological activities and toxicological effects of Mimosa pudica- A review, Pharmacological Research - Modern Chinese Medicine 7 (2023) 100241. https://doi.org/10.1016/j.prmcm.2023.100241

[6] Q.-W. Zhang, L.-G. Lin, W.-C. Ye, Techniques for extraction and isolation of natural products: a comprehensive review, Chin Med 13 (2018) 20. https://doi.org/10.1186/s13020-018-0177-x

[7] W. Wulan, A. Yudistira, H. Rotinsulu, UJI AKTIVITAS ANTIOKSIDAN DARI EKSTRAK ETANOL DAUN Mimosa pudica Linn. MENGGUNAKAN METODE DPPH, PHARMACON 8 (2019) 106. https://doi.org/10.35799/pha.8.2019.29243

[8] K.O. Fagbemi, D.A. Aina, O.O. Olajuyigbe, Soxhlet Extraction versus Hydrodistillation Using the Clevenger Apparatus: A Comparative Study on the Extraction of a Volatile Compound from Tamarindus indica Seeds, The Scientific World Journal 2021 (2021) 1–8. https://doi.org/10.1155/2021/5961586

[9] S.T. Tasnuva, U.A. Qamar, I.S.M. Zaidul, Mimosa pudica L.: A comparative study via in vitro analysis and GC Q-TOF MS profiling on conventional and supercritical fluid extraction using food grade ethanol, 2017.

[10] S. Madan Mohan, B. Pandey, S.G. Rao, Phytochemical Analysis and Uses of Mimosa pudica Linn. in Chhattisgarh, n.d. www.iosrjournals.org

Separation Technology - ICoST 2025   Materials Research Forum LLC
Materials Research Proceedings 59 (2026) 154-161   https://doi.org/10.21741/9781644903957-20

[11] R.L.L. Pambi, P. Musonge, Application of response surface methodology (RSM) in the treatment of final effluent from the sugar industry using Chitosan, in: 2016: pp. 209–219. https://doi.org/10.2495/WP160191

[12] W. Huang, H. Zhang, Convergence analysis of deep residual networks, Analysis and Applications 22 (2024) 351–382. https://doi.org/10.1142/S021953052350029X

[13] J. Fang, Z. He, Parameter Estimation in RSM Taking into Account Errors in Independent Variables, in: 2010 3rd International Conference on Information Management, Innovation Management and Industrial Engineering, IEEE, 2010: pp. 183–186. https://doi.org/10.1109/ICIII.2010.366

[14] J. Assunção, H.M. Amaro, F.X. Malcata, A.C. Guedes, Factorial Optimization of Ultrasound-Assisted Extraction of Phycocyanin from Synechocystis salina: Towards a Biorefinery Approach, Life 12 (2022) 1389. https://doi.org/10.3390/life12091389

[15] Z. Ye, D. Ouyang, Prediction of small-molecule compound solubility in organic solvents by machine learning algorithms, J Cheminform 13 (2021) 98. https://doi.org/10.1186/s13321-021-00575-3

[16] H. Hu, J. Wu, M. Zhang, Microcalorimetry Techniques for Studying Interactions at Solid–Liquid Interface: A Review, Surfaces 7 (2024) 265–282. https://doi.org/10.3390/surfaces7020018

[17] V. Sangu, T. Yamuna, G. Anu Preethi, & A. Sineha, Green synthesis and characterization of silver nanoparticles using ethanolic extract of Mimosa Pudica linn leaves, 2021.

[18] A.A. Ahuchaogu, G.I. Ogbuehi, P.O. Ukaogo, Ifeanyi.E. Otuokere, Gas Chromatography Mass Spectrometry and Fourier transform Infrared Spectroscopy analysis of methanolic extract of Mimosa pudica L. leaves, Journal of Drugs and Pharmaceutical Science 4 (2020) 1–9. https://doi.org/10.31248/JDPS2020.031

[19] R.M. Dickey, A.M. Forti, A.M. Kunjapur, Advances in engineering microbial biosynthesis of aromatic compounds and related compounds, Bioresour Bioprocess 8 (2021) 91. https://doi.org/10.1186/s40643-021-00434-x

[20] J. de Moraes, R.N. de Oliveira, J.P. Costa, A.L.G. Junior, D.P. de Sousa, R.M. Freitas, S.M. Allegretti, P.L.S. Pinto, Phytol, a Diterpene Alcohol from Chlorophyll, as a Drug against Neglected Tropical Disease Schistosomiasis Mansoni, PLoS Negl Trop Dis 8 (2014) e2617. https://doi.org/10.1371/journal.pntd.0002617

[21] K. Ahn, D.S. Johnson, B.F. Cravatt, Fatty acid amide hydrolase as a potential therapeutic target for the treatment of pain and CNS disorders, Expert Opin Drug Discov 4 (2009) 763–784. https://doi.org/10.1517/17460440903018857

Separation Technology - ICoST 2025
Materials Research Proceedings 59 (2026) 162-169

Materials Research Forum LLC
https://doi.org/10.21741/9781644903957-21

Prediction of Solvent Component and Composition for Absorption-Based Acid Gas Removal Unit using Optuna-LightGBM and K-means

Rafi Jusar WISHNUWARDANA[1,a*], Madiah OMAR[2,b], Haslinda ZABIRI[1,c], Kishore BINGI[3,d], Rosdiazli IBRAHIM[3,e]

[1]Chemical Engineering, Universiti Teknologi PETRONAS, Perak, Malaysia

[2]Integrated Engineering Department, Universiti Teknologi PETRONAS, Perak, Malaysia

[3]Electrical and Electronic Engineering Department, Universiti Teknologi PETRONAS, Perak, Malaysia

[a]rafi_24002236@utp.edu.my, [b]madiah.omar@utp.edu.my, [c]haslindazabiri@utp.edu.my, [d]bingi.kishore@utp.edu.my, [e]rosdiazli@utp.edu.my

Keywords: Acid Gas Removal Unit, Solvent Component, Solvent Composition, LightGBM, Optuna, K-Means

Abstract. Acid gas removal unit (AGRU) is a pivotal component of a natural gas processing plant. The primary purpose of acid gas removal is to reach the industrial pipeline standard of H_2S below four ppm and CO_2 below 2% per volume for pipeline quality. The most widely used technique is an absorption-based AGRU using amine as a solvent. MDEA is the most utilized solvent but has the drawback of absorbing CO_2. The mixture of other amine and physical solvents is necessary to assist the absorption of CO_2. However, the main problem of mixing solvents is parameter complexity. The machine learning method is utilized to find the most optimal solvent based on its operational parameters. LightGBM tuned with Optuna are used to classify the solvent component, followed by K-means to identify solvent composition. The algorithm is applied to six different solvent blends and two feed gas compositions, resulting in 37,786 data points. The LightGBM model tuned with Optuna performed excellently with accuracy of 0.98 and training time under 0.2 seconds. K-means showed the silhouette score averaging 0.5, showing that the data is not well clustered. This model demonstrates reliable capability in analyzing and distinguishing the solvent component and its respective composition.

Introduction

Natural gas was presented as the most environmentally friendly fossil fuel, with the CO_2 emission 23.2% and 57.4% lower than oil and coal, respectively [1]. The purification of natural gas from dangerous gas such as H_2S and CO_2 is critical in industrial chemistry to guarantee the quality of product gas. Hence, it is essential to fulfil the industrial pipeline standard of H_2S below 4 ppm and CO_2 below 2% per volume for pipeline quality [2,3]. To achieve this feat, the removal of H_2S and CO_2 from natural gas is essential.

In recent developments, several techniques have been implemented in the AGRU process, including physical and chemical solvent absorption, adsorption, membrane, etc. However, among all the techniques, absorption-based AGRU using amine as a chemical solvent is widely applied in the chemical industry [4-6]. Methyldiethanolamine (MDEA) is the most used amine considering its selectivity in absorbing H_2S [7]. Despite its selectivity over H_2S, it has a drawback in MDEA reacting with the CO_2. To tackle that drawback, MDEA needs to be blended with an amine that can assist the absorption of CO_2. Besides using a chemical solvent, the blending of MDEA with sulfolane (SFL) shows promising capability in enhancing the absorption of H_2S and energy saving. Also, MDEA, SFL, and (piperazine) PZ have been discovered to have the same ability as the SFL blend but have advantages at the CO_2 absorption [8]. Based on all previously mentioned research,

Separation Technology - ICoST 2025
Materials Research Forum LLC
Materials Research Proceedings 59 (2026) 162-169
https://doi.org/10.21741/9781644903957-21

the major problem of solvent blending requires careful optimization of operational conditions to achieve the best performance. Hence, the data-driven approach is growing in interest.

This paper will develop a more general model for determining solvent components as well as composition on six different solvents blending such as MDEA + SFL + PZ, MDEA +PZ, MDEA+ SFL, MDEA + (diglycoamine) DGA, MDEA + (methylethanolamine) MEA, and MDEA + Diisopropylamine (DIPA). At the same time, two different feed gas compositions are applied to the model, providing a large amount of the dataset for the model. The model employed in this paper is LightGBM for classifying the solvent component, and K means to determine the solvent composition ranges. LightGBM will be tuned with the Optuna Hyperparameter optimization framework to ensure high accuracy and rapid training on the model.

Methodology
In this paper, solvent components and composition of absorption-based AGRU were determined by employing Optuna-LightGBM model integrated with K-means models. The main pipeline (see Fig. 1).

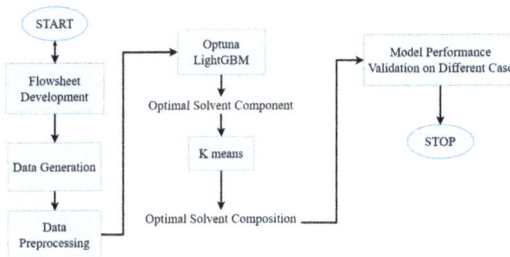

Figure 1 Research flowchart.

Flowsheet development
As mentioned beforehand, absorption-based AGRU using amine is extensively utilized in industry. To recreate the actual operation of the AGRU process, flowsheets were developed in Aspen HYSYS software as portrayed in Fig. 2, which has been validated in previous work [9] by using a case study from real plant data, resulting in an acceptable average error of 2%. Therefore, the developed flowsheet is suitable for generating data.

Data generation
Data can be retrieved from the validated flowsheet by simulating different values of a variable using the Aspen Simulation workbook. Several variables that influence the AGRU process are selected based on the previous literature. Large datasets can improve the model accuracy [10], and better data generalization [11]. Table 1 shows the variables that are going to be generated in this research based on the previous literature [9]. The parameters are defined using a specific framework with the Case study tool embedded in Aspen HYSYS software, with each case study generating 201 samples. To develop a generalized model, different datasets will be generated. It is critical to test the generalization ability of the model on different datasets by observing its performance error [15]. The dataset will be generated based on the feed gas composition that is summarized in Table 2. Case 1 will be utilized as the base dataset for model development.

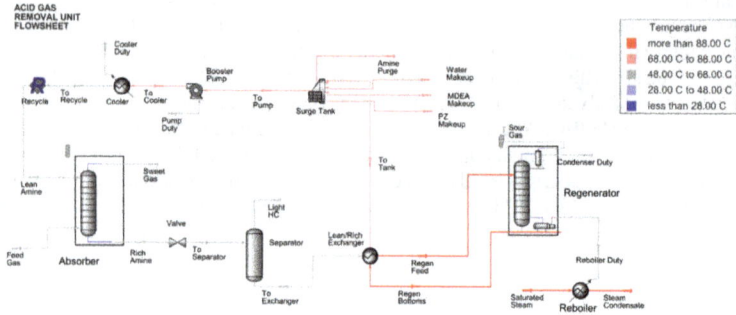

Figure 2 Developed Aspen HYSYS flowsheet.

Table 1 Generated parameters in simulation.

Variables		Minimum	Base	Maximum	Reference
MDEA + PZ	MDEA (wt%)	35	50	45	[9]
	PZ (wt%)	0	5	10	
MDEA + SFL	MDEA (wt%)	30.8	44	57.2	
	SFL (wt%)	1.05	1.5	1.95	
MDEA + DGA	MDEA (wt%)	30.8	44	57.2	[8]
	DGA (wt%)	1.05	1.5	1.95	
MDEA + SFL + PZ	MDEA (wt%)	30.8	44	57.2	
	SFL (wt%)	0.525	0.75	0.975	
	PZ (wt%)	0.525	0.75	0.975	
MDEA + MEA	MDEA (wt%)	8.4	12	15.6	[12]
	MEA (wt%)	19.6	28	36.4	
MDEA + DIPA	MDEA (wt%)	0	15	30	[13]
	DIPA (wt%)	30	15	0	
Temperature (°C)		38.50	55.00	71.50	
Absorber Pressure (bar)	Feed	38.12	54.02	69.93	
	Lean	38.33	54.32	70.32	[14]
	Top	37.3	52.85	68.41	
	Bottom	37.64	53.33	69.03	

Data pre-processing

In this research, data pre-processing begins by defining the model's input and output. Lean amine temperature, absorber pressure, H_2S and CO_2 composition on sweet gas are the inputs, while the solvent component and its composition are the targeted outputs. The next stage is changing the category for the solvent component into a numerical form. Furthermore, the data is moved to standardization to prevent bias because of inevitable feature domination [17]. The subsequent step involves data splitting into training and testing datasets. This research will apply 80% training data and 20% test data since training data should significantly outweigh testing data, ideally by three to four times [18]. The last step is balancing data using Synthetic minority oversampling technique (SMOTE) [19]. The purpose of data balancing is to prevent model to lean toward the majority class and performing poorly in the minority class

Table 2 *Sour gas composition.*

Component	Mole fraction	
	Case 1 [14]	Case 2 [16]
H2O	0.00234	0
i-Pentane	0.0052	0
n-Pentane	0.0056	0.0024
i-Butane	0.01243	0
n-Butane	0.0215	0.0018
CO_2	0.03001	0.0535
H_2S	0.00024	0.0005
Methane	0.72583	0.8086
n-Ethane	0.11717	0.0163
i-Ethane	0.07377	0
n-Propane	0.00234	0.0035
N2	0	0.1134

Model development

This research employed LightGBM to classify solvent components on AGRU. The key strength of LightGBM is its capability to achieve high accuracy while operating on less memory, making it compatible with regression and classification tasks [20]. Moreover, a leaf-wise strategy is utilized in this algorithm resulting in fewer nodes and enhanced computational performance, diverging from conventional methods [21]. Furthermore, Optuna hyperparameter tuning framework will be applied on LightGBM by following the steps below,

1. Determining the optimization goal and selecting hyperparameter boundaries. Two target function are defined, which are accuracy of more than 0.98 and a training time of less than 0.7 seconds. The chosen hyperparameters for LightGBM are summarized in Table 3.

Table 3 *LightGBM Hyperparameter ranges.*

Hyperparameter	Range
learning_rate	0.001 - 0.4
num_boost_round	3 to 100
max_depth	1 - 300
num_leaves	2 - 300
lambda_l1 and lambda_l2	$1e^{-12}$ - 0.03

2. Perform the optimization. Optuna was configure for having 2,000 optimization trial.
3. Implementing the final model and analyzing its performance. The best-performing hyperparameters from the trial data were extracted, and a LightGBM model was latter trained using these optimal values.

Lastly, K-means is applied to determine the solvent composition. The output will be the optimal concentration range of amine solvent. K-means is one of the unsupervised learning methods that has been proven by its capability and extensive application in current research and industry [22]. K-means works by initialising k cluster centres through random selection, defining the difference between each object and the initial clustering centre, and assigning it to the nearest clustering centre [23].

Model performance metrics evaluation

The performance of the LightGBM model was evaluated using metrics such as accuracy and training duration. Accuracy is calculated as follows,

$$Accuracy = \frac{(TP+TN)}{(TP+TN+FP+FN)} \tag{1}$$

Separation Technology - ICoST 2025

Materials Research Proceedings 59 (2026) 162-169

Materials Research Forum LLC

https://doi.org/10.21741/9781644903957-21

where TP is true positive, TN is true negative, FP is false positive and FN is false negative prediction of the model. To evaluate the performance of the K-means model, evaluation metrics such as the Silhouette Score and Davies Bouldin Score were used to assess the performance of the unsupervised algorithm. These performance metrics are calculated as follows,

$$Silhoutte\ Score = \frac{b(i)-a(i)}{max\{a(i),b(i)\}} \tag{2}$$

where b(i) is the smallest average distance between i and some other clusters and a(i) is the average distance between i and all other points of its cluster,

$$Davies - Bouldin\ Score = \frac{1}{k}\sum_{i=1}^{k}\left(\max_{i\neq j}\frac{s_i+s_j}{d_{ij}}\right) \tag{3}$$

where k Is the number of the clusters, si is the average distance between each point of the cluster i and its centroid (same for cluster j and sj), and dij Is the distance between the i'th and j'th cluster centroid.

Results and Discussion
Analysis of generated data
The correlation heatmap of solvent composition with the operating parameters such as temperature of the amine solvent, absorber pressure, and the composition of H_2S and CO_2 in the product gas\sweet gas are illustrated in Fig. 3. Generally, there is a significant correlation between the solvent and the temperature and pressure. The correlation heatmap shows the distinctive trait of each solvent blend on its capability to absorb H_2S and CO_2. For instance, there is a negative relationship between MDEA and all solvent blending with the H_2S, indicating that MDEA with a higher mass fraction tends to absorb H_2S than CO_2. This is why secondary amines such as PZ, SFL, DGA, MEA, and DIPA are needed to assist in the absorption of CO_2 from sweet gas. It can be observed that most of the secondary amines have a negative relationship with CO_2, such as PZ, SFL, DGA, and MEA, except DIPA. Given the personal characteristics of each blend, the classification of solvent components is necessary, followed by the determination of solvent composition.

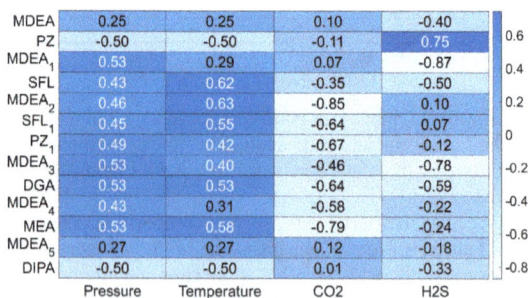

	Pressure	Temperature	CO2	H2S
MDEA	0.25	0.25	0.10	-0.40
PZ	-0.50	-0.50	-0.11	0.75
MDEA$_1$	0.53	0.29	0.07	-0.87
SFL	0.43	0.62	-0.35	-0.50
MDEA$_2$	0.46	0.63	-0.85	0.10
SFL$_1$	0.45	0.55	-0.64	0.07
PZ$_1$	0.49	0.42	-0.67	-0.12
MDEA$_3$	0.53	0.40	-0.46	-0.78
DGA	0.53	0.53	-0.64	-0.59
MDEA$_4$	0.43	0.31	-0.58	-0.22
MEA	0.53	0.58	-0.79	-0.24
MDEA$_5$	0.27	0.27	0.12	-0.18
DIPA	-0.50	-0.50	0.01	-0.33

Figure 3 Correlation matrix of solvent composition and operating parameters.

Model tuning and validation using different dataset
Before discussing the model validation first, LightGBM model hyperparameter tuning with the Optuna was performed, resulting in a high accuracy of 0.98 and a training time of 0.17 seconds after performing the iteration of 2000 combinations of hyperparameters. Hyperparameter tuning was done on Scenario 1 as the base case before being validated in another case for generalization. Hyperparameters are required to achieve high accuracy as long as a fast model. Table 4 shows the

Separation Technology - ICoST 2025
Materials Research Proceedings 59 (2026) 162-169

Materials Research Forum LLC
https://doi.org/10.21741/9781644903957-21

model's performance metrics applied to different scenarios. The machine learning models used in this research are LightGBM and K-means with Accuracy and training time as the performance metrics of LightGBM and Silhouette, and Davies Bouldin as the performance metrics of the means as the unsupervised learning methods.

Table 4 Model validation on different case study.

Model	Metric	1	2
LightGBM	Accuracy	0.98	0.989
	Training time (s)	0.17	0.193
K-means	Silhouette	0.481616	0.598307
	Davies-Bouldin	0.825007	0.599234

Overall, the generalization process was successfully conducted, as indicated by the minimum difference between the performance metrics for each scenario. LightGBM accuracy shows a very high score above 0.98 across all the scenarios, with case 2 being the highest one. The training time is considerably rapid, with the maximum training time of 0.193 seconds, occurring in scenario 2. The K-means performance metrics, show how the model can recognize the cluster of data. Basically, the output of the K-means model will determine the range composition of the solvent. It shows that the silhouette score doesn't have any difference between scenarios, with the difference not until the 0.5 score. The Davies Bouldin score has the same pattern as the Silhoutte score. From this discussion, it can be concluded that the generalization process for this model was successfully conducted.

Conclusion

This research presents the technique to define the solvent component and composition for absorption-based AGRU using six different solvent blends. In this paper, a general model of component and composition AGRU has been developed. The proposed model has a consistent accuracy of 0.98 across all scenarios, while the clustering performance shows adequate performance with the range of 0.48-0.59. This model successfully determines the solvent component and composition across six different solvent blends on two different feed gas compositions with a total of 37,786 data points.

Acknowledgement

The authors gratefully acknowledge the support provided by Universiti Teknologi Petronas for this research, funded through grant Yayasan Universiti Teknologi PETRONAS (YUTP), with grant number YUTP-PRG 015PBC-037 and YUTP-FRG 015LCO-362.

References

[1] R. Faiz, M. Al-Marzouqi, Insights on natural gas purification: Simultaneous absorption of CO_2 and H_2S using membrane contactors, Sep. Purif. Technol. 76 (2011) 351–361. https://doi.org/10.1016/j.seppur.2010.11.005

[2] S. Sheikh, Enhancements in modeling gas sweetening, in: Proc. Abu Dhabi Int. Petroleum Exhibition and Conference (ADIPEC), Society of Petroleum Engineers, 2015. https://doi.org/10.2118/177656-MS

[4] A.M. Tamidi, N.H. Yasin, CO_2 and H_2S co-removal from natural gas using membrane contactor technology, in: Proc. SPE Asia Pacific Oil and Gas Conference and Exhibition (APOG), Society of Petroleum Engineers, 2020.

[5] L.A. Pellegrini, S. Moioli, F.M. Munari, P. Vergani, B. Picutti, A. Uccelletti, The acid gas removal unit at Gasco's Habshan 5: Simulation and comparison with field data, in: Proc. Offshore Mediterranean Conference and Exhibition (OMC), 2015.

[6] M. Faqih, M.B. Omar, R.J. Wishnuwardana, N.I. Ismail, M.H.M. Zaid, K. Bingi, Prediction of solvent composition for absorption-based acid gas removal unit on gas sweetening process, Molecules 29 (19) (2024) 4591. https://doi.org/10.3390/molecules29194591

[7] W. Zhu, H. Ye, X. Zou, Y. Yang, H. Dong, Analysis and optimization for chemical absorption of H_2S/CO_2 system: Applied in a multiple gas feeds sweetening process, Sep. Purif. Technol. 276 (2021) 119301. https://doi.org/10.1016/j.seppur.2021.119301

[8] A. Esmaeili, T. Yoon, T.A. Atsbha, C.J. Lee, Rate-based modeling and energy optimization of acid gas removal from natural gas stream using various amine solutions, Process Saf. Environ. Prot. 177 (2023) 643–663. https://doi.org/10.1016/j.psep.2023.07.030

[9] M. Hakimi, M.B. Omar, R. Ibrahim, Application of a neural network for predicting H_2S from an acid gas removal unit (AGRU) with different compositions of solvents, Sensors 23 (2023) 1020. https://doi.org/10.3390/s23021020

[10] I. Izonin, R. Muzyka, R. Tkachenko, I. Dronyuk, K. Yemets, S.-A. Mitoulis, A method for reducing training time of ML-based cascade scheme for large-volume data analysis, Sensors 24 (15) (2024) 4762. https://doi.org/10.3390/s24154762

[11] P. Anandan, A. Manju, M.R. Reddy, Classification of massive data sets using a revolutionary grey wolf optimization algorithm and a deep learning model in a cloud-based setting, in: Proc. 2023 Int. Conf. Data Sci., Agents & Artif. Intell. (ICDSAAI), 2023, pp. 1–6.

[12] L.C. Law, N.Y. Azudin, S.R. Syamsul, Optimization and economic analysis of amine-based acid gas capture unit using monoethanolamine/methyl diethanolamine, Clean Technol. Environ. Policy 20 (2018) 451–461. https://doi.org/10.1007/s10098-017-1430-1

[13] B.K. Choi, S.M. Kim, K.M. Kim, U. Lee, J.H. Choi, J.S. Lee, I.H. Baek, S.C. Nam, J.H. Moon, Amine blending optimization for maximizing CO_2 absorption capacity in a diisopropanolamine–methyldiethanolamine–H_2O system using the electrolyte UNIQUAC model, Chem. Eng. J. 419 (2021) 129517. https://doi.org/10.1016/j.cej.2021.129517

[14] A. Sharifi, E.O. Amiri, Effect of the tower type on the gas sweetening process, Oil Gas Sci. Technol. 72 (2017). https://doi.org/10.2516/ogst/2017018

[15] S. Zhang, The study of model generalization ability for spam classification based on machine learning models, in: Proc. 2024 9th Int. Symp. Comput. Inf. Process. Technol. (ISCIPT), 2024, pp. 43–46.

[16] H.A. Behrooz, Robust synthesis of amine-based natural gas sweetening plants, Gas Sci. Eng. 113 (2023) 204970.

[17] Y. Li, Y. Cao, J. Yang, M. Wu, A. Yang, J. Li, Optuna-DFNN: An Optuna framework driven deep fuzzy neural network for predicting sintering performance in big data, Alexandria Eng. J. 97 (2024) 100–113. https://doi.org/10.1016/j.aej.2024.04.026

[18] L. Zhou, D. Garg, Y. Qiu, S.M. Kim, I. Mudawar, C.R. Kharangate, Machine learning algorithms to predict flow condensation heat transfer coefficient in mini/micro-channel utilizing universal data, Int. J. Heat Mass Transf. 162 (2020) 120351. https://doi.org/10.1016/j.ijheatmasstransfer.2020.120351

[19] S. Banerjee, K. Bhavna, T. Raychoudhury, Prediction of transport behavior of nanoparticles using machine learning algorithm: Physical significance of important features, J. Contam. Hydrol. 258 (2023) 104237. https://doi.org/10.1016/j.jconhyd.2023.104237

Separation Technology - ICoST 2025
Materials Research Proceedings 59 (2026) 162-169

Materials Research Forum LLC
https://doi.org/10.21741/9781644903957-21

[20]   O. Alshboul, G. Almasabha, A. Shehadeh, K. Al-Shboul, A comparative study of LightGBM, XGBoost, and GEP models in shear strength management of SFRC-SBWS, Structures 61 (2024) 106009. https://doi.org/10.1016/j.istruc.2024.106009

[21]   R. Bakır, C. Orak, A. Yüksel, Optimizing hydrogen evolution prediction: A unified approach using random forests, LightGBM, and Bagging Regressor ensemble model, Int. J. Hydrogen Energy 67 (2024) 101–110. https://doi.org/10.1016/j.ijhydene.2024.04.173

[22] K. Yuan, G. Chi, Y. Zhou, H. Yin, A novel two-stage hybrid default prediction model with k-means clustering and support vector domain description, Res. Int. Bus. Finance 59 (2022) 101536. https://doi.org/10.1016/j.ribaf.2021.101536

[23] G. Niu, Y. Ji, Z. Zhang, W. Wang, J. Chen, P. Yu, Clustering analysis of typical scenarios of island power supply system by using cohesive hierarchical clustering based K-means clustering method, Energy Rep. 7 (2021) 250–256. https://doi.org/10.1016/j.egyr.2021.08.049

Separation Technology - ICoST 2025

Materials Research Proceedings 59 (2026) 170-176

Materials Research Forum LLC

https://doi.org/10.21741/9781644903957-22

Physicochemical Analysis of Neem Extract Loaded Polyvinylpyrrolidone/2-Hydroxypropyl-ß-Cyclodextrin Nanofibers for Wound Dressing Application

Crystal Hui Man TIONG[1,a], Sharol CHIA[1,b],
Nur Iman Batrisyia MOHD SHAHRUL NIZAM[1,c], Nurizzati MOHD DAUD[1,2,d*]

[1]Department of Biomedical Engineering and Health Sciences, Faculty of Electrical Engineering, Universiti Teknologi Malaysia, 81310 Johor Bahru, Johor

[2]Centre of Lipids Engineering & Applied Research, Ibnu Sina Institute for Scientific and Industrial Research, Universiti Teknologi Malaysia, 81310 Johor Bahru, Johor, Malaysia

[a]tionghui@graduate.utm.my, [b]chiasharol@graduate.utm.my,
[c]nurimanbatrisyia@graduate.utm.my, [d]nurizzati.md@utm.my

Keywords: Neem Extract, Polyvinylpyrrolidone, 2-Hydroxypropyl-ß-Cyclodextrin, Electrospinning, Nanofiber, Wound Dressing

Abstract. Neem (*Azadirachta indica*), a medicinal plant known for its antimicrobial and anti-inflammatory has emerged as a promising therapeutic agent for wound dressing. However, incorporating its bioactive compounds into stable and mechanically robust dressing matrices remains challenging due to compatibility and structural integrity issues. This study explores the fabrication and characterization of neem extract (NE)-loaded electrospun nanofibers composed of polyvinylpyrrolidone (PVP) and 2-hydroxypropyl-beta-cyclodextrin (HPßCD), focusing on the effect of polymer ratios on their physicochemical and mechanical properties. The presence of hydrogen bonding and host–guest complex formation among NE, PVP, and HPßCD was confirmed, indicating enhanced polymer compatibility and structural stability. Tensile testing revealed that increasing PVP concentration enhanced tensile strength, elongation at break, and Young's modulus, indicating improved flexibility and durability of the nanofibers. The optimized sample NE/PVP/HPßCD formulation (1:15:5) exhibited superior mechanical resilience and strong molecular compatibility, making it suitable for advanced wound dressing applications. Optimizing PVP concentrations enables tunable structural stability and mechanical properties, offering a promising plant-based nanofiber for advanced wound dressings.

Introduction

The advancement of drug delivery systems over the past several decades has led to significant improvements in therapeutic outcomes, particularly in the management of wounds. Nanofiber-based delivery platforms, especially electrospun ones, have attracted growing interest due to their high porosity, large surface area-to-volume ratio, and ability to mimic the extracellular matrix (ECM) which collectively enhance wound healing [1]. Electrospinning is further recognized for its simplicity and ability to encapsulate various therapeutic agents into polymeric matrices [2].

The use of natural plant-derived compounds in biomedical applications has gained increasing attention due to their pharmacological potential and reduced likelihood of adverse effects. *Azadirachta indica* (neem) has long been utilized in traditional medicine systems such as Ayurveda and Unani for its therapeutic versatility [3]. Neem has been reported to exhibit antibacterial, anti-inflammatory, antioxidant, and wound-healing properties relevant to wound management. Its leaves contain diverse bioactive constituents, including azadirachtin, nimbin, nimbolide, quercetin, and flavonoids, which contribute to its medicinal benefits [4].

Separation Technology - ICoST 2025   Materials Research Forum LLC
Materials Research Proceedings 59 (2026) 170-176   https://doi.org/10.21741/9781644903957-22

However, a key obstacle in translating the bioactive agents such as neem extract into wound dressing applications lies in their poor compatibility with polymer matrices and the reduced structural integrity of the resulting fibers, which compromises their mechanical strength and stability during application [5]. To address these issues, this study had incorporated neem extract into polymer-based electrospun nanofiber systems to enhance the compatibility and fiber structural integrity.

Among the polymers used, polyvinylpyrrolidone (PVP) is a synthetic, water-soluble polymer known for its hydrophilicity, non-toxicity, and biocompatibility [6]. It also exhibits excellent film-forming properties, chemical stability, and mechanical strength, which make it an ideal matrix for electrospun nanofibers in wound dressing applications. Complementing PVP, hydroxypropyl-β-cyclodextrin (HPβCD) is a cyclic oligosaccharide derivative widely applied to stabilize hydrophobic bioactive compounds and enhance their solubility and compatibility by forming inclusion complexes within its hydrophobic cavity [7].

The combination of PVP and HPβCD in neem extract-loaded nanofibers for the development of plant-based wound dressing materials has not yet been explored. This study aims to fabricate and characterize neem extract-loaded PVP/HPβCD nanofibers with varying PVP concentrations using the electrospinning technique. The focus is to investigate how different PVP polymer ratios affect the physicochemical properties of the nanofibers, such as chemical functionalities and tensile strength. The goal is to develop a plant-based wound dressing that effectively integrates traditional therapeutic agents with modern nanofiber nanotechnology to deliver robust, mechanically stable, and compatible bioactive scaffolds for wound care.

Materials and Methods

Neem leaves were collected from local farmer at Johor, Malaysia and its identification was authenticated by Forest Research Institute Malaysia (FRIM) (sample ID: PID 151124-08). Polyvinylpyrrolidone (PVP, average Mw ~1,300,000) was purchased from Chemiz, Malaysia. 2-Hydroxypropyl-β-cyclodextrin (HPβCD, average Mw ~1460) and phosphate buffer saline (PBS, pH 7.4) were obtained from Sigma Aldrich, USA. Ultrapure water was prepared using a Direct-Q 3 UV Merck Millipore purification system. All reagents were of analytical grade and were used without further purification.

Preparation of NE/PVP/HPβCD solution for electrospinning

40 mg of NE were weighed and dissolved in 2 mL of ultrapure water in a glass vial. Subsequently, HPβCD was firstly added in 1:5 (NE: HPβCD ratio) as shown in Table 1 and stirred using a magnetic stirrer on a hot plate until fully dissolved and then followed by adding PVP with varying concentration. A control solution consisting only of PVP without NE and HPβCD was also prepared. All prepared solutions were subsequently sonicated using a digital ultrasonic cleaner to remove air bubbles and ensure homogeneity.

Table 1 Ratio composition of NE, PVP and HPβCD for each sample.

Sample	NE	PVP	HPβCD
Control PVP	-	10	-
1	1	10	5
2	1	15	5

Fabrication of NE/PVP/HPβCD nanofibers

The prepared NE/PVP/HPβCD solutions, along with the control formulations, were electrospun to fabricate nanofiber mats using an electrospinning setup located at the Material Preparation Lab, V01 UTM [8]. The system consisted of a high-voltage power supply (PS35-PCL, Nanolab Instruments), a syringe pump (NE-300, New Era Pump Systems), and a flat plate collector

Separation Technology - ICoST 2025
Materials Research Proceedings 59 (2026) 170-176

Materials Research Forum LLC
https://doi.org/10.21741/9781644903957-22

measuring 15 cm × 15 cm × 7 cm. Each solution was transferred into a 5 mL syringe fitted with a 23G blunt-end metal needle and mounted onto the syringe pump. The collector surface was covered with aluminium foil to provide a clean and detachable surface for fiber deposition. A consistent set of electrospinning parameters was maintained for all samples, including an applied voltage of 20 kV between the needle tip and the collector, a flow rate of 1 mL/h, and a fixed distance of 12 cm between the needle tip and the collector. Electrospinning was carried out for 1 hour for each sample to allow sufficient nanofiber accumulation. Upon completion, the nanofiber mats were carefully peeled off the aluminium foil and stored in sealed ziplock bags for further analysis.

Chemical functionalities analysis

Fourier Transform Infrared (FTIR) spectroscopy was employed to investigate the chemical structure and functional groups present in NE/PVP/HPβCD nanofiber samples according to the method from Lin et al. [9] with some modifications. Small sections of each electrospun nanofiber mat were cut into squares and placed in direct contact with the Attenuated Total Reflectance (ATR) crystal of the FTIR spectrophotometer (Spectrum Two, PerkinElmer, USA). Spectral data were collected in the mid-infrared range from 4000 to 400 cm^{-1} using an average of 32 scans with a resolution of 4 cm^{-1} to ensure high precision. The resulting ATR-FTIR spectra were then imported into OriginPro 2024 software (OriginLab Corporation, USA) for detailed analysis, allowing for the identification and comparison of characteristic peaks corresponding to the neem extract and the polymeric components (PVP and HPβCD) within the nanofiber matrix.

Tensile testing

Tensile testing was conducted to evaluate the mechanical properties of the electrospun NE/PVP/HPβCD nanofiber mats based on the method from Caloian et al. [10] with some modifications. To avoid damage, the nanofiber mats were carefully peeled from the aluminum foil collector. The thickness of each mat was measured using a digital micrometer, while the average width was determined by taking measurements at three different points across the sample. Each sample was mounted between two square paper frames, with the top and bottom edges secured using cellophane tape to prevent slippage during testing. The framed samples were then positioned in a Shimadzu Universal Testing Machine (Model AGX-V), Japan, and the initial gauge length was measured as the distance between the grips. A uniaxial tensile force was applied at a constant crosshead speed of 2 mm/min until the sample fractured. Force-elongation data collected during the test were converted into stress-strain curves, which were then used to determine the tensile strength, elongation at break, and Young's modulus of the nanofiber mats.

Results and Discussion

Chemical functionalities analysis

The FTIR spectra of the control formulation (PVP only) and selected composite sample NE/15 PVP/HPβCD were analyzed in the spectral range of 400–4000 cm^{-1} as presented in Table 2 and illustrated in Fig. 1 to identify the presence and interaction of characteristic functional groups. The FTIR spectrum of sample NE/15 PVP/HPβCD exhibited broader O–H stretching bands in the range of 3384 cm^{-1}, suggesting stronger intermolecular hydrogen bonding interactions between the hydroxyl groups of HPβCD and the hydrophilic segments of PVP. The peaks at 2923 cm^{-1} (C–H stretching), 1652 cm^{-1} (C=O stretching), 1422 cm^{-1} (C–H bending), and 1291 cm^{-1} (C–N stretching) were preserved, indicating that the polymer structure remained intact. A notable peak at 1033 cm^{-1} confirmed the presence of HPβCD through its ether linkages. The broader O–H bands than control sample is reported similar spectral behaviour in PVP-based blends incorporating bioactive compounds reflected improved polymer compatibility and hydrogen bonding with the neem constituents [11].

Separation Technology - ICoST 2025   Materials Research Forum LLC
Materials Research Proceedings 59 (2026) 170-176   https://doi.org/10.21741/9781644903957-22

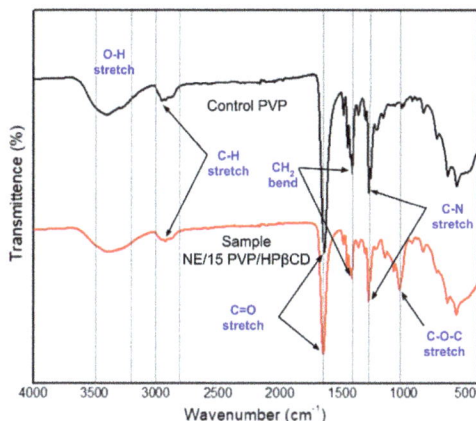

Figure 1 Chemical functionalities analysis of electrospun NE/PVP/HPβCD nanofibers: sample NE/15 PVP/HPβCD with its control PVP.

Table 2 Spectroscopic data of the samples.

Control PVP	Sample NE/15 PVP/HPβCD	Correlation
3404 cm^{-1}	3384 cm^{-1}	O-H stretching
2956 cm^{-1}	2923 cm^{-1}	C-H stretching
1653 cm^{-1}	1652 cm^{-1}	C=O stretching
1421 cm^{-1}	1422 cm^{-1}	C-H bending
1290 cm^{-1}	1291 cm^{-1}	C-N stretching
-	1033 cm^{-1}	C-O-C stretching

Meanwhile, the control formulation composed solely of PVP displayed sharp peaks at 1421 cm^{-1} (C–H bending), 1290 cm^{-1} (C–N stretching), 1653 cm^{-1} (C=O stretching), 2956 cm^{-1} (C–H stretching), and a relatively narrow O–H region (3404 cm^{-1} (stretching)), indicating the absence of significant hydrogen bonding due to absence of NE and HPβCD. Similarly, FTIR evidence of neem extract incorporation in electrospun alginate/chitosan/PEO fibers, confirming polymer with extract molecular interactions while preserving the fibrous structure [12].

Tensile Testing
Tensile testing was conducted to evaluate the mechanical strength and flexibility of the electrospun nanofiber mats (NE/10 PVP/HPβCD and NE/15 PVP/HPβCD). These properties are crucial for wound dressing applications, where materials must maintain adequate tensile strength for structural integrity during handling and application, while also possessing sufficient flexibility to conform to the wound surface. The stress–strain profiles of both samples are presented in Fig. 2.

The tensile stress–strain curves were obtained by calculating the force per unit displacement for each sample. As shown in Fig. 2, the curve for the sample NE/10 PVP/HPβCD displayed a steep linear increase, indicating a proportional relationship between stress and strain. This region reflects the elastic deformation phase, where the nanofiber can return to its original length upon stress release [13]. In contrast, the curve for the sample NE/15 PVP/HPβCD showed extended strain and higher tensile stress values, suggesting improved tensile strength and flexibility. Such behaviour indicates that the fibers can endure higher forces before fracture and elongate further, making them more durable and stretchable under mechanical stress.

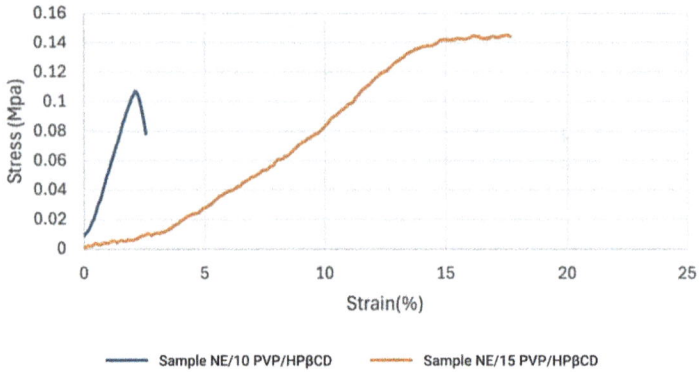

Figure 2 *Stress-strain graph of electrospun NE/PVP/HPβCD nanofibers with two varying PVP concentrations (1:10:5 and 1:15:5).*

Sample NE/10 PVP/HPβCD reached its maximum stress at a lower strain value and fractured earlier, demonstrating poorer mechanical performance and reduced elasticity compared to the sample NE/15 PVP/HPβCD. These findings indicate that increasing the PVP concentration from ratio of 10 parts to 15 parts in the NE:PVP:HPβCD system enhances both tensile strength and elongation at break. This improvement can be linked to the role of increment of PVP content in promoting molecular entanglement within the nanofiber matrix, which strengthens the overall structural network [14].

The mechanical parameters derived from the tensile tests are summarized in Table 3, including the ultimate tensile strength and Young's modulus. Young's modulus, representing material stiffness, was calculated as the ratio of stress (σ) to strain (ε) in the linear elastic region of the stress–strain curve [15].

Table 3 *Mechanical properties of each nanofiber sample.*

Sample	NE/10 PVP/HPβCD	NE/15 PVP/HPβCD
Ultimate tensile strength (MPa)	0.1052	0.1448
Young's modulus (MPa)	0.0539	0.01142

Sample NE/15 PVP/HPβCD exhibited a higher ultimate tensile strength (0.1448 MPa vs. 0.1052 MPa) and a lower Young's modulus (0.01142 MPa vs. 0.0539 MPa) compared to the sample NE/10 PVP/HPβCD, indicating greater strength with increased compliance. This means less stress is required to achieve the same strain in NE/15 PVP/HPβCD, while its higher ultimate tensile strength supports handling durability. Both samples lie within the commercial monolayer range (ultimate stress ≤ 0.77 MPa; Young's modulus < 1 MPa), with NE/15 offering the more favorable balance for dimensional stability during handling and conformability on wound surfaces [16].

Conclusion

This study successfully demonstrated the fabrication of neem extract-loaded PVP/HPβCD nanofibers using electrospinning and highlighted their potential as wound dressing materials. The FTIR analysis confirmed the successful incorporation of neem extract and HPβCD into the PVP

Separation Technology - ICoST 2025
Materials Research Proceedings 59 (2026) 170-176

Materials Research Forum LLC
https://doi.org/10.21741/9781644903957-22

matrix through characteristic functional group interactions, particularly the broadening of the O-H stretching band, which signified enhanced hydrogen bonding and improved polymer compatibility. The preservation of key spectral peaks further indicated that the polymeric network remained structurally intact, confirming the maintenance of structural integrity upon addition of bioactives. Furthermore, tensile testing revealed that increasing the PVP concentration from 10 to 15 parts markedly enhanced both tensile strength and elongation at break, reflecting improved flexibility and mechanical durability which are critical attributes for wound dressing applications. The optimized sample NE/15 PVP/HPβCD nanofiber, confirmed by FTIR to possess strong intermolecular compatibility and by tensile testing to exhibit superior mechanical resilience, demonstrated enhanced structural stability and elasticity, making it a promising candidate for advanced bioactive wound dressings with improved strength and durability. Future research should focus on in vitro and in vivo biocompatibility studies to validate the safety and therapeutic efficacy of these nanofibers, as well as antimicrobial evaluations against clinically relevant pathogens. Long-term stability testing and clinical assessments of scalability are recommended to ensure practicality for commercial applications, while the incorporation of additional natural bioactives could further extend their applicability to a wider range of wounds.

Acknowledgment

This research was supported by the Ministry of Higher Education Malaysia under Fundamental Research Grant Scheme FRGS/1/2023/TK05/UTM/02/1. The authors would also like to express their gratitude to Dr Khong Wui Gan and Dr Azam Ahmad Bakir from the University of Southampton Malaysia for their assistance in conducting the tensile tests.

References

[1] X. Lu, L. Zhou, W. Song, Recent Progress of Electrospun Nanofiber Dressing in the Promotion of Wound Healing, Polymers (Basel) 16 (2024) 2596. https://doi.org/10.3390/polym16182596

[2] M. Ahmadi Bonakdar, D. Rodrigue, Electrospinning: Processes, Structures, and Materials, Macromol 4 (2024) 58–103. https://doi.org/10.3390/macromol4010004

[3] T. Tufail, H. Bader Ul Ain, A. Ijaz, M.A. Nasir, A. Ikram, S. Noreen, M.T. Arshad, M.A. Abdullahi, Neem (Azadirachta indica): A Miracle Herb; Panacea for All Ailments, Food Sci Nutr 13 (2025). https://doi.org/10.1002/fsn3.70820

[4] P. Mohanasundaram, M.S. Antoneyraj, A systematic review of neem flower (Azadirachta indica): a promising source of bioactive compounds with pharmacological and immunomodulating properties, Traditional Medicine Research 10 (2025) 41. https://doi.org/10.53388/TMR20241025001

[5] S. Alven, S. Peter, Z. Mbese, B.A. Aderibigbe, Polymer-Based Wound Dressing Materials Loaded with Bioactive Agents: Potential Materials for the Treatment of Diabetic Wounds, Polymers (Basel) 14 (2022) 724. https://doi.org/10.3390/polym14040724

[6] B.K. Borji, M. Pourmadadi, A. Tajiki, M. Abdouss, A. Rahdar, A.M. Díez-Pascual, Polyvinyl pyrrolidone/starch/hydroxyapatite nanocomposite: A promising approach for controlled release of doxorubicin in cancer therapy, J Drug Deliv Sci Technol 95 (2024) 105516. https://doi.org/10.1016/j.jddst.2024.105516

[7] Á. Sarabia-Vallejo, M. del M. Caja, A.I. Olives, M.A. Martín, J.C. Menéndez, Cyclodextrin Inclusion Complexes for Improved Drug Bioavailability and Activity: Synthetic and Analytical Aspects, Pharmaceutics 15 (2023) 2345. https://doi.org/10.3390/pharmaceutics15092345

[8] M. Paczkowska-Walendowska, A. Miklaszewski, J. Cielecka-Piontek, Is It Possible to Improve the Bioavailability of Resveratrol and Polydatin Derived from Polygoni cuspidati Radix as a Result of Preparing Electrospun Nanofibers Based on Polyvinylpyrrolidone/Cyclodextrin?, Nutrients 14 (2022) 3897. https://doi.org/10.3390/nu14193897

[9] T.-C. Lin, C.-Y. Yang, T.-H. Wu, C.-H. Tseng, F.-L. Yen, Myricetin Nanofibers Enhanced Water Solubility and Skin Penetration for Increasing Antioxidant and Photoprotective Activities, Pharmaceutics 15 (2023) 906. https://doi.org/10.3390/pharmaceutics15030906

[10] I. Caloian, J. Trapp, B.Y. Kantepalle, P. Latimer, T.J. Lawton, C. Tang, Mechanical Properties of Dual-Layer Electrospun Fiber Mats, Polymers (Basel) 17 (2025) 1777. https://doi.org/10.3390/polym17131777

[11] Danushika.C. Manatunga, J.A.B. Jayasinghe, C. Sandaruwan, R.M. De Silva, K.M.N. De Silva, Enhancement of Release and Solubility of Curcumin from Electrospun PEO–EC–PVP Tripolymer-Based Nanofibers: A Study on the Effect of Hydrogenated Castor Oil, ACS Omega 7 (2022) 37264–37278. https://doi.org/10.1021/acsomega.2c03495

[12] A. Hameed, T.U. Rehman, Z.A. Rehan, R. Noreen, S. Iqbal, S. Batool, M.A. Qayyum, T. Ahmed, T. Farooq, Development of polymeric nanofibers blended with extract of neem (Azadirachta indica), for potential biomedical applications, Front. Mater. 9 (2022). https://doi.org/10.3389/fmats.2022.1042304

[13] H. Zhang, Y. Jin, C. Chi, G. Han, W. Jiang, Z. Wang, H. Cheng, C. Zhang, G. Wang, C. Sun, Y. Chen, Y. Xi, M. Liu, X. Gao, X. Lin, L. Lv, J. Zhou, Y. Ding, Sponge particulates for biomedical applications: Biofunctionalization, multi-drug shielding, and theranostic applications, Biomaterials 273 (2021) 120824. https://doi.org/10.1016/j.biomaterials.2021.120824

[14] M. Rajeev, C.C. Helms, A Study of the Relationship between Polymer Solution Entanglement and Electrospun PCL Fiber Mechanics, Polymers (Basel) 15 (2023) 4555. https://doi.org/10.3390/polym15234555

[15] R.H. Alasfar, S. Ahzi, N. Barth, V. Kochkodan, M. Khraisheh, M. Koç, A Review on the Modeling of the Elastic Modulus and Yield Stress of Polymers and Polymer Nanocomposites: Effect of Temperature, Loading Rate and Porosity, Polymers (Basel) 14 (2022) 360. https://doi.org/10.3390/polym14030360

[16] M. Minsart, S. Van Vlierberghe, P. Dubruel, A. Mignon, Commercial wound dressings for the treatment of exuding wounds: an in-depth physico-chemical comparative study, Burns Trauma 10 (2022). https://doi.org/10.1093/burnst/tkac024

Separation Technology - ICoST 2025
Materials Research Proceedings 59 (2026) 177-184

Materials Research Forum LLC
https://doi.org/10.21741/9781644903957-23

Removal of Cadmium from Simulated Wastewater through Synergistic Reactive Extraction

Fadzlin Qistina FAUZAN[1,a], Izzat Naim SHAMSUL KAHAR[1,b],
Norisya Balqis MOHD AMIN[1,c], Norasikin OTHMAN[1,2,d],
Shuhada A. IDRUS-SAIDI[1,2,e*],

[1]Faculty of Chemical and Energy Engineering, Universiti Teknologi Malaysia, 81310 Skudai, Johor, Malaysia

[2]Centre of Lipids Engineering & Applied Research (CLEAR), Ibnu Sina Institute for Scientific and Industrial Research, Universiti Teknologi Malaysia, 81310 Skudai, Johor, Malaysia

[a]fadzlinqistina@graduate.utm.my, [b]izzatnaim.sk@utm.my, [c]norisyabalqis@graduate.utm.my, [d]norasikin@cheme.utm.my, [e]shuhada.atika@utm.my

Keywords: Cadmium, Synergistic Extractants, Reactive Extraction, Stoichiometric, Cyanex 302

Abstract. Cadmium (Cd) contamination in water poses serious environmental risks due to its high toxicity, making effective removal strategies essential. Among various methods, reactive extraction has emerged as a promising and greener approach for Cd removal from wastewater. The use of a single extractant has shown potential in Cd removal from wastewater. However, its efficiency remains limited, highlighting the need for improved formulations. In this study, the effects of extractant type, concentration, and synergistic combinations were investigated to determine the most effective system for Cd extraction. The results revealed that the synergistic pairing of phosphinothioic acid (Cyanex 302) as the base extractant, with tributyl phosphate (TBP) as a synergist, using palm oil as diluent, achieved high efficiency. At the optimized conditions of 0.03 M Cyanex 302 and 0.001 M TBP, Cd extraction reached 99.47%. Stoichiometric analysis conducted indicated that 0.5 mole of Cyanex 302 interacted with one mole of Cd during complex formation, confirming the extraction mechanism. For Cd stripping, 0.15 M H_2SO_4 exhibited the highest stripping efficiency, with 96.73% Cd successfully stripped from the organic phase. Stoichiometric analysis revealed that 1.5 moles of H_2SO_4 are required to complex with one mole of Cd, validating the stripping mechanism.

Introduction

Cadmium (Cd) is a hazardous heavy metal prevalent in the environment due to human activities such as mining, industrial waste, and agriculture. In wastewater, Cd concentrations can reach levels of several milligrams per liter, exceeding safe discharge limits. Despite its toxicity, Cd is valuable in industries for batteries, pigments, coatings, and plastics. Recovering Cd from wastewater reduces environmental and health risks and recycles this valuable resource, contributing to sustainable industrial practices. Cd contamination in water poses significant health risks, such as kidney damage, skeletal damage, and respiratory issues. It also adversely affects aquatic life, leading to bioaccumulation in the food chain, impacting both ecosystems and human health [1].

Existing methods for reducing Cd contamination face challenges, including low efficacy, high costs, and negative environmental effects. Conventional extraction methods like solvent extraction, membrane filtering, adsorption, and precipitation have limitations, such as cost, fouling, and secondary waste generation. Reactive extraction has been proposed for its simplicity, efficiency, and selectivity. However, the system based on a single extractant typically lacks efficiency and selectivity, leading to high chemical consumption and environmental concerns,

Separation Technology - ICoST 2025 | Materials Research Forum LLC
Materials Research Proceedings 59 (2026) 177-184 | https://doi.org/10.21741/9781644903957-23

especially from petroleum-based diluents [2]. Common petroleum diluents, such as kerosene and toluene, are unsustainable due to their toxicity, high volatility, and non-biodegradability. To overcome these limitations, research has shifted towards green diluents, including vegetable oils, ionic liquids, and deep eutectic solvents, which offer advantages such as low toxicity, renewability, and biodegradability. Among these, vegetable oils have gained significant attention due to their wide availability and low cost. These vegetable oils are proven to have good compatibility with various extractants, showing promising results in metal removal [3]. Novel approaches using synergistic extractants have also emerged as an effective strategy for metal removal. By combining two extractants, synergism can significantly improve selectivity and reduce the chemical consumption [4].

This study aims to develop a sustainable and cost-effective method for Cd removal from aqueous phases using a synergistic reactive extraction process. By optimizing the key parameters such as extractant combinations and stripping agents, the research seeks to enhance environmental remediation strategies while safeguarding ecosystem balance and human health in Cd-contaminated areas. To achieve these objectives, the study focuses on three core aspects: first, the formulation of a synergistic extractant system for the selective removal of Cd through reactive extraction, second, the evaluation of extraction and stripping efficiency under varying concentrations of extractants and stripping agents, and third, the elucidation of the underlying mechanisms involved in Cd extraction and stripping through slope of analysis method [2,5]. By advancing both the efficiency of Cd removal, this research contributes to sustainable remediation practices and supports environmentally responsible industrial applications.

Materials and Methods

The simulated feed phase was prepared by dissolving cadmium sulfate ($CdSO_4$, 99% purity, Sigma-Alrich) in distilled water. The chemicals for the extraction step included di-(2-ethylhexyl) phosphoric acid (D2EHPA, 97% purity, Sigma-Aldrich), phosphinothioic acid (Cyanex 302, 97% purity, Sigma-Aldrich), triethanolamine (TEA, 99% purity, Merck), and tributyl phosphate (TBP, 99% purity, Acros Organics). Palm oil (100% purity, BURUH) purchased from a local supermarket was used as the green diluent. For the stripping step, sulphuric acid (H_2SO_4), sodium hydroxide (NaOH), and hydrochloric acid (HCl) were evaluated as agents for Cd stripping.

Reactive extraction

The reactive extraction of Cd was carried out by mixing equal volumes (10 mL each) of the simulated feed phase (47.56 ppm Cd) and the organic phase (extractant dissolved in palm oil) in a conical flask. The mixture was agitated in an incubator shaker at 320 rpm for 18 h to reach equilibrium. Following extraction, the phase was transferred to a separating funnel and allowed to settle for 30 min to ensure complete phase separation by gravity settling. The Cd concentration in the aqueous phase was determined using atomic absorption spectroscopy (AAS). The extractant that exhibited the highest Cd extraction efficiency was designated as the base extractant.

To develop the synergistic formulation, mixtures of the selected base extractant with potential synergistic extractants were evaluated for Cd extraction. For each experiment, the organic phase (10 mL) was prepared by diluting the base and synergistic extractants in palm oil. The extraction procedure and Cd concentration analysis followed the methods described previously. The effect of the synergistic extractant on the extraction efficiency and the synergistic coefficient (SC) was then assessed, and the combination that demonstrated the highest extraction efficiency and SC was selected for further investigation. The experimental data for reactive extraction experiments were reported as triplicate with standard deviation rounded to three decimal places.

The stripping of Cd was carried out by mixing equal volumes (10 mL each) of the loaded organic phase (containing Cd after extraction) and the stripping phase (H_2SO_4, NaOH, or HCl) in a conical flask. The mixture was agitated in an incubator shaker at 320 rpm for 18 h to achieve

Materials Research Forum LLC
https://doi.org/10.21741/9781644903957-23

equilibrium. Following stripping, the phase was transferred to a separating funnel and allowed to settle for 30 min to ensure complete phase separation. Similar to extraction, the Cd concentration in the stripping phase was determined using AAS. The experimental data for stripping experiments were reported as triplicate with standard deviation rounded to three decimal places.

Data analysis

The extraction (E), distribution ratio (D), SC, and stripping (S) were determined using the following equations.

$$E\ (\%) = \frac{C_i - C_f}{C_i} \times 100\% \qquad (1)$$

$$D = \frac{C_{organic}}{C_{aqueous}} \qquad (2)$$

$$SC = \frac{D_{mix}}{(D_{extractant} + D_{synergist})} \qquad (3)$$

$$S\ (\%) = \frac{C_{stripping}}{C_{organic}} \times 100\% \qquad (4)$$

Results and Discussion

Effect of single extractant

Four types of extractants (TEA, TBP, D2EHPA and Cyanex 302) were evaluated for Cd extraction (Fig. 1). Among these, Cyanex 302 achieved the highest Cd extraction (97.31%) at 0.05 M. This is mainly due to the presence of sulfur donor atoms in its thiophosphinic acid structure (P=S) [6]. According to the hard-soft-acid-base (HSAB) principle, a soft acid, such as Cd^{2+}, has a stronger preference for soft donor atoms, such as sulfur, leading to stable Cd-S complexes in the organic phase [6,7]. In contrast, D2EHPA contains only oxygen donor atoms (P=O, hard bases), which interact less strongly with Cd^{2+}, resulting in a lower extraction efficiency (69.63%). TEA, a basic extractant, demonstrated a comparatively high efficiency of 89.28%, although still lower than Cyanex 302. This may be due to a relatively weaker interaction between Cd and the amine group compared to the stronger interaction between Cd and the dithiophosphonic acid of Cyanex 302. TBP, primarily functioned as a solvating extractant, exhibited the lowest efficiency (11.00%) due to its limited complexation ability, arising from its hard base characteristic. It is generally more effective for metals with different coordination chemistry. Other than that, TBP tends to form complexes with the green diluent used, further limiting its extraction activity [5].

The percentage of Cd extracted from the feed phase was obtained using Eq. 1, where C_i is the concentration of Cd in the aqueous phase before extraction, and C_f is the concentration of Cd in the aqueous phase after extraction. The distribution ratio of Cd ion, in which the ratio of Cd ion concentration being transported to the organic phase from the aqueous feed phase is represented in Eq. 2, where $C_{organic}$ is the concentration of Cd ion in the organic phase and C_{aqeous} is the concentration of Cd ion in the aqueous phase. Eq. 3 was used to determine the SC of extractants, where D_{mix} is the distribution ratio of mixed extractants (base and synergist), $D_{extractant}$ and $D_{synergist}$ are the individual distribution ratios of the base and synergist extractant, respectively. The percentage of Cd recovered from the loaded organic phase was obtained using Eq. 4, where $C_{stripping}$ is the Cd concentration in the stripping phase after the stripping process.

Separation Technology - ICoST 2025
Materials Research Proceedings 59 (2026) 177-184

Materials Research Forum LLC
https://doi.org/10.21741/9781644903957-23

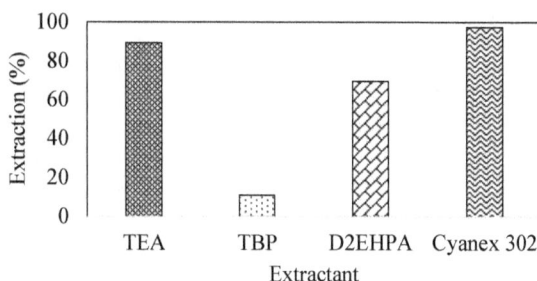

Figure 1 *Effect of several different types of extractants at 0.05 M towards Cd extraction (Experimental conditions: [Cd]: 47.56 ppm; diluent: palm oil; aqueous: organic phase: 1:1; agitation speed: 320 rpm; agitation time: 18 hours; temperature: 26 °C).*

A similar trend was reported in a recent study where Cd ion adsorption was facilitated by donor atoms capable of electron donation, as validated by spectroscopic analysis and HSAB principle. The findings support that Cd ions preferentially interact with softer donor sites, consistent with the observation that donor atom softness strongly influences extraction efficiency [7]. Since Cyanex 302 provided the highest extraction among the four extractants tested, it was selected as the base extractant for further formulation studies.

Effect of mixed extractants

Mixtures of extractants with equal concentrations were investigated for their synergistic effect. Among the tested systems, the combination of Cyanex 302 + TEA exhibited the highest extraction efficiency for Cd (97.58%). Both extractants were individually effective, with Cyanex 302 acting as a chelating agent and TEA serving as a basic extractant capable of forming ion pairs with Cd. This observation was quite similar to the initial observation, where both provided high Cd extraction. However, the high extraction observed in the mixture was attributed to their individual activities rather than a true synergistic effect. This is supported by the fact that all combinations yielded SC of less than one (Table 1), confirming antagonistic interactions despite the high extraction efficiency. The combination of Cyanex 302 with D2EHPA achieved only moderate efficiency, which was lower than the Cd extraction by D2EHPA alone. As evidenced by the SC, this observation confirmed that the Cyanex 302-D2EHPA combination exhibited the lowest synergistic effect among all extractant mixtures, confirming their low interaction contribution. The mixture of Cyanex 302 and TBP showed the lowest extraction efficiency (52.99%). Since TBP functioned primarily as a solvating extractant, its ability to form stable complexes with Cd was limited. Its interaction with Cyanex 302 was also weak due to a low SC value, resulting in poor extraction efficiency.

Table 1 *Effect of mixed extractants at 0.025 M towards Cd extraction (Experimental conditions: [Cd]: 47.56 ppm; diluent: palm oil; aqueous: organic phase: 1:1; agitation speed: 320 rpm; agitation time: 18 hours; temperature: 26°C).*

Base extractant	Synergist Extractant	Extraction (%)	SC
Cyanex 302	TBP	52.99 ± 0.016	0.028
	TEA	97.58 ± 0.005	0.556
	D2EHPA	65.30 ± 0.032	0.016

Separation Technology - ICoST 2025   Materials Research Forum LLC
Materials Research Proceedings 59 (2026) 177-184   https://doi.org/10.21741/9781644903957-23

Effect of TEA and Cyanex 302 with Synergist TBP. To enhance the Cd extraction efficiency, TBP was introduced as a modifier in mixtures with TEA and Cyanex 302 as base extractants. The study examined the combined effect of TBP with each base extractant at concentrations of 0.02 M and 0.03 M. The inclusion of TBP was intended to improve the solubility of the extractant mixtures, thereby enhancing Cd removal efficiency (Table 2). At very low concentrations, TBP typically acts as a phase modifier without interfering antagonistically with the base extractant in extracting the target solute [5].

In the TEA-TBP system, the extraction efficiency was lower than that of Cyanex 302-TBP, and varying TEA concentration from 0.02 M to 0.03 M slightly reduced the extraction efficiency and SC. The results of the TEA-TBP system for Cd extraction indicated that this system is moderately effective and is less suitable compared to the Cyanex 302-TBP system under the tested conditions (Table 2). The TEA-TBP system showed that TBP concentration significantly affects Cd extraction. At 0.02 M TEA and 0.001 M TBP, the extraction efficiency was 76.81%, but it dropped to 70.20% when TBP concentration increased to 0.002 M. At 0.03 M TEA, efficiencies were 73.57% and 73.74% for 0.001 M and 0.002 M TBP, respectively. These results suggest that the optimal condition was 0.02 M TEA with 0.001 M TBP. The high SC (Table 2) indicated that the interaction between TEA and TBP plays a key role in Cd extraction. In contrast, the Cyanex 302-TBP system achieved higher Cd extraction efficiency compared to the TEA-TBP system, although its SC values were lower. This system showed that TBP concentration did not significantly affect Cd extraction. At 0.02 M Cyanex 302 and 0.001 M TBP, the efficiency was 95.04%, increasing slightly to 95.31% at 0.002 M TBP. At 0.03 M Cyanex 302, the efficiency reached 99.47% and 99.02% for 0.001 M and 0.002 M TBP, respectively, indicating nearly complete Cd extraction (Table 2).

Table 2 Effect of mixed extractant concentration on Cd extraction (Experimental conditions: [Cd]: 47.56 ppm; mixture extractant: TEA + TBP and Cyanex 302 + TBP; diluent: palm oil; aqueous: organic phase: 1:1; agitation speed: 320 rpm; agitation time: 18 hours; temperature: 26°C).

Base Extractant (M)	TBP (M)	SC	Extraction (%)
TEA			
0.02	0.001	2.876	76.81 ± 0.014
	0.002	2.261	70.20 ± 0.022
0.03	0.001	1.422	73.57 ± 0.045
	0.002	1.520	73.74 ± 0.022
Cyanex 302			
0.02	0.001	0.661	95.04 ± 0.013
	0.002	0.703	95.31 ± 0.001
0.03	0.001	1.221	99.47 ± 0.001
	0.002	0.652	99.02 ± 0.001

Based on this study, TBP successfully enhanced the efficiency of both systems, with the Cyanex 302-TBP mixture consistently outperformed the TEA-TBP mixture. This observation aligned with the initial single-extractant experiment (Fig. 1), where Cyanex 302 demonstrated a higher Cd extraction efficiency than TEA. Although TBP addition relatively increased the potential of TEA, its overall extraction efficiency remained lower than that of Cyanex 302, likely due to the presence of sulfur donor atoms in Cyanex 302 as discussed previously.

The highest extraction efficiency was achieved at 0.03 M Cyanex 302 with 0.001 M TBP, yielding almost complete Cd extraction. Notably, TBP significantly enhanced the Cyanex 302 extraction efficiency at this specific concentration, producing a true synergistic effect (SC>1). At other concentrations, the high extraction efficiencies were primarily attributed to the inherent activity of Cyanex 302, rather than the synergistic interaction with TBP (SC<1). These results highlight that the Cyanex 302-TBP system can exhibit true synergistic behaviour under specific concentrations. Therefore, this study provides valuable insight into optimizing extractant formulations for more efficient Cd extraction from aqueous phases, which could be applied to the design of more selective and cost-effective remediation strategies [21].

Effect of Different Stripping Agents. Cd was first extracted from the aqueous phase using 0.03 M Cyanex 302 with 0.001 M TBP before assessing the stripping efficiency of the selected agents. The results showed significant variation (Table 3). Among them, 0.1 M H_2SO_4 achieved the highest Cd stripping (92.84%), indicating its strong ability to dissociate the Cd-Cyanex 302 complex and promote Cd stripping into the aqueous phase. 0.1 M HCl showed moderate efficiency (62.75%), while 0.1 M NaOH was significantly ineffective (8.61%), suggesting that alkaline conditions are not favorable for breaking the Cd-Cyanex 302 interaction. These results are consistent with the reports that acidic agents are generally more effective for metal recovery due to their ability to create a low-pH environment that facilitates ion dissociation. For instance, a study demonstrated selective stripping of zinc using H_2SO_4. However, the effectiveness of a particular acid depends on the type of metal ion, as different metal ions might exhibit varying affinities towards specific acids [9]. Effect of Stripping Agent Concentration. To enhance the stripping efficiency of Cd ions, a series of H_2SO_4 concentrations was further tested (Table 4).

The 0.1 M H_2SO_4 served as the baseline, where the concentrations tested range from 0.01 M to 0.2 M. This study found Cd stripping with 0.01 M and 0.05 M H_2SO_4 was lower (78.54% to 91.61%) than the baseline at 0.1 M (92.84%), indicating that limited protonation of the extractant and destabilization of the Cd-Cyanex 302 complex. As the acid concentration increased to 0.15 M, the Cd stripping peaked (96.73%), consistent with a stronger driving force for complex dissociation. At this concentration, the higher proton availability effectively displaced Cd^{2+} from the Cyanex 302 complex and facilitated its transfer into the aqueous phase. However, further increasing the H_2SO_4 concentration to 0.2 M resulted in a decline in stripping efficiency (85.98%). This can be attributed to the elevated ionic strength of the aqueous phase, which hinder the mass transfer across the interface.

Table 3 *Effect of different stripping agents on Cd recovery at 0.1 M concentration. (Experimental conditions: [Cd]: 47.56 ppm; aqueous: organic ratio = 1:1; agitation speed: 320 rpm, agitation time: 18 hours; temperature: 26℃).*

Stripping Agent	Stripping (%)
NaOH	8.61 ± 0.026
H_2SO_4	92.84 ± 0.085
HCl	62.75 ± 0.081

Table 4 *Effect of different concentrations of H_2SO_4 on Cd recovery. (Experimental conditions: [Cd]: 47.56 ppm; aqueous: organic ratio = 1:1; agitation speed: 320 rpm, agitation time: 18 hours; temperature: 26℃).*

H_2SO_4 (M)	Stripping (%)
0.01	78.54 ± 0.135
0.05	91.61 ± 0.195
0.10	92.84 ± 0.085
0.15	96.73 ± 0.085
0.20	85.98 ± 0.156

Mechanistic study of extraction and stripping

A stoichiometry analysis was carried out to elucidate the reaction between the extractants present in the organic phase and Cd in the aqueous phase. This stoichiometry analysis is used to determine

Separation Technology - ICoST 2025
Materials Research Proceedings 59 (2026) 177-184

Materials Research Forum LLC
https://doi.org/10.21741/9781644903957-23

the number of moles of Cyanex 302 required for Cd extraction. TBP was not included in the equation as it was involved in the reaction as the phase modifier rather than a reactant. Based on Eq. 5, the extraction of Cd occurs through the formation of 1 mol of Cd ion and n mol of Cyanex 302, resulting in a single complex.

$$M^{m+}_{(aq)} + \frac{n+m}{2}(HR)_{2\,(org)} \rightleftharpoons MR_m(HR)_{n\,(org)} + mH^+_{(aq)} \tag{5}$$

Here, (HR) denotes the Cyanex 302 molecule, and n represents the stoichiometric coefficient indicating the number of extractant molecules involved in the complexation with one Cd ion. To determine the stoichiometric coefficient n, a plot of log D versus log [Cyanex302] was constructed. From the regression equation in Fig. 2a (y = 0.482x + 2.5431, R^2=0.9115), the slope was found to be 0.48 (n=0.5). This result indicates an apparent half-order dependence on Cyanex 302 concentration, suggesting that only a fraction of the extractant molecules participate in the complexation process (Eq. 6). Accordingly, the extraction can be described by the foolowing general reaction.

$$Cd^{2+}_{(aq)} + \frac{5}{2}HR_{(org)} \rightleftharpoons CdR_2(HR)_{1/2\,(org)} + 2H^+_{(aq)} \tag{6}$$

To determine the stoichiometric coefficient n of the stripping equation, a plot of log D versus log [H₂SO₄] was constructed. From the regression equation in Fig. 2b (y = 1.3398x + 2.6832, R^2=0.932), the slope was found to be 1.34, which is in good agreement with the value of n=1.5. This result indicates a fractional order dependence on [H₂SO₄], suggesting that 1.5 molecules of H₂SO₄ were involved in stripping one Cd^{2+} ion from the organic phase (Eq. 7). In this stripping process, the three protons (3H⁺) released from 1.5 mol H₂SO₄ displace the Cd^{2+} ion bound to the Cyanex 302 in the organic phase. The overall stripping reaction can therefore be represented as:

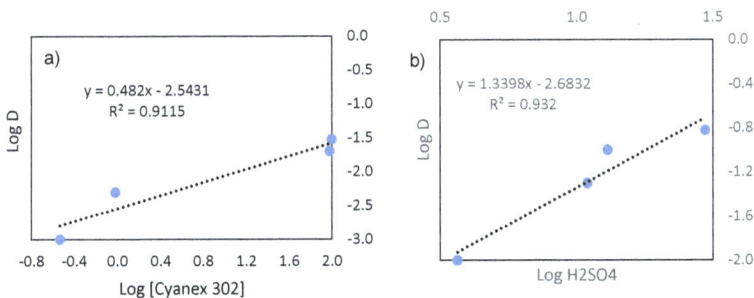

$$CdR_2(HR)_{1/2\,(org)} + 3H^+_{(aq)} \rightleftharpoons Cd^{2+}_{(aq)} + \frac{5}{2}(HR)_{(org)} + H^+_{(aq)} \tag{7}$$

Figure 2 Stoichiometric plot, (a) Cd extraction using Cyanex 302-TBP and (b) Cd stripping using H₂SO₄.

Conclusion

In conclusion, this study demonstrated the significant impact of introducing TBP as a modifier in synergistic extractant mixtures for Cd extraction. The combination of Cyanex 302 and TBP at concentrations of 0.03 M and 0.001 M, respectively, successfully demonstrates synergistic effects in Cd extraction from aqueous phase. The highest extraction efficiency of 99.47% highlighted the strong interaction between Cyanex 302 and TBP, with the mixture showing higher stability and selectivity in the organic phase, resulting in near-complete Cd removal. The stripping process achieved 96.73% stripping efficiency using 0.15 M H_2SO_4 as the stripping agent. These findings provide valuable insights into optimizing extractant mixtures for improved heavy metal extraction, contributing to sustainable industrial practices and effective environmental remediation.

Acknowledgment

The authors gratefully acknowledge the Ministry of Higher Education Malaysia for financial support under the Fundamental Research Grant Scheme (FRGS/1/2023/TK05/UTM/02/3).

References

[1] M. Irfan, X. Liu, K. Hussain, S. Mushtaq, J. Cabrera, P. Zhang, The global research trend on cadmium in freshwater: a bibliometric review, Environ. Sci. Pollut. Res. 30 (2021) 71585–71598. https://doi.org/10.1007/s11356-021-13894-7

[2] Ankush, Ritambhara, S. Lamba, Deepika, R. Prakash, Cadmium in Environment—An Overview, in: 2024: pp. 3–20. https://doi.org/10.1007/978-3-031-54005-9_1

[3] N. Jusoh, N. Othman, R.N.R. Sulaiman, N.F.M. Noah, K.S.N. Kamarudin, M.A.A. Zaini, D.A.B. Sidik, Development of palm oil-based synergist liquid membrane formulation for silver recovery from aqueous solution, J. Membr. Sci. Res. 7 (2021) 59–63. https://doi.org/10.22079/JMSR.2020.120174.1327

[4] I. N. S. Kahar, S.A. Idrus-saidi, M.A. Musa, N. Othman, A. Rosli, Removal of phosphate from aqueous solution through synergistic extractive extraction, 53 (2025) 136–145.

[5] I.N.S. Kahar, S.A. Idrus-Saidi, N. Othman, M.A.A. Mohd Akta, M.A. Ahmad Zaini, A. Rosli, Removal of Tetracycline Antibiotic Using Green Synergistic Extractive Extraction from Aqueous Solution, Malaysian J. Fundam. Appl. Sci. 21 (2025) 1874–1882. https://doi.org/10.11113/mjfas.v21n2.3754

[6] J. Hu, D.L. Kuang, H.Q. Zhou, L.H. Chung, J. He, Sulphur-enriched Metal-Organic Frameworks: Design, Manipulation, Strategies, Recent Advances and Perspectives, Coord. Chem. Rev. 540 (2025) 216759. https://doi.org/10.1016/j.ccr.2025.216759

[7] C. Bulin, T. Guo, Ultra fast and highly efficient recovery of cadmium with graphene oxide-chitosan grafted by nickel ferrite as a recyclable adsorbent and atomic scale mechanism, Sci. Total Environ. 991 (2025) 179974. https://doi.org/10.1016/j.scitotenv.2025.179974

[8] N.F.M. Noah, N. Othman, I.N.S. Kahar, S.S. Suliman, Potential use of synergist D2EHPA/Cyanex 302 in kerosene system for reactive extraction: Zinc recovery and organic phase regeneration, Chem. Eng. Process. - Process Intensif. 176 (2022) 108976. https://doi.org/10.1016/j.cep.2022.108976

[9] P. Liu, X. Wang, W. Zhang, Impact of organic acids on extraction of rare earth elements: Mechanisms and optimization, J. Rare Earths (2025).https://doi.org/10.1016/j.jre.2025.02.012

Separation Technology - ICoST 2025
Materials Research Proceedings 59 (2026) 185-192

Materials Research Forum LLC
https://doi.org/10.21741/9781644903957-24

Microalgae-Mediated Biological Synthesis of Silver Nanoparticles: Optimization and Morphological Characterization

Hui Ying TEH[1,a], Man Kee LAM[1,b*], Voon-Loong WONG[2,c], Wai Hong LEONG[3,d], Inn Shi TAN[4,e], Henry Chee Yew FOO[4,f]

[1]Chemical Engineering Department, HICoE-Centre for Biofuel and Biochemical Research, Universiti Teknologi PETRONAS, 32610 Seri Iskandar, Perak, Malaysia

[2]School of Energy and Chemical Engineering, Xiamen University Malaysia, 43900 Sepang, Selangor Darul Ehsan, Malaysia

[3]Algal Bio Co. Ltd, Todai-Kashiwa Venture Plaza, 5-4-19 Kashiwanoha, Kashiwa, Chiba, 277-0082, Japan

[4]Department of Chemical and Energy Engineering, Faculty of Engineering and Science, Curtin University Malaysia, CDT250, 98009 Miri, Sarawak, Malaysia

[a]hui_22009859@utp.edu.my, [b]lam.mankee@utp.edu.my, [c]voonloong.wong@xmu.edu.my, [d]waihong@algalbio.co.jp, [e]tan.s@curtin.edu.my, [f]henry.foo@curtin.edu.my

Keywords: Microalgae, Silver Nanoparticles, Green Synthesis, Plasmon Resonance, Photoinduction

Abstract. Silver nanoparticles (AgNPs) are essential in biomedical and water treatment applications due to their superior localized surface plasmon resonance (LSPR) property. This study adopted *C. vulgaris* as a bio-reducing agent for eco-friendly AgNP synthesis from silver nitrate ($AgNO_3$) via photo-induced reduction. The reaction rate was found to be highly dependent on the illumination source. To optimize the process, the volumetric ratio of *C. vulgaris*-to-precursor solution was varied alongside the illumination wavelengths from different LED colors (white, blue, red, and green). The optimal *C. vulgaris*-to-precursor ratio was 1:5, achieving a LSPR increment of 0.407 a.u. within 35 min. Blue light accelerated the reaction, reducing the time to 30 min and yielding a higher LSPR increment of 0.445 a.u.. Particle size analysis further confirmed the enhanced synthesis under blue light, with a narrower particle size distribution peak in the 39 - 47 nm range. FESEM revealed larger particle sizes due to aggregation in drying. However, upon re-suspension of the dried AgNPs, smaller particles were recovered with HRTEM showing individual AgNPs as small as ~16 nm. The results highlight the positive role of *C. vulgaris* in mediating competitive nanosized silver production through a greener, more sustainable approach.

Introduction

Since as early as 1954, silver nanoparticles (AgNPs) have been recognized for their antimicrobial potential, although silver itself has been utilized for various purposes for centuries [1]. Progress across multiple scientific and industrial disciplines has broadened their use to diverse fields, including biomedicine [2], water and air purification [3], cosmetic formulations [4], and food packaging technologies [5]. When silver is engineered at the nanoscale ranging from 1 to 100 nanometers, its properties outstand with enhanced surface-area-to-volume ratio. This size reduction amplifies its localized surface plasmon resonance (LSPR) as well, underpin its remarkable optical performance [6].

To date, the existing synthesis methods of AgNPs can be classified as top-down and bottom-up approaches. Top-down method refers to the mechanical grinding of bulk metal to produce nano-structured atomic silver. In contrast, bottom-up approach adopts chemical reduction among the particles of precursor [7,8]. A widely employed early chemical method for producing AgNPs is citrate reduction documented by M. C. Lea in 1889 [9]. In the era of sustainability, researchers have intensively explored the integration of environmentally friendly feedstock to achieve a holistic and environmentally friendly biological synthesis of AgNPs. The biological synthesis of metal nanoparticles is being carried out by a vast range of macro-microscopic organisms in nature, including plants [10], viruses [11], bacteria [12] and algae [13]. This is achieved by harnessing the biomolecules or metabolites of these organisms, which may include proteins [14], lipids [15] and carbohydrates [16].

Among the various biological sources of reducing agents for AgNPs fabrication, algae-mediated green synthesis stands out due to the organism's strong potential as an industrial feedstock. Algae exhibit exceptionally fast growth rates and require far less fertile soil, freshwater, and pesticide input compared to most terrestrial plants [17]. Consequently, large-scale algae cultivation exerts minimal pressure on vital human resources and does not compete significantly with the food supply [18]. In addition, algae, particularly *Chlorella* species, are non-pathogenic [19], making them a safe and promising option for environmentally friendly AgNP synthesis.

Recent studies have demonstrated that both intracellular and extracellular metabolites of microalgae possess remarkable reducing abilities to produce AgNPs with excellent stability and LSPR bands [20]. The reduction rate of silver ions is often enhanced through the incorporation of electromagnetic radiation [21]. Compared to other sources of electromagnetic radiation, visible light is more abundant in the solar spectrum that reaches the Earth's surface. Visible light refers to the portion of the electromagnetic spectrum that stimulates the human visual system. The specific range of wavelengths can vary, depending on the amount of radiant energy that reaches the retina. Typically, the lower limit of visible light is around 360–400 nm, while the upper limit is around 760–830 nm [22].

In this study, the green synthesis of AgNPs involves visible light induction. The mechanism of photo-induced or light-assisted biosynthesis of AgNPs begins with exposing the mixture of precursor and extract to light, which excites the biomolecules in the microalgae extract. When biomolecules are exposed to radiation, light energy is absorbed, exciting biomolecules to a higher energy state. This promotes the donation of electrons from reactive biomolecules to silver ions [23]. Therefore, this study aims to optimize the photo-induced synthesis of AgNPs by manipulating the volumetric ratio of *C. vulgaris*-to-precursor, as well as varying the illumination wavelengths sourced from different colours of light-emitting diodes (white, blue, red, and green).

Material and Methods

Cultivation of microalgae

The *C. vulgaris* green microalgae strain donated by Prof. Lee Keat Teong, School of Chemical Engineering, USM, was used in this study. The *C. vulgaris* strain was conserved in Bold's Basal Medium (BBM) culture medium at pH 6.8. The nutrient source for the growth of microalgae was TANI brand granulate chicken compost. To prepare a nutrient medium, 5 g of chicken compost was immersed in 300 mL of tap water and stirred for 24 h using a magnetic stirrer at 600 rpm. The compost solution was then filtered using filter paper (Whatman Filter Paper, Grade 1) to remove non-soluble particulates. The filtrate served as the nutrient source of cultivation and could be stored up to 3 days at ~ 4 °C. A 1 L Erlenmeyer flask was filled with 80 mL of the filtered compost solution with 840 mL of tap water as the culture medium. Subsequently, 80 mL of microalgae suspension with an initial cell concentration of 0.3×10^6 cells was prepared from the seed culture and introduced into the photobioreactor. The pH of the medium was then adjusted to 3 - 3.5. Subsequently, the photobioreactor was continuously aerated with air at 4.0 L/min [24]. The microalgae culture was illuminated under the optimized intensity of 250 μmol/m^2.s (light-emitting diodes, cool-day light, 6500 K), for 14 days.

Separation Technology - ICoST 2025
Materials Research Proceedings 59 (2026) 185-192

Materials Research Forum LLC
https://doi.org/10.21741/9781644903957-24

Synthesis of silver nanoparticle via photo-induction

On day 14 of cultivation, the living culture of *C. vulgaris* and 1mM silver nitrate (AgNO₃) were mixed in 5 different volumetric ratios: 1:1, 1:2, 1:5, 1:10, and 1:50 (v/v), to determine the optimal ratio of *C. vulgaris*-to-precursor. The reaction mixture was illuminated with white LED light to initiate the reduction process. After determining the optimal ratio from the spectrophotometric results, the illumination wavelength was varied among white (400 - 700 nm), blue (450 - 490 nm), green (490 - 570 nm), and red (630 - 750 nm) to study the effects of photoinduction wavelength on the green synthesis of AgNPs. Absorbance data of the reaction suspension were collected every 5 min using a UV-Vis spectrophotometer (UV-2600, Shimadzu).

Characterization

AgNPs were recovered from the suspension by washing three times using a centrifuge (HERMLE Z 206A) at 5000 rpm with distilled water. Subsequently, sonicator (Quasar 405) was used to redisperse the washed AgNPs in distilled water at 70 kHz for 15 min, employing a cold-water bath to prevent overheating of the sonicator. The dispersion was then analyzed for particle size using a Malvern Zetasizer Nano-ZSP. After the analysis, the samples were oven-dried at 40 °C for 24 h for future analysis. The dried AgNPs were stored in the dark to protect them from light.

The morphological characteristics of the synthesized AgNPs were examined using Field Emission Scanning Electron Microscopy (FE-SEM, TESCAN CLARA). Images were captured at magnifications of 50 - 100 kx with an accelerating voltage of 5 keV, low voltage was selected to minimize charging effects while preserving surface details.

The dried AgNPs were re-suspended in hexane solvent for analysis. The structural and morphological analysis of the AgNPs was conducted using High Resolution Transmission Electron Microscopy (HRTEM, Hitachi HT8730). A drop of the AgNP suspension was carefully placed on a carbon-coated copper grid and left to air-dry for 30 min. The sample was then examined under HRTEM at an accelerating voltage of 110 kV. The images were captured to determine the nanoparticle shape and size.

Results and Discussion

Spectrophotometry results reflect both the intensity of the absorbance peak and the size of nanoparticles. The intensity of the absorbance peak defines the performance of the LSPR effect of the assayed nanoparticles, while the wavelength of the absorbance peak qualitatively indicates the size of the produced nanoparticles (larger or smaller). Fig. 1(a) illustrates the effect of the *C. vulgaris*-to-precursor volumetric ratio on the reaction time, as well as the increment in absorbance intensity ($Abs_{time=t} - Abs_{time=0}$). Spectrophotometry results were recorded at 5-min intervals up to 60 min of incubation time.

For the experimental sets with 1:1 (v/v) and 1:2 (v/v) ratios, no optimum incubation time was observed within the first 60 min. Absorbance values continued to increase without showing a bell shape, which would indicate an optimum value. However, when the volume of *C. vulgaris* was further decreased while keeping the precursor volume constant, the results showed a promising bell curve, as seen in the 1:5, 1:10 and 1:50 (v/v) ratios in Fig. 1(a). The optimum reaction times were 35 min with an absorbance increment of 0.407 a.u. at λ_{max}, 30 min with an increment of 0.307 a.u. at λ_{max}, and 30 min with an increment of 0.161 a.u. at λ_{max}, for 1:5, 1:10 and 1:50 (v/v) ratios, respectively. It can be observed that further reducing the *C. vulgaris* volume accelerated the reaction by 5 min, but the absorbance increment decreased significantly.

Fig. 1(b) further ascertains that the optimal reactants ratio was 1:5 (v/v). The maximum absorbance increased as the ratio changed from 1:1 to 1:5 (v/v), but decreased beyond this point, with a drastic reduction in absorbance observed at 1:50 (v/v). Furthermore, Fig. 1(b) shows the changes in λ_{max} with different volumetric ratios. λ_{max} is defined as the wavelength at which the maximum LSPR value is observed. A shift to longer wavelengths was seen with an increasing

Separation Technology - ICoST 2025   Materials Research Forum LLC
Materials Research Proceedings 59 (2026) 185-192   https://doi.org/10.21741/9781644903957-24

precursor volume, showing a linear trend across ratios from 1:1 to 1:50 (v/v). This red shift phenomenon indicates that the increasing $AgNO_3$ volume in the reactant mixture has shifted the λ_{max} value from 415.5 to 527.75 nm, as illustrated by Fig. 1(b), with ratios ranging from 1:1 to 1:50 (v/v). As reported by [10,12,20], a longer wavelength of the absorbance peak indicates larger nanoparticle size, providing a qualitative explanation for the red shift in λ_{max} observed. Additionally, it is noteworthy that the λ_{max} values for 1:10 (v/v) (506.45 nm) and 1:50 (v/v) (527.75 nm) fall outside the typical range of AgNPs LSPR λ_{max} values (380 - 500 nm) reported in the literature [25]. Therefore, based on the absorbance increment value, the optimal ratio was 1:5 (v/v), with a reaction time of 35 min. The spectrophotometry results suggest that the reaction should be terminated at 35 min. Prolonged exposure beyond the optimal reaction time would degrade the LSPR properties and performance of the AgNPs produced, as indicated by decreasing absorbance values.

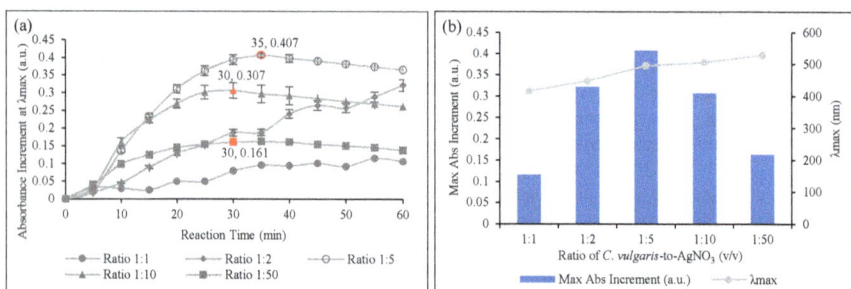

Figure 1 (a) Absorbance increment of AgNPs in reaction suspension against reaction time, (b) maximum absorbance increment of AgNPs and maximum wavelength (λ_{max}) of AgNPs at 1:1, 1:2, 1:5, 1:10 and 1:50 (v/v) of C. vulgaris-to-AgNO₃.

Effects of illumination wavelength in photo-induced green synthesis of AgNPs

Visible light consists of different wavelengths, which are perceived as different colours [26]. Based on Fig. 2(a), the effectiveness of enhancing the reaction with different colours of light follows the order: Blue > White > Green > Red light-emitting diodes. The wavelength ranges for blue, green, and red light are approximately 450 - 490 nm, 490 - 570 nm, and 630 - 750 nm, respectively. This trend aligns with the fact that electromagnetic radiation energy is inversely proportional to wavelength [27]. As a result, less energy is absorbed by the biomolecules in microalgae at longer wavelengths, leading to slower reaction rates in the experimental sets illuminated with lower-energy wavelengths of visible light.

Red light possesses the lowest energy profile, provides insufficient energy to be absorbed by the biomolecules in microalgae, thereby preventing electron excitation. The negative absorbance increment observed suggests that no significant reaction occurred under red light illumination, indicating that red light failed to induce the reduction of silver ions. When comparing white light to blue light, blue-light-induced AgNPs synthesis demonstrated a shorter reaction time of 30 min and a higher absorbance increment (0.445 a.u.) compared to white-light-induced AgNPs (0.407 a.u.). White light, which is a mixture of various colours, has a broad spectral range from 500 to 780 nm, with a narrower and more intense peak in the blue wavelength range of 450 to 490 nm [28]. This explains why white light induces a faster reaction and higher absorbance increment than green and red light, but lower than blue light. However, AgNPs induced by blue light degraded more rapidly beyond the optimal incubation time compared to those mediated by white light.

Separation Technology - ICoST 2025
Materials Research Proceedings 59 (2026) 185-192

Materials Research Forum LLC
https://doi.org/10.21741/9781644903957-24

The superiority of blue light in enhancing the reaction is further supported by the particle size distribution results shown in Fig. 2(b). The number percent distribution illustrates the size distribution of particles in AgNP samples induced by blue and white light. A higher percentage of smaller particles was observed in the blue-light-mediated AgNPs, as evidenced by the narrower and more intense peak in the 39 - 47 nm range. In contrast, the white-light-assisted AgNPs displayed a broader and less intense particle size distribution, ranging from 41 - 53 nm. Both the spectrophotometry and particle size analysis results ascertain that blue light was the most effective wavelength in the synthesis process, while green and red light had significantly less influence.

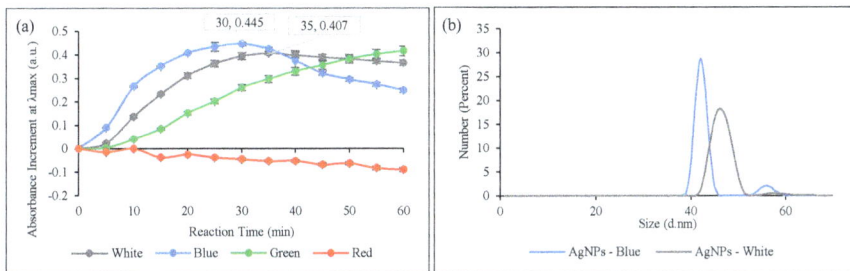

Figure 2 (a) Absorbance increment of AgNPs in reaction suspension against reaction time at different colours of light-emitting diodes, (b) number percent distribution of AgNPs incubated under blue and white light-emitting diodes analyzed using Malvern Zetasizer Nano-ZSP.

Morphological characteristics

The FESEM analysis revealed a mixture of AgNPs with sizes ranging from 55.95 to 68.34 nm, as shown in Fig. 3(a). Larger AgNPs exceeding 100 nm were also observed, which can be attributed to aggregation during the drying process. These larger particles likely resulted from particle coalescence prompted by thermal energy. However, the particle sizes observed in FESEM were larger compared to the PSA results, primarily due to differences in sample preparation and the state of the sample. While PSA examines a colloidal sample before drying, FESEM characterizes the sample in its dried state. As a result, the larger particle sizes in FESEM are expected, as thermal drying can induce agglomeration.

HRTEM was used to examine the sample of dried AgNPs re-suspended in solvent. The HRTEM analysis as represented by Fig. 3(b) revealed a range of 16 - 38 nm quasi-spherical AgNPs, which is smaller than the results obtained from PSA. This demonstrates the higher sensitivity and resolution of HRTEM. Discrepancies between PSA and HRTEM measurements are expected, as PSA measured nanoparticles numerically, whereas HRTEM measured nanoparticles through direct imaging. Additionally, HRTEM analysis of the re-suspended AgNPs revealed even smaller particles, with sizes approaching ~16 nm. This confirms the recoverability of smaller nanoparticle sizes in thermally dried AgNPs through resuspension, as the size observed is smaller than that measured in the FESEM analysis of the dried sample. The AgNPs exhibited roughly spherical shapes, likely influenced by their crystalline planes. The repeating pattern of fringes observed in the image represents the crystal lattice structure of the particles. The presence of a denser zone indicates thicker particles, reflecting Ostwald ripening, where larger nanoparticles grow at the expense of smaller ones through dissolution–reprecipitation [29].

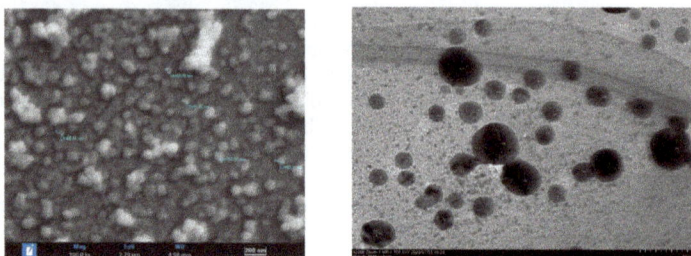

Figure 3 (a) FESEM image of AgNPs synthesized under the illumination of blue light with 100.0 kx magnification, (b) HRTEM image of re-suspended AgNPs colloidal with 200k magnification, with particles sizes in the range of 16 – 38nm.

Conclusion

The optimal *C. vulgaris*-to-precursor ratio for the synthesis of AgNPs was determined to be 1:5 (v/v), with a reaction time of 35 min and an absorbance increment of 0.407 a.u. A red shift in λ_{max} was observed as the volume of *C. vulgaris* decreased. The effectiveness of different light wavelengths in enhancing the reaction followed the order: Blue > White > Green > Red. Blue light proved to be the most efficient, reducing the optimal incubation time to 30 min and achieving a higher absorbance increment of 0.445 a.u. Furthermore, PSA confirmed the superior performance of blue light, with a narrower and more intense particle size distribution in the 39 - 47 nm range. These findings highlight the positive impact in the integration of blue light to induce the green synthesis of AgNPs with *C. vulgaris*. FESEM analysis showed that thermal heating led to aggregation of AgNPs, resulting in an increased particle size range. However, upon re-suspending the dried AgNPs in hexane solvent, smaller particles of silver were recovered. HRTEM analysis further confirmed this by showing smaller particle sizes than those observed by FESEM, with its higher magnification capability detecting quasi-spherical AgNPs in the size range of 16 - 38 nm.

Acknowledgement

The authors would like to acknowledge the funding provided by Ministry of Higher Education (MoHE) Malaysia through Fundamental Research Grant Scheme with project number FRGS/1/2024/TK08/UTP/ 02/2 and cost centre 015MA0-166.

References

[1] A. Uthaman, H. M. Lal, and S. Thomas, Fundamentals of Silver Nanoparticles and Their Toxicological Aspects, in: H. M. Lal, S. Thomas, T. Li, and H. J. Maria (Eds.), Polymer Nanocomposites Based on Silver Nanoparticles Synthesis, Characterization and Applications, Springer Nature Switzerland AG, Switzerland, 2021, pp. 1-24.

[2] T. Bruna, F. Maldonado-Bravo, P. Jara, and N. Caro, Silver Nanoparticles and Their Antibacterial Applications, International Journal of Molecular Sciences, 22(2021). https://doi.org/10.3390/ijms22137202.

[3] Y. Yu, Z. Zhou, G. Huang, H. Cheng, L. Han, S. Zhao, Y. Chen, F. Meng, Purifying water with silver nanoparticles (AgNPs)-incorporated membranes: Recent advancements and critical challenges, Water Research. 222 (2022) 118901.

[4] Information on https://www.nanotechproject.tech/cpi/products/solefreshtm-socks/

[5] Information on https://product.statnano.com/product/6858/silver-nanoparticle-food-containers

Separation Technology - ICoST 2025
Materials Research Proceedings 59 (2026) 185-192

Materials Research Forum LLC
https://doi.org/10.21741/9781644903957-24

[6] G. Sunil, H. Aditya, D. Shradhey, M. Omkar, K. Pranav, N. Supriya, and K. Suresh, Silver Nanoparticles: Properties, Synthesis, Characterization, Applications and Future Trends, IntechOpen, (2021). https://doi.org/10.5772/intechopen.99173.

[7] N. Baig, I. Kammakakam, and W. Falath, Nanomaterials: A review of synthesis, properties, recent progress, and challenges, Materials Advances, 2(2021). https://doi.org/10.1039/D0MA00807A.

[8] G. Habibullah, J. Viktorova, and T. Ruml, Current Strategies for Noble Metal Nanoparticle Synthesis, Nanoscale Research Letters, 16(2021). https://doi.org/10.1186/s11671-021-03480-8.

[9] E. Casals, M.F. Gusta, N. Bastus, J. Rello, V. Puntes, Silver Nanoparticles and Antibiotics: A Promising Synergistic Approach to Multidrug-Resistant Infections, Microorganisms. 13(4) (2025) 952.

[10] M. Ansari, S. Ahmed, A. Abbasi, M.T. Khan, M. Subhan, N.A. Bukhari, A.A. Hatamleh, N.R. Abdelsalam, Plant mediated fabrication of silver nanoparticles, process optimization, and impact on tomato plant, Scientific Reports. 13(1) (2023) 18048.

[11] N. Telkapalliwar, and P. S. Deharkar, A Comprehensive Overview of Virus-Mediated Synthesis of Silver Nanoparticles, Journal of Advanced Scientific Research. 16(4) (2025) 1-4.

[12] E. Akdaşçi, F. Eker, H. Duman, M. Bechelany, S. Karav, Microbial-Based Green Synthesis of Silver Nanoparticles: A Comparative Review of Bacteria- and Fungi-Mediated Approaches, International Journal of Molecular Sciences. 26(20) (2025) 10163.

[13] R. S. Hamida, M. A. Ali, Z. N. Almohawes, H. Alahdal, M. A.-O. Momenah, and M. M. Bin-Meferij, Green Synthesis of Hexagonal Silver Nanoparticles Using a Novel Microalgae Coelastrella aeroterrestrica Strain BA_Chlo4 and Resulting Anticancer, Antibacterial, and Antioxidant Activities, Pharmaceutics, 14(2022) 2002. doi: 10.3390/pharmaceutics14102002.

[14] R. Anjali, Fabrication of silver nanoparticles from marine macro algae Caulerpa sertularioides: Characterization, antioxidant and antimicrobial activity, Process Biochemistry, 121(2022) 601-618. https://doi.org/https://doi.org/10.1016/j.procbio.2022.07.027.

[15] S. Mora-Godínez, F.F. Contreras-Torres, A. Pacheco, Characterization of Silver Nanoparticle Systems from Microalgae Acclimated to Different CO2 Atmospheres, ACS Omega. 8(24) (2023) 21969-21982.

[16] Abdullah, N. S. Al-Radadi, T. Hussain, S. Faisal, and S. Ali Raza Shah, Novel biosynthesis, characterization and bio-catalytic potential of green algae (Spirogyra hyalina) mediated silver nanomaterials, Saudi Journal of Biological Sciences, 29(2022) 411-419. https://doi.org/https://doi.org/10.1016/j.sjbs.2021.09.013.

[17] A. Abdelfattah et al., Microalgae-based wastewater treatment: Mechanisms, challenges, recent advances, and future prospects, Environmental Science and Ecotechnology, 13(2023) 100205. https://doi.org/https://doi.org/10.1016/j.ese.2022.100205.

[18] P. Sharma, L.K.S. Gujjala, S. Varjani, S. Kumar, Emerging microalgae-based technologies in biorefinery and risk assessment issues: Bioeconomy for sustainable development, Science of The Total Environment. 813 (2022) 152417.

[19] S.M. Aly, N.I. ElBanna, M. Fathi, Chlorella in aquaculture: challenges, opportunities, and disease prevention for sustainable development, Aquaculture International. 32(2) (2024) 1559-1586.

[20] D. Chugh, V. S. Viswamalya, and B. Das, Green synthesis of silver nanoparticles with algae and the importance of capping agents in the process, Journal of Genetic Engineering and Biotechnology, 19(2021) 126. https://doi.org/https://doi.org/10.1186/s43141-021-00228-w.

[21] N. Jara, N.S. Milán, A. Rahman, L. Mouheb, D.C. Boffito, C. Jeffryes, S.A. Dahoumane, Photochemical Synthesis of Gold and Silver Nanoparticles—A Review, Molecules. 26(15) (2021) 4585.

[22] S. Gorman, The inhibitory and inactivating effects of visible light on SARS-CoV-2: A narrative update, Journal of Photochemistry and Photobiology. 15 (2023) 100187.

[23] M. Fahim, A. Shahzaib, N. Nishat, A. Jahan, T.A. Bhat, A. Inam, Green synthesis of silver nanoparticles: A comprehensive review of methods, influencing factors, and applications, JCIS Open. 16 (2024) 100125.

[24] U. Suparmaniam, C.Y. Li, M.K. Lam, N.T. Sahrin, H. Rawindran, C.S. Liew, J.W. Lim, I.S. Tan, S.Y. Lau, B.L.F. Chin, Enhancing lipid yield in freshwater microalgae through synergistic abiotic stressors for sustainable biodiesel production, Biomass and Bioenergy. 200 (2025) 107978.

[25] K. Ahmad et al., Green synthesis and characterization of silver nanoparticles through the Piper cubeba ethanolic extract and their enzyme inhibitory activities, Sec. Medicinal and Pharmaceutical Chemistry, 11(2023) 2296-2646.

[26] P.U.P.A. Gilbert, Chapter 6 - Color and color vision, in: P.U.P.A. Gilbert (Ed.), Physics in the Arts (Third Edition), Academic Press, 2022, pp. 101-125.

[27] A.T. Black, Ultraviolet A (UVA), in: P. Wexler (Ed.), Encyclopedia of Toxicology (Fourth Edition), Academic Press, Oxford, 2024, pp. 647-654.

[28] A. Sharakshane, An easy estimate of the PFDD for a plant illuminated with white LEDs: 1000 lx = 15 µmol/s/m<sup>2</sup&gt, bioRxiv. (2018) 289280.

[29] C.N. Nanev, Thermodynamic and molecular-kinetic considerations of the initial growth of newly born crystals; crystal size distribution; Dissolution of small crystals during Ostwald ripening due to temperature changes, Progress in Crystal Growth and Characterization of Materials. 69(2) (2023) 100604.

Separation Technology - ICoST 2025
Materials Research Proceedings 59 (2026) 193-200

Materials Research Forum LLC
https://doi.org/10.21741/9781644903957-25

Green Valorization of Palm Oil Solid Condensate using Supercritical CO$_2$ Ethanol Extraction for Rich-Bioactive Compounds

Noor Azwani MOHD RASIDEK[1,a*], Liza MD SALLEH[1,b*],
Nurizzati MOHD DAUD[3,c], Noor Sabariah MAHAT[1,d], Zuhaili IDHAM[2,e],
Mohammad Lokman HILMI[1,f], Muhammad Abbas AHMAD ZAINI[1,g]

[1]Centre of Lipids Engineering & Applied Research (CLEAR), IbnuSina Institute for Scientific and Industrial Research, Universiti Teknologi Malaysia, 81310 UTM Johor Bahru, Johor, Malaysia

[2]Department of Deputy Vice Chancellor (Research & Innovation), Universiti Teknologi Malaysia, 81310 UTM Johor Bahru, Johor, Malaysia

[3]Department of Biomedical Engineering and Health Sciences, Faculty of Electrical Engineering, Universiti Teknologi Malaysia, 81310 UTM Johor Bahru, Johor, Malaysia

[a]noor.azwani@utm.my, [b]r-liza@utm.my, [c]nurizzati.md@utm.my, [d]noorsabariah@utm.my, [e]zuhailiidham@utm.my, [f]m.lokmanhilmi@gmail.com, [g]r-abbas@utm.my

Keywords: Palm Oil Sterilization Condensate, Supercritical Fluid Extraction, Phenolic, Antioxidant Activity, Anti-Inflammatory Potential

Abstract. Palm oil sterilization condensate (POSteC), a milling by-product, is rich in bioactive compounds with therapeutic potential. This study valorized its solid fraction (POSC) using supercritical fluid extraction (SFE) with CO$_2$–ethanol across 40–80°C and 10–30 MPa. Extraction yield, β-carotene, α-tocopherol, total phenolic content (TPC), total flavonoid content (TFC), and antioxidant activity (AA) were assessed, alongside LC-MS/MS profiling of anti-inflammatory constituents. TPC increased with harsher conditions, peaking at 52.214 mg/g GAE DW (80°C, 30 MPa), whereas TFC was highest under mild conditions (155.384 mg/g QE DW at 40°C, 10 MPa) due to thermal sensitivity. AA remained consistently high (>95%), indicating preserved radical-scavenging potential. Key bioactives, including pseudobrucine, fawcettiine, and eclalbasaponin V, were linked to NF-κB and COX pathway modulation. Overall, mid SFE conditions (60°C, 20 MPa) provided the most balanced phytochemical profile. These findings demonstrate SFE as a green strategy to convert POSC into antioxidant and anti-inflammatory-rich extracts for nutraceutical, cosmetic, and pharmaceutical applications.

Introduction

The palm oil industry is a major contributor to global vegetable oil production, with Malaysia accounting for approximately 25% of the global supply [1]. Although annual yields of palm oil, particularly crude palm oil (CPO), are substantial; however, precise global production figures are best obtained from industry reports. Alongside this output, significant environmental challenges arise, most notably from palm oil mill effluent (POME). Generated during the milling process, POME is characterized by high biochemical oxygen demand (BOD), chemical oxygen demand (COD), and suspended solids, which can severely impact wastewater management systems and surrounding ecosystems if left untreated [2].

On average, processing one tonne of fresh fruit bunches (FFB) produces approximately 0.67 tonnes of POME nearly three times the volume of CPO extracted [3]. Additionally, palm oil sterilization condensate (POSteC), generated during sterilization, contributes a further 0.2 to 0.3 tonnes per tonne of FFB, making Malaysia annual POME yield particularly significant [2]. Therefore, given that POME is classified as a high-strength but biodegradable wastewater, effective treatment and valorization strategies are essential to minimize its environmental footprint

[1]. Conventional management typically involves anaerobic digestion systems that facilitate methane recovery. However, alternative strategies include nutrient recycling through conversion into fertilizers, animal feed, or other biodegradable products. More recently, research has highlighted the valorization potential of POME fractions, particularly POSteC, which is rich in bioactive compounds such as polyphenols and flavonoids. These compounds exhibit strong antioxidant and anti-inflammatory properties and have been increasingly studied [4].

Supercritical fluid extraction (SFE) has emerged as a sustainable alternative to conventional solvent-based extraction methods. It offers several advantages, including high selectivity, low toxicity, and mild processing conditions that help preserve sensitive bioactive. SFE has proven effective in extracting antioxidant compounds such as tocopherols, tocotrienols, carotenoids, phenolic acids, and flavonoids, which act as radical scavengers, lipid peroxidation inhibitors, and anti-inflammatory agents through mechanisms like NF-κB suppression, COX-2 inhibition, and cytokine regulation [2,3]. The extraction conditions during SFE are crucial for optimizing solubility and yield. Compound solubility, which strongly influences extraction efficiency, depends on the pressure and temperature of SFE. Increasing pressure raises CO_2 density, thereby enhancing its ability to dissolve solutes and extract bioactive effectively. Studies have demonstrated that higher CO_2 density correlates with increased recovery of total phenolics and flavonoids, validating the importance of fine-tuning extraction parameters [5]. Additionally, the incorporation of food-grade co-solvents such as ethanol has been shown to improve the recovery of polar compounds by adjusting the polarity of the extraction medium, providing a boost to extraction effectiveness [4]. Notably, SFE allows for selective extraction under mild conditions, preserving the stability and therapeutic efficacy of heat-sensitive compounds. For example, tocopherols and carotenoids, which are often degraded under high-temperature solvent extractions, remain intact under SFE, ensuring higher-quality extracts [6].

Recent studies have expanded the understanding of the anti-inflammatory properties of bioactive compounds extracted via SFE. For instance, flavonoids and tocopherols have been identified as effective modulators of inflammatory pathways, demonstrating significant promise for therapeutic applications. Similarly, phenolic compounds extracted from various plant sources have shown potential to inhibit the production of pro-inflammatory cytokines, providing evidence for their role in chronic disease prevention. As the field advances, exploring optimal extraction conditions specific to various bioactive compounds remains essential. The selection of parameters such as temperature, pressure, and the use of co-solvents can significantly influence the bioactivity of the resulting extracts, reinforcing the need for further research in this area [7]. Hence, SFE has been reported to enhance the recovery of bioactive compounds while offering potential applications in sustainable waste valorization and the development of products.

In contrast to conventional solvent extraction, SFE-derived extracts are generally superior in both purity and stability. The absence of residual toxic solvents enhances their safety for food and pharmaceutical applications, while the lower processing temperatures reduce thermal degradation of heat-sensitive compounds such as carotenoids and tocopherols. Additionally, the tuneable solvation power of supercritical CO_2 allows selective targeting of desired antioxidant fractions, improving both yield and bioactivity. As a result, SFE extracts generally demonstrate greater functional stability and bioefficacy, positioning them as more desirable for incorporation into high-value health-related products [8]. Therefore, in this study seeks to employ SFE to recover the phytochemicals from solid form of POSC, while assessing their antioxidant properties. By integrating green technologies into POME valorization, this approach not only strengthens sustainable waste management practices but also supports the circular bio economy. Transforming palm oil by-products into value-added functional ingredients reduces environmental burdens while creating new economic opportunities for agro-industrial residues [1].

Separation Technology - ICoST 2025   Materials Research Forum LLC
Materials Research Proceedings 59 (2026) 193-200   https://doi.org/10.21741/9781644903957-25

Methodology

Fresh palm oil sterilization condensate (POSTeC) was collected from a local PPNJ Kahang palm oil mill in Kluang, Johor Malaysia, and kept stored at 4°C until further use. Analytical grade carbon dioxide (99.9%) was purchased from a local supplier. Food grade ethanol ($\geq$99.5%), Folin–Ciocalteu reagent, sodium carbonate, aluminum chloride, DPPH, gallic acid, quercetin, ascorbic acid, and other analytical reagents were obtained from Sigma-Aldrich (USA).

The POSTeC sample was first stirred thoroughly to ensure homogeneity. Immediately filtered using muslin cloth before it was then centrifuged at 4000 rpm for 40 min. The resulting solid phase known as palm oil solid condensate (POSC) was collected by filtered through Whatman No. 1 filter paper. This solid POSC was stored for subsequent SFE extraction process and analysis.

Supercritical CO₂–ethanol extraction

The SFE extraction was carried out using ethanol as a co-solvent. The chiller temperature was maintained at 2°C, and the back-pressure regulator heater at 60°C. About 5 ± 0.005 g of raw POSC was loaded into a 30 mL extraction vessel. Liquid CO_2 was pumped at 4 mL/min, followed by ethanol at 0.4 mL/min. Extracts were collected every 30 min in sealed vials and stored at 4°C. Excess ethanol was removed using a MIVAC concentrator at 40°C for approximately 5 h to yield a concentrated extract (~13 mL). A design of experiment (DOE) used to optimize key parameters temperature (40 to 80°C) and pressure (10 to 30 MPa) with CO_2 and ethanol flow rates fixed at 4.0 mL/min and 0.4 mL/min, respectively. The responses analyzed were oil yield (%), β-carotene (mg/kg), and α-tocopherol (mg/kg) to assess extraction efficiency and product quality.

Determination of α-Tocopherol and β-Carotene and structural characterization

Approximately 1 mg of POSC extract was mixed with 3 mL of methanol in a glass tube, vortexed, and sonicated for 15 min to dissolve carotenoids and tocopherols. The mixture was left to stand at room temperature for 10 min before analysis using a UV-Vis spectrophotometer (Genesys 10S, Thermo Scientific, Japan) at 450 nm for β-carotene and 293 nm for α-tocopherol. All measurements were conducted in triplicate, and concentrations were calculated from standard curves, expressed as β-carotene and α-tocopherol equivalents (mg/kg sample).

About 1 mg of POSC extract was mixed with 100 mg of dry KBr, ground in an agate mortar, and pressed into a 13 mm pellet. The pellet was kept dry and analyzed immediately to avoid moisture interference. FTIR spectra were recorded from 4000 to 400 cm^{-1} at 4 cm^{-1} resolution, using a pure KBr pellet as background. All spectra were baseline-corrected, normalized, and examined for characteristic functional groups, with replicate pellets prepared for reproducibility.

Determination of TPC and TFC

Total phenolic content (TPC) was determined using the Folin–Ciocalteu method. Briefly, 0.5 mL of POSC extract (1 mg/mL) was mixed with 2.5 mL of Folin–Ciocalteu reagent (1:10 dilution) and 2.0 mL of 7.5% sodium carbonate, followed by incubation at room temperature for 30 min in the dark. Absorbance was measured at 765 nm using a UV–Vis spectrophotometer, and results were expressed as mg gallic acid equivalents (GAE)/g extract. Total flavonoid content (TFC) was determined by the aluminum chloride colorimetric method, in which 0.5 mL of extract was mixed with 1.5 mL of ethanol, 0.1 mL of 10% aluminum chloride, 0.1 mL of 1 M potassium acetate, and 2.8 mL of distilled water. After incubation for 30 min at room temperature, absorbance was recorded at 415 nm, and results were expressed as mg quercetin equivalents (QE)/g extract.

DPPH free-radical scavenging activity

The antioxidant activity of the POSC extracts was evaluated using the DPPH (2,2-diphenyl-1-picrylhydrazyl) radical scavenging method. Briefly, a 0.1 mM DPPH solution was prepared in methanol and stored in the dark until use. For the assay, 100 μL of extract solution at various concentrations was added to 3.9 mL of freshly prepared DPPH solution. The mixture was vortexed gently and incubated at room temperature in the dark for 30 min to allow the reaction to proceed.

Separation Technology - ICoST 2025

Materials Research Proceedings 59 (2026) 193-200

Materials Research Forum LLC

https://doi.org/10.21741/9781644903957-25

The decrease in absorbance was measured at 517 nm using a UV–Vis spectrophotometer against methanol as blank. Ascorbic acid (or Trolox, if preferred) was used as a positive control. The radical scavenging activity was calculated using Eq. (1) [1],

$$\text{DPPH scavenging activity (\%)} = \frac{A_{control} - A_{sample}}{A_{control}} \times 100 \qquad (1)$$

where $A_{control}$ is the absorbance of the DPPH solution without extract and A_{sample} is the absorbance with extract and results were expressed as percentage inhibition.

Results and Discussion

Effect of SFE conditions on POSC

The 13 runs of content of extract yield, β-carotene and α-tocopherol in the POSC extracts varied significantly with changes in temperature and pressure as shown in (Table 1) suggesting that the extraction of these bioactive is highly sensitive to the processing conditions.

Table 1 SFE of POSC within vary temperature and pressure.

Run	Factors		Responses		
	A: Temp. (°C)	B: Pressure (MPa)	Extract yield (g)	β-carotene (mg/kg)	α-tocopherol (mg/kg)
1	80	30	3.1559	6.9835	10.2082
2	60	30	1.8815	8.6788	9.9164
3	60	10	2.3425	8.0353	9.9327
4	60	20	2.7282	7.6435	11.0864
5	60	20	2.8641	6.3412	13.1918
6	60	20	2.4818	6.3412	12.3055
7	40	30	3.2249	6.0529	10.9609
8	80	20	2.5697	8.4294	14.0209
9	40	10	1.7593	7.6506	9.5127
10	60	20	2.614	7.9647	10.9718
11	40	20	3.3879	5.9424	11.6782
12	60	20	2.4169	8.7353	12.5345
13	80	10	2.5628	8.3306	15.6764

The extract yield of POSC obtained from SFE ranged from 1.76–3.39 g, mainly influenced by the combined effects of temperature and pressure. Generally, increasing pressure improved extraction performance due to the higher CO_2 density and solvent strength. For example, Run 9 (40°C, 10 MPa) produced the lowest yield (1.76 g), while Run 11 (40°C, 20 MPa) produced the highest (3.39 g) [9]. However, further increasing pressure to 30 MPa did not enhance yield and instead caused a decline, likely due to solute saturation or restricted mass transfer at high fluid viscosity [10]. Temperature showed a non-linear effect. At moderate levels (60°C), yields remained relatively stable due to a good balance between solute volatility and CO_2 density (Run 3: 2.34 g; Run 5: 2.86 g). At high temperature with lower pressure, yield dropped (Run 13: 2.56 g), indicating that reduced CO_2 density outweighs the benefits of increased solute vapor pressure [11].

The β-Carotene content ranged from 5.94–8.74 mg/kg. The highest concentration was observed under moderate extraction conditions (60°C, 20 MPa, Run 12: 8.74 mg/kg), confirming the need to prevent both thermal degradation and low solvent density. Very low pressure (Run 9) or excessively high temperature (Run 8) reduced β-carotene recovery, which is aligned with previous findings showing optimum conditions around 50–65°C and 15–25 MPa [13]. The α-Tocopherol content ranged from 9.51–15.68 mg/kg, indicating that it is more temperature-sensitive. The highest recovery occurred at 80°C and 10 MPa (Run 13: 15.68 mg/kg), suggesting enhanced solubility at elevated temperatures. However, applying both high temperature and high pressure

Separation Technology - ICoST 2025   Materials Research Forum LLC
Materials Research Proceedings 59 (2026) 193-200   https://doi.org/10.21741/9781644903957-25

reduced extraction efficiency (Run 2: 9.92 mg/kg), demonstrating that excessive pressure may hinder compound selectivity [14]. Overall, moderate pressure with balanced temperature conditions achieved the best combination of extract yield and phytochemical recovery, particularly for β-carotene and α-tocopherol.

FTIR structural profiling of POSC

Fig. 1 shows the FTIR spectra of POSC extracts obtained under low (40°C, 10 MPa), mid (60°C, 20 MPa), and high (80°C, 30 MPa) SFE conditions. The variations in peak intensities reflect how different temperature–pressure settings affect the solubility and stability of phytochemicals in POSC. β-Carotene is identified by strong C–H stretching peaks at 2065–2098 cm^{-1} from its long hydrocarbon chain, and a conjugated C=C stretch around 1468 cm^{-1}, which represents the polyene structure typical of carotenoids. α-Tocopherol shows a distinctive O–H stretch near 3300 cm^{-1} (hydroxyl group), C–H stretches at 2800–3000 cm^{-1} (alkyl chains), an aromatic C=C stretch near 1600 cm^{-1}, and C–O stretching peaks at 1100–1200 cm^{-1} corresponding to ether/alcohol groups (Ramos-Hernández et al., 2018). Overall, the spectra confirm characteristic signals of both compounds: β-carotene is dominated by conjugated C=C and C–H vibrations, whereas α-tocopherol is marked by O–H, aromatic C=C, and C–O functional group peaks.

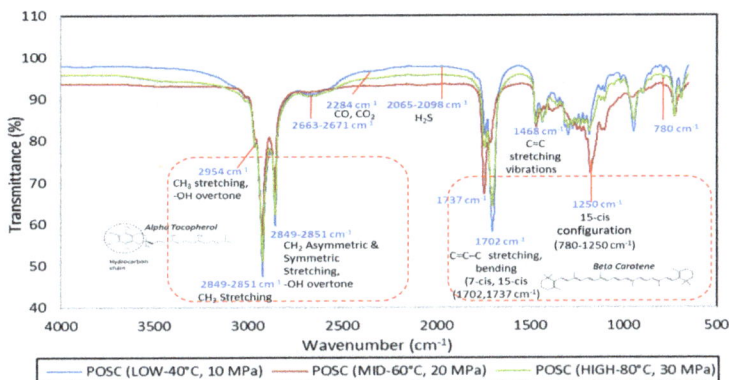

Figure 1 *Structural profiling of POSC at low, mid and high SFE conditions.*

Phytochemical composition of POSC

The total phenolic content (TPC), total flavonoid content (TFC), and antioxidant activity (AA) of POSC extracts under different SFE conditions were presented in (Table 2). The total phenolic content (TPC) increased with greater extraction severity, from 10.438 mg/g GAE DW at 40 °C, 10 MPa to 52.214 mg/g GAE DW at 80°C, 30 MPa, due to enhanced solvent power and ethanol-induced polarity modification at elevated pressure and temperature [17–19]. Conversely, total flavonoid content (TFC) was highest under mild conditions (155.384 mg/g QE DW), then declined at 60°C, 20 MPa and further at 80°C, 30 MPa, confirming the heat-sensitivity and reduced selectivity toward flavonoids under harsher conditions [18,20].

Table 2 *Phytochemical of POSC extracts at different SFE conditions.*

Phytochemicals compositions	POSC (low) 40°C, 10 MPa	POSC (mid) 60°C, 20 MPa	POSC (high) 80°C, 30 MPa
Total phenolic content (TPC, mg/g GAE DW)	10.438 ± 0.52^c	15.624 ± 0.37^b	52.214 ± 1.15 [a]
Total flavonoid content (TFC, mg/g QE DW)	155.384 ± 2.41^a	116.729 ± 1.98^b	64.462 ± 1.22^c
Antioxidant activity (AA, %)	96.43 ± 0.56^a	95.02 ± 0.42^b	95.75 ± 0.35^{ab}

Values are means ± standard deviation (n = 3). Different superscript letters (a–c) within the same row indicate significant differences (p < 0.05) according to one-way ANOVA followed by Tukey's HSD test.

Despite this opposing behavior, antioxidant activity remained consistently high (≈95–96%), indicating that multiple compound groups contribute to radical scavenging. The absence of proportional increases in antioxidant response at high TPC levels suggests dilution by non-antioxidant components co-extracted under severe conditions, similar to prior observations in palm oil-based systems [21].

Conclusion

This study demonstrates that palm oil solid condensate (POSC), a fraction of palm oil sterilization condensate typically considered waste, can be effectively valorized into high-value antioxidant extracts using supercritical fluid extraction (SFE). Extraction parameters strongly influenced bioactive recovery: phenolic content was maximized under high pressure–temperature conditions, flavonoids were better preserved under mild conditions, and antioxidant activity remained consistently high across all settings. Among the tested conditions, mid SFE (60°C, 20 MPa) offered the most versatile phytochemical balance, whereas low SFE favored heat-sensitive flavonoids and high SFE enriched potent alkaloids. These outcomes establish SFE as a green and efficient platform for converting palm oil by-products into bioactive-rich extracts, advancing both sustainable waste management and the development of functional ingredients for health-related industries.

Acknowledgment

The authors gratefully acknowledge the support of *Jabatan Tanah dan Survei Sarawak* for funding this project through the Government Agency Grant (R.J130000.7346.1R038). The authors also wish to express their sincere appreciation to the Centre of Lipid Engineering and Applied Research (CLEAR), Universiti Teknologi Malaysia (UTM), for providing the instrumentation and research facilities.

References

[1] V. Vijay, S.L. Pimm, C.N. Jenkins, and S.J. Smith, The impacts of oil palm on recent deforestation and biodiversity loss, PLoS One 11 (2016) e0159668. https://doi.org/10.1371/journal.pone.0159668

[2] T. Naidua, D. Qadir, R. Nasir, H. Mannan, H. Mukhtar, K. Maqsood, A.Ali, and A. Abdulrahman, Utilization of *Moringa oleifera* and nanofiltration membrane to treat palm oil mill effluent (POME), *Materials Science & Engineering Technology.* 52 (2021) 346–356. https://doi.org/10.1002/mawe.202000084

[3] N.M.F.M. Yasin, M.S. Hossain, H.P.S. Abdul Khalil, M. Zulkifli, A. Al-Gheethi, A.J. Asis, and A.N.A. Yahaya, Treatment of palm oil refinery effluent using tannin as a polymeric coagulant: isotherm, kinetics, and thermodynamics analyses, *Polymers* 12 (2020) 2353. https://doi.org/10.3390/polym12102353

Separation Technology - ICoST 2025   Materials Research Forum LLC
Materials Research Proceedings 59 (2026) 193-200   https://doi.org/10.21741/9781644903957-25

[4] J. A. Pinem, I. Tumanggor, and E. Saputra, The application of nanofiltration membrane for palm oil mill effluent treatment by adding polyaluminium chloride (PAC) as coagulant, *J. Rekayasa Kim. Lingkung.* 15 (2020) 1–9. https://doi.org/10.23955/rkl.v15i1.13952

[5] W.Zhang, D.N. Rizkiyah, and N.R. Putra, Innovative Techniques in Sandalwood Oil Extraction: Optimizing Phenolic and Flavonoid Yields with Subcritical Ethanol. *Separations.* 11 (2024) 201. https://doi.org/10.3390/separations11070201

[6] A. Meléndez-Martínez, A.I. Mandić, F. Bantis, V. Böhm, G.I.A. Borge, M. Brnčić, et al, A comprehensive review on carotenoids in foods and feeds: status quo, applications, patents, and research needs. *Critical Reviews In Food Science And Nutrition* 62. (2021) 1999–2049. https://doi.org/10.1080/10408398.2020.1867959

[7] V.A. Cruz, N.J. Ferreira, E. Le Roux, E. Destandau, and A.L. Oliveira, Intensification of the SFE using ethanol as a cosolvent and integration of the SFE process with SC-CO₂ followed by PLE using pressurized ethanol of black soldier fly (*Hermetia illucens L.*) larvae meal extract yields and characterization. *Foods* 13 (2024) 1620. https://doi.org/10.3390/foods13111620

[8] I. Usman, H. Saif, A. Imran, M. Afzaal, F. Saeed, I. Azam, A. Afzal, H. Ateeq, F. Islam, Y.A. Shah, and M.A. Shah, Innovative applications and therapeutic potential of oilseeds and their by-products: An eco-friendly and sustainable approach, *Food Sci. Nutr.* 11 (2023) 2599–2609. https://doi.org/10.1002/fsn3.3322

[9] A. He, N.R. Putra, and L. Qomariyah, Enhanced extraction of flaxseed oil, tocopherols, and fatty acids using supercritical carbon dioxide with ethanol: Process optimization and modelling, *Can. J. Chem. Eng.* 103 (2025) 1234–1246. https://doi.org/10.1002/cjce.25685

[10] V. Kitrytė, A. Laurinavičienė, M. Syrpas, A. Pukalskas, and P.R. Venskutonis, Modeling and optimization of supercritical carbon dioxide extraction for isolation of valuable lipophilic constituents from elderberry (*Sambucus nigra* L.) pomace, *J. CO₂ Util.* 35 (2020) 225–235. https://doi.org/10.1016/j.jcou.2019.09.020

[11] N. Nastić, J.A. Mazumder, and F. Banat, Supercritical CO₂ extraction of oil from fruit seed by-product: advances, challenges, and pathways to commercial viability, *Crit. Rev. Food Sci. Nutr.* (2025) 1–18. https://doi.org/10.1080/10408398.2025.2527946

[12] D. Kostrzewa, A. Dobrzyńska-Inger, B. Mazurek, and M. Kostrzewa, Pilot-scale optimization of supercritical CO₂ extraction of dry paprika (*Capsicum annuum*): influence of operational conditions and storage on extract composition, *Molecules* 27 (2022) 2090. https://doi.org/10.3390/molecules27072090

[13] M. Xu, and J. Watson, Microencapsulated vitamin A palmitate degradation mechanism study to improve the product stability, *J. Agric. Food Chem.* 69 (2021) 15681–15690. https://doi.org/10.1021/acs.jafc.1c06087

[14] A.A. Gigi, U. Praveena, P.S. Pillai, K.V. Ragavan, and C. Anandharamakrishnan, Advances and challenges in the fractionation of edible oils and fats through supercritical fluid processing, *Compr. Rev. Food Sci. Food Saf.* 23 (2024) e70017. https://doi.org/10.1111/1541-4337.70017

[15] M.Y. Kim, E.J. Kim, Y.-N. Kim, C. Choi, and B.-H. Lee, Comparison of the chemical compositions and nutritive values of various pumpkin (*Cucurbitaceae*) species and parts, *Nutr. Res. Pract.* 6 (2012) 21–27. https://doi.org/10.4162/nrp.2012.6.1.21

[16] N. Takatani, F. Beppu, Y. Yamano, T. Maoka, and M. Hosokawa, Seco-type β-apocarotenoid generated by β-carotene oxidation exerts anti-inflammatory effects against activated macrophages, *J. Oleo Sci.* 70 (2021) 549–558. https://doi.org/10.5650/jos.ess20329

[17] H.P. Tai, C.T.T. Hong, T.N. Huu, and T.N. Thi, Extraction of custard apple (*Annona squamosa* L.) peel with supercritical CO_2 and ethanol as co-solvent, *J. Food Process. Preserv.* 46 (2022) e17040. https://doi.org/10.1111/jfpp.17040

[18] S.A. Radzali, M. Markom, and N. Md Saleh, Co-solvent selection for supercritical fluid extraction (SFE) of phenolic compounds from *Labisia pumila*, *Molecules* 25 (2020) 5859. https://doi.org/10.3390/molecules25245859

[19] L.M. Buelvas-Puello, G. Franco-Arnedo, and H.A. Martínez-Correa, Supercritical fluid extraction of phenolic compounds from mango (Mangifera indica L.) seed kernels and their application as an antioxidant in an edible oil, *Molecules* (2021). https://doi.org/10.3390/molecules26247516

[20] N. Ahmad, Z.A.A. Hasan, H. Muhamad, S.H. Bilal, N.Z. Yusof, and Z. Idris, Determination of total phenol, flavonoid, antioxidant activity of oil palm leaves extracts and their application in transparent soap, *J. Oil Palm Res.* (2018). https://doi.org/10.21894/jopr.2018.0010

[21] A.C. Hui, C.S. Foon, and C.C. Hock, Antioxidant activities of *Elaeis guineensis* leaves, *J. Oil Palm Res.* 29 (2017) 343–351. https://doi.org/10.21894/jopr.2017.2903.06

Separation Technology - ICoST 2025
Materials Research Proceedings 59 (2026) 201-207

Materials Research Forum LLC
https://doi.org/10.21741/9781644903957-26

Computer-Aided Design and Fabrication of Advanced Membranes for Nitrogen/Methane Separation

Jimoh K. ADEWOLE[1,a*], Mohammed S. AL-AJMI[1,b], Amna S. AL-JABRI[1,c],
Faisal R. AL MARZUQI[1,d], Habeebllah B. OLADIPO[1,e], Faruq B. Owoyale[1,2,f]

[1]Membrane Science and Engineering Lab, Department of Process Engineering, National University of Science & Technology, Oman

[2]Department of Chemical Engineering, University of Ilorin, Ilorin, Nigeria

[a]jimohadewole@nu.edu.om, [b]004392-17@imco.edu.om, [c]004356-17@imco.edu.om,
[d]faisal.almarzuqi@imco.edu.om, [e]habeebllah@imco.edu.om, [f]faruqbelloowoyale@gmail.com

Keywords: Membrane Engineering, Process Intensification, Driving Forces/Resistances, Computer-Aided Molecular Design, Print-Assisted Membrane Fabrication, Hydraulic Resistance

Abstract. One of the effective strategies for advancing the goals of process intensification is the purposeful introduction of reproducible structures through a combination of computer-aided molecular design and print-assisted fabrication techniques. In this study, molecular models of two polymers—MDA and ODA—were developed using Avogadro software. These models were subsequently fabricated into membranes using a print-assisted method, specifically employing a LaserJet printer to produce structured flat-sheet composite membranes. The membranes were evaluated for their performance in N_2/CH_4 gas separation. Characterization of the membranes was conducted using microscopy, while gas transport properties and separation performance were assessed via constant-pressure gas permeation tests, measuring both permeability and hydraulic resistance. Although the same substrate was used for printing both polymers, the hydraulic resistance varied significantly between the materials and the gases tested. For MDA, resistance to N_2 and CH_4 ranged from 38,560.78 to 95,697.10 $kPa \cdot s/m^3$, whereas ODA exhibited a range of 28,132.33 to 88,124.16 $kPa \cdot s/m^3$. In terms of selectivity for N_2/CH_4, ODA demonstrated values of 2.34 and 3.10 for single-layer and seven-layer membranes, respectively, compared to 1.71 and 1.92 for MDA under the same conditions. Morphological analysis revealed that the printed membranes possessed well-defined structures with evenly distributed voids, contributing to enhanced reproducibility and predictability in performance. Overall, the fabricated membranes exhibit promising characteristics for intensifying membrane-based separation processes, particularly in the context of N_2/CH_4 separation.

Introduction

The demand for engineered membrane materials has risen rapidly owing to a drastic increase in the demand for membrane separation equipment and processes for gas separation. The growing interest in the industrial application of membrane technology is due to its intrinsic beneficial features, such as high energy efficiency, operation simplicity, modularity, lower footprint, no phase change involved, low weight, low maintenance, and ease of scaling up [1]. Despite some initial successes in various membrane materials fabrication methods, many challenges remain in developing membrane materials that are suitable for separating a variety of gas mixtures. The performance of a membrane separation system is dictated by the semi-permeable membrane materials employed for the separation. In most cases, the materials are made with polymers with known chemical structures. Therefore, majority believe that molecular design and engineering methods that allow the tailoring of the physical, chemical and micro-structural properties of polymers is the most successful method to enhance the separation performance [2]. Much research effort has been published on chemical modifications of polymers which resulted in controlling the gas separation performance of the membrane [3,4,5,6,7,8]. The practice of chemical modification of polymer followed by subsequent synthesis and fabrication of these membranes involves series of chemical processes and the use of hazardous volatile solvents and chemicals which are dangerous to both researchers and the environment at large. In addition, the processes are also time-consuming and may involve the use of expensive chemicals and reagents. It is therefore imperative to devise a means of simulating the performance of these polymers as membranes before fabrication. The simulation can be used as a quick screening tool to reduce waste, cost and time, save the environment and safeguard the health and safety of researchers. A computer-aided molecular design and print-assisted fabrication is thereby proposed as a tool for the design and fabrication of membranes for gas separation.

In this study, the incorporation of the inkjet coating procedure into the membrane fabrication process was demonstrated using a complex mixture of nanoparticles, polymers and dyes. The capabilities of the produced membranes for the separation of various gas mixtures were investigated. The application of inkjet printing in membrane fabrication could facilitate the production of mixed-matrix membranes with uniformly distributed nanoparticles. Thus, the central aim of this study is to evaluate the fundamental physio-chemical properties of a membrane that is designed by molecular design software and then fabricated using inkjet printing as a deposition and patterning tool. Specifically, molecular design of MDA and ODA was carried out using Avogadro software. The designed polymer was then printed using an inkjet printer. The printed membrane was evaluated in terms of the gas permeability and selectivity of the membrane.

It is expected that this work will open a new frontier in the approach used for membrane design and fabrication. It will also provide some insights into how membrane performance can be simulated and predicted before carrying out the practical routes of its synthesis in the lab. Membranes produced with this type of method are expected to possess a uniform and repeatable morphology, which will make it easier to understand the membrane physico-chemical behaviours and predict their performance. The understanding of the physio-chemical behaviours can ultimately provide insight required for accurate prediction of membrane performance.

Methodology and Materials
The materials used in this research include an inkjet printer, ink, lignocellulosic paper-based gases (CH, and N₂). Molecular design of two polymers, 4,4'-Oxydianiline (ODA) and methylenedianiline (MDA) was done using Avogadro's software. The optimized geometry was then duplicated to many molecular chains. The chains were arranged side by side. The two polymers were then printed using ink-jet printer [9,10,11,12]. The flat sheet membrane was printed using word processing software and a laser jet printer.

The morphology of the fabricated membrane samples was studied using SEM and Digital Microscope USB X1000 RoHS to capture an optical image of the sample surfaces and cross-section.

Hydraulic resistance and gas permeability were measured by a constant pressure/variable volume apparatus. Detailed information about this apparatus has been published elsewhere [13]. The apparatus is composed of the permeation cell and a gas flowmeter on the downstream side. At steady-state condition, gas permeability was calculated using,

$$P = \frac{22.414}{A} \frac{l}{(p_2 - p_1)} \frac{p_1}{RT} \frac{dV}{dt} \tag{1}$$

where A is the membrane area (cm²), p_2 and p_1 are feed or upstream and permeate or downstream pressures, respectively, R is the universal gas constant (6236.56 cm³ cm Hg/mol.K), T is the

Separation Technology - ICoST 2025
Materials Research Proceedings 59 (2026) 201-207

Materials Research Forum LLC
https://doi.org/10.21741/9781644903957-26

absolute temperature (K), dV/dt is the volumetric flow rate obtained from the flowmeter (cm³/s) and 22.414 is the number of cm³ STP of penetrant per mole [14,15].

Results and Discussion

Figs. 1 and 2 provide information about the uniformity of deposition and dispersion of particles in the substrate. Fig. 1(a) shows a clear distinction between the deposited layer and the underlying substrate, with the former being a black layer overlaying the whitish cellulosic paper support. This optical comparison validates the effective formation of the layers through the LaserJet printing-assisted method. Fig. 1(b)-(f) shows the structural development in terms of more layers. As an example, the distribution of the active phase in Fig. 1(e) of the single-layer printed membrane is relatively homogeneous, similar to that of the seven-layer samples. The uniformity of the microstructural arrangement of different thicknesses indicates that the print-assisted fabrication method can support the formation of mixed-matrix membranes with dispersed phases across the entire continuous matrix.

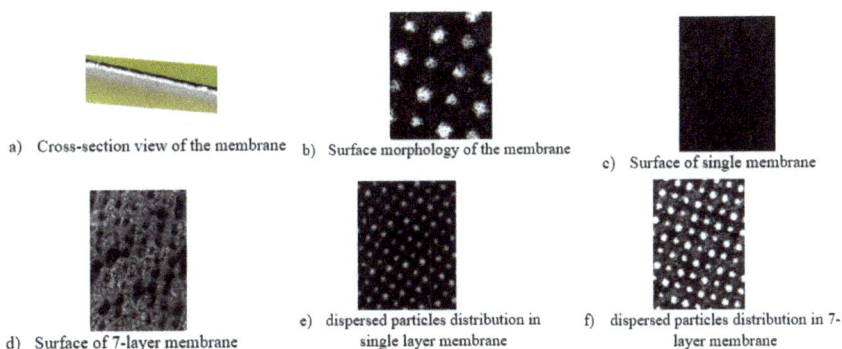

a) Cross-section view of the membrane

b) Surface morphology of the membrane

c) Surface of single membrane

d) Surface of 7-layer membrane

e) dispersed particles distribution in single layer membrane

f) dispersed particles distribution in 7-layer membrane

Figure 1 Optical microscopic images of different layers of printed membranes.

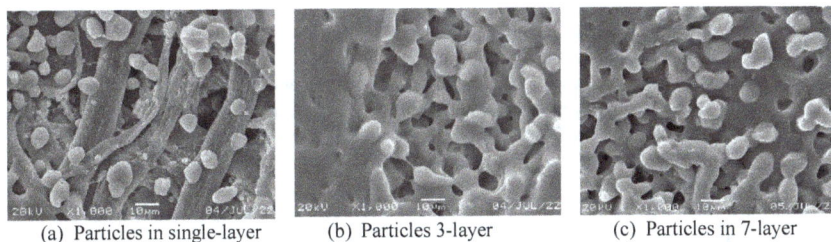

(a) Particles in single-layer (b) Particles 3-layer (c) Particles in 7-layer

Figure 2 SEM images of membrane samples for different layers of printed membranes.

Optical microscopy observations (Fig. 1) are also supported by SEM analysis (Fig. 2), which gives evidence of the microstructural features at a higher resolution. SEM images show that the particles of ink are not only deposited on the surface but also embedded into the fibrous structure of the cellulosic substrate and create a cohesive interface between the active and support layers. Additionally, the lack of large agglomeration or void formation at the interface indicates a stable fabrication process that can be used to generate defect-reduced mixed-matrix membranes that can be used in gas separation.

Separation Technology - ICoST 2025   Materials Research Forum LLC
Materials Research Proceedings 59 (2026) 201-207   https://doi.org/10.21741/9781644903957-26

Gas transport properties

As shown in Table 1, the hydraulic resistance measurements are in close agreement with what was seen in the SEM images. Both MDA and ODA single-layer membranes exhibited a high number of very fine pores, but the ones in ODA were slightly bigger, albeit by a very small margin. This slight variation assisted in lowering the flow resistance, particularly in the case of methane (CH_4). As an example, the CH_4 resistance of the single-layer ODA membrane was approximately 28,132 kPa s/m³, which was significantly lower than that of MDA, 38,561 kPa s/m³. In the case of nitrogen (N_2), the two membranes showed nearly identical resistance (65,827 and 65,933 kPa s/m³ for MDA and ODA, respectively), indicating that the slight difference in pore size between the two polymers was not significant to this gas. Though the exact reason for this behaviour needs to be further investigated, N_2 is known to be less condensable and has weaker interactions with the polymer matrix and is more dependent on overall pore connectivity rather than small changes in pore size. At a membrane thickness of seven layers, the SEM images revealed that the surfaces became rougher and some of the pore mouths were partially blocked or misaligned between layers. This stratification effect automatically increased resistance to gas flow, since gases had to travel through more interfaces and through a more tortuous path. In the case of MDA, there was a definite rise in resistance to both gases: CH_4 to approximately 49,868 kPa s/m³ and N_2 to 95,697 kPa s/m³. The greater rise in N_2 indicates that the added roughness and narrower passages in the layered structure were more difficult for the smaller gas to flow through.

Table 1 Hydraulic Resistance of Various Layer Membranes

Polymer	Layer	Gas	Hydraulic resistance
MDA-1-layer	Single	N_2	65826.79
ODA-1-layer	Single	N_2	65933.37
MDA-7-layer	7-layer	N_2	95697.10
ODA7-layer	7-layer	N2	88124.16
MDA-1-layer	Single	CH_4	38560.78
ODA-1-layer	Single	CH_4	28132.33
MDA-7-layer	7-layer	CH_4	49867.72
ODA-7-layer	7-layer	CH_4	28425.523

In the case of ODA, the resistance of N_2 increased significantly (from about 65,933 to 88,124 kPa s/m³) when more layers were added, but the resistance of CH_4 remained very small (28,132 to 28,426 kPa s/m³). This indicates that the slightly larger pores and improved orientation of layers in ODA allowed the preservation of open CH_4 pathways, even after stacking. The ODA layered structure thus dealt with the increased complexity more effectively than MDA, probably because the pores were more continuous throughout the stack and less likely to be choked off at the interfaces. Comparing the two polymers, the findings indicate that although both MDA and ODA single-layer membranes have similar performance with N_2, ODA has a lower resistance to CH_4 and keeps it after layering. However, DA is penalized more when stacked, especially on N_2. This is the direct translation of what is observed in the SEM: MDA-stacked membranes appear rougher and more blocked on the surface, whereas ODA are more open and better connected.

In summary, the microstructure has a strong effect on the hydraulic resistance trends. Single layers have the advantage of a high density of fine pores, but even a slight difference in pore size can lower resistance to larger, more condensable gases such as CH_4. The more layers added, the more resistance is added in both cases, but the surface roughness and layer alignment are important factors to avoid excessive pressure drops. Of the two, ODA was more stable in structure when layered, whereas MDA was more susceptible to stacking effects.

Separation Technology - ICoST 2025
Materials Research Proceedings 59 (2026) 201-207

Materials Research Forum LLC
https://doi.org/10.21741/9781644903957-26

Permeability and selectivity

The gas permeability measurements of single-layer and septuple-layer membranes produced with MDA and ODA inks indicate a clear change in gas transport behaviour, which is caused by the structural change brought about by the consecutive deposition of the layers and by the nature of the ink-substrate interface. As shown in Fig. 3(a) the permeability of methane decreased to 0.0219 (seven layers) and 0.0257 cm^3 (STP) cm/(cm^2 scmHg) (single layer) in the case of MDA-based membranes, and so did the permeability of nitrogen to 0.0114 and 0.0151 (cm^3 (STP) cm/(cm^2 scmHg). A similar tendency was observed in ODA-based membranes, where the nitrogen permeability decreased to 0.0124 cm^3 (STP) cm/(cm^2 scmHg); interestingly, the permeability of methane slightly increased to 0.0384 with each additional layer.

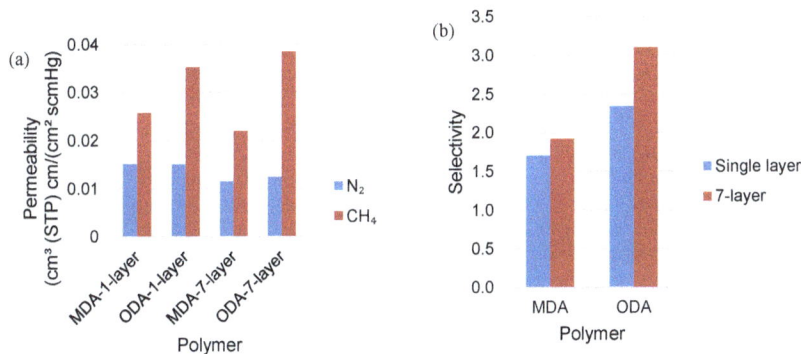

Figure 3 (a) Nitrogen (N_2) permeability of single-layer and seven-layer MDA and ODA membranes, showing the impact of membrane layering and polymer type on gas transport, and (b) CH_4/N_2 selectivity of single-layer and seven-layer MDA and ODA membranes, illustrating the enhancement in gas separation efficiency upon increasing the number of layers.

These differences in permeability indicate that there is a layer-dependent modulation of the transportation of gases: MDA membranes are more likely to become densified with extra layers, whereas ODA membranes are more likely to exhibit increased methane permeability, which is likely to be caused by minor changes in the interfacial packing and the mobility of polymer chains. CH_4/N_2 selectivity increased significantly as the number of deposited layers increased in both membrane systems. The selectivity of MDA-based membranes increased by 1.71 (single layer) to 1.92 (seven layers), and ODA-based membranes increased by a more significant margin of 2.34 to 3.10 as in Fig. 3(b). This observation suggests that the extra layers enhance the molecular-sieving effect and probably reduce the occurrence of non-selective defects that would otherwise promote the diffusion of nitrogen. These observations are in line with the data of morphological and structural characterization. SEM micrographs (Figs. 1 and 2) indicate that there is a relatively homogenous distribution of ink particles in the cellulosic matrix, and the seven-layered membranes have a more continuous and compact deposition as compared to the single-layered specimens. This densification effect is probably one of the causes of the reported decrease in gas permeability, especially in the case of the MDA membranes, as it reduces the effective transport pathways and raises tortuosity. The slight rise in CH_4 permeability after multilayer deposition in the case of ODA membranes can be explained by the fact that interfacial compatibility between the ink and substrate was improved due to the more integrated and smooth deposition patterns in the SEM micrographs.

In the case of the ODA membranes, the higher CH_4 permeability of the seven-layer sample can be attributed to the fact that the rigid amorphous fraction was reduced slightly, thus facilitating the diffusion of CH_4 without significantly reducing N_2 rejection. Generally, the gas permeation properties presented herein demonstrate a subtle interaction between layer thickness, particle distribution, crystallinity and interaction of functional groups. Single-layer membranes had a relatively high permeability, and seven-layered structures provided better selectivity, especially in ODA-based systems. These results demonstrate the urgent need to optimize the number of deposited layers to balance flux and selectivity in specific gas-separation processes.

Conclusion

This study successfully demonstrated the feasibility of print-assisted fabrication of membranes with controlled layer deposition ranging from 0 to 10 layers. Physicochemical characterization results showed a clear difference in both hydraulic and gas separation performance (in terms of permeability and selectivity between N_2 and CH_4), uniform distribution of the particles throughout the layers; and the presence of sulphate, phenolic, and carbohydrate-related functional groups that provided structural framework. The single-layered membranes were expectedly relatively more permeable whereas seven-layered membranes were more selective. These results indicate that print-assisted methods have the potential to produce scalable membranes with reproducible microstructures for gas separation. It can also serve as a tool for predicting the performance of a new polymer for a membrane before synthesis. Further research is needed to investigate more polymer structures and create a database for structure-property relationships for a number of polymers for membrane fabrication. Moreover, the print-assisted method can serve as a strong tool for evaluating the performance of mixed-matrix membranes for gas separation.

Acknowledgement

The authors would like to thank the Membrane Engineering Lab and Research and Graduate Studies Office of the International Maritime College Campus, National University of Science & Technology, Oman, under the grant NUFRG/23/IM/0027. We would also like to thank Mr Khalid S. Al Sa'idi for his help in drawing and Dr Muna Al Hinai for her help in facilitating the FTIR and XRD characterization.

References

[1] J. K. Adewole, A. L. Ahmad, S. Ismail, and C. P. Leo, 'International Journal of Greenhouse Gas Control Current challenges in membrane separation of CO 2 from natural gas : A review', *International Journal of Greenhouse Gas Control*, vol. 17, pp. 46–65, 2013. https://doi.org/10.1016/j.ijggc.2013.04.012.

[2] H. Sanaeepur, A. Ebadi Amooghin, S. Bandehali, A. Moghadassi, T. Matsuura, and B. Van der Bruggen, 'Polyimides in membrane gas separation: Monomer's molecular design and structural engineering', *Progress in Polymer Science*, vol. 91, pp. 80–125, 2019. https://doi.org/10.1016/j.progpolymsci.2019.02.001.

[3] Y. Zhuang, J. G. Seong, and Y. M. Lee, 'Polyimides containing aliphatic/alicyclic segments in the main chains', *Progress in Polymer Science*, vol. 92, pp. 35–88, 2019. https://doi.org/10.1016/j.progpolymsci.2019.01.004.

[4] J. Zou and W. S. W. Ho, 'CO 2 -selective polymeric membranes containing amines in crosslinked poly (vinyl alcohol)', vol. 286, pp. 310–321, 2006. https://doi.org/10.1016/j.memsci.2006.10.013.

[5] C. Staudt-bickel and W. J. Koros, 'Improvement of CO 2 / CH 4 separation characteristics of polyimides by chemical crosslinking', vol. 155, pp. 145–154, 1999.

[6] M. L. Jue and R. P. Lively, 'Targeted gas separations through polymer membrane functionalization', *Reactive and Functional Polymers*, vol. 86, pp. 88–110, 2015. https://doi.org/https://doi.org/10.1016/j.reactfunctpolym.2014.09.002.

[7] J. K. Adewole and A. S. Sultan, 'Polymeric Membranes for Natural Gas Processing: Polymer Synthesis and Membrane Gas Transport Properties BT - Functional Polymers', M. A. Jafar Mazumder, H. Sheardown, and A. Al-Ahmed, Eds., Cham: Springer International Publishing, 2019, pp. 941–976. doi: 10.1007/978-3-319-95987-0_26.

[8] J. Hao, P. A. Rice, and S. A. Stern, 'Upgrading low-quality natural gas with H 2 S- and CO 2 -selective polymer membranes Part I . Process design and economics of membrane stages without recycle streams', vol. 209, pp. 177–206, 2002.

[9] N. Verma, R. Kumar, and V. Sharma, 'Analysis of laser printer and photocopier toners by spectral properties and chemometrics', *Spectrochimica Acta Part A: Molecular and Biomolecular Spectroscopy*, vol. 196, pp. 40–48, 2018. https://doi.org/https://doi.org/10.1016/j.saa.2018.02.001.

[10] M. I. Szynkowska, K. Czerski, T. Paryjczak, and A. Parczewski, 'Ablative analysis of black and colored toners using LA-ICP-TOF-MS for the forensic discrimination of photocopy and printer toners', *Surface and Interface Analysis*, vol. 42, no. 5, pp. 429–437, May 2010. https://doi.org/https://doi.org/10.1002/sia.3194.

[11] A. Metzinger, R. Rajkó, and G. Galbács, 'Discrimination of paper and print types based on their laser induced breakdown spectra', *Spectrochimica Acta Part B: Atomic Spectroscopy*, vol. 94–95, pp. 48–57, 2014. https://doi.org/https://doi.org/10.1016/j.sab.2014.03.006.

[12] S. Badalov and C. J. Arnusch, 'Ink-jet printing assisted fabrication of thin film composite membranes', *Journal of Membrane Science*, vol. 515, pp. 79–85, 2016. https://doi.org/https://doi.org/10.1016/j.memsci.2016.05.046.

[13] A. L. Ahmad, J. K. Adewole, C. P. Leo, A. S. Sultan, and S. Ismail, 'Preparation and gas transport properties of dual-layer polysulfone membranes for high pressure CO<inf>2</inf>removal from natural gas', *Journal of Applied Polymer Science*, vol. 131, no. 20, 2014. https://doi.org/10.1002/app.40924.

[14] J. K. Adewole, 'Transport properties of gases through integrally skinned asymmetric composite membranes prepared from date pit powder and polysulfone', *Journal of Applied Polymer Science*, vol. 133, no. 28, 2016. https://doi.org/10.1002/app.43606.

[15] J. K. Adewole, A. L. Ahmad, S. Ismail, C. P. Leo, and A. S. Sultan, 'Comparative studies on the effects of casting solvent on physico-chemical and gas transport properties of dense polysulfone membrane used for CO2/CH4 separation', *Journal of Applied Polymer Science*, vol. 132, no. 27, 2015. https://doi.org/10.1002/app.42205.

Keyword Index

2-Hydroxypropyl-ß-Cyclodextrin 178

Acid Gas Removal Unit 162
Active Packaging 64
Adsorption 25, 33
Aerogel 113
Alginate 113
Aluminium Electrode 48
Amygdalin 87
Anaerobic Digestion 41
Anti-Inflammatory Potential 193
Anti-Inflammatory 87
Antioxidant Activity 193
Anti-Oxidant 87
Applied Voltage 48

Biochar 33
Biogas Upgrading 106
Biogas 41
Biomass Utilization 17
Bio-Oil 10
Biorefining 10
Box-Behnken Design 1

C/N Ratio 41
Cadmium 177
CBD 56
Chemical Absorption 106
Chemical Activator 98
CO_2 Removal 106
Computer-Aided Molecular Design 201
Cyanex 302 177

Defatted Microalgae Biomass 10
Diffusion 80
Driving Forces/Resistances 201

Electrocoagulation 48
Electronic Waste 1
Electrospinning 178
Energy Consumption 121
Eugenol 138
Extractive Extraction 130

Factorial Design 72
Feed Temperature 121
Freezing Dynamics 121

Gold Recovery 1
Green Extraction 138
Green Synergistic Formulation 130
Green Synthesis 185

Heavy Metal Adsorption 17
Hydraulic Resistance 201
Hydroxychavicol 138

Iron Oxide Nanoparticles 25

K-Means 162

Leaching 1
LightGBM 162
Lipid Condensate 98
Livestock Waste 41

Magnesium Sulphate 121
Mass Transfer 80
Membrane Engineering 201
Methylene Blue Adsorption 98
Methylene Blue 25
Microalgae 10, 185
Mimosa Pudica Linn 154
Molybdenum Disulfide 56

Nanofiber	178	Sirukam Dairy Farm	41
Neem Extract	178	Solute Recovery	121
		Solvent Component	162
Oleoresin	72	Solvent Composition	162
Optical Ammonia Sensing	56	Soxhlet Extraction	154
Optimization	145	Stoichiometric	177
Optuna	162	Subcritical Water Extraction (SWE)	72, 138, 145
Organic Content	48	Supercritical Carbon Dioxide Extraction	80
Palm Oil Mill Effluent (POME)	48	Supercritical Drying	113
Palm Oil Sterilization Condensate	193	Supercritical Fluid Extraction	193
Palm Oil Sterilizer Condensate	130	Surface Charge	25
Pectin Functionalization	25	Swietenia Macrophylla	80
Phenolic Compounds	130	Synergistic Extractants	177
Phenolic	193	Syzigium Aromaticum	72
Phosphoric Acid	33		
Photoinduction	185	Tetracycline	33
Piper Betel Leaves	138	Thermal Insulation	113
Plasmon Resonance	185	Thermochemical Process	10
POF	56	Thin Film	64
Polyvinylpyrrolidone	178	Thiourea Leachate	1
Pomegranate Peel	33	Trichanthera Gigantea	145
Porous Adsorbent	98	Two-Level Factorial	64
Print-Assisted Membrane Fabrication	201	Ultrasonic Assisted Extraction	87
Process Intensification	201		
Progressive Freeze Concentration	121	Vertical Finned Crystallizer	121
Protein	145		
Prunus Armeniaca	87	Waste Printed Circuit Boards	1
Reactive Extraction	177	Wastewater Treatment	98
Recovery	130	Water Scrubbing	106
Renewable Energy	41	Watermelon Rind Extract	64
Response Surface Methodology (RSM)	154	Wound Dressing	170
Response Surface Methodology	145		
		Zinc Chloride Recovery	98
Silver Nanoparticles	185	Zirconia	113
Single Sphere Model	80		

About the Editors

Muhammad Abbas AHMAD ZAINI is an associate professor of chemical engineering at Universiti Teknologi Malaysia (UTM). He received a PhD from Chiba University, Japan in 2010. His research revolves around activated carbon manufacture from agricultural and industrial wastes for water pollutants removal. He is a chartered engineer with UK Engineering Council, professional engineer with Board of Engineers Malaysia and professional technologist with Malaysia Board of Technologists.

Dr. Syed Anuar Faua'ad SYED MUHAMMAD is a Lecturer at UTM since 1997, specializing in Chemical and Bioprocess Engineering. He teaches courses in chemical plant safety, fluid mechanics, microbiology for engineer, green energy engineering, and laboratory of bioprocess engineering. Recently, his research focused on supercritical carbon dioxide extraction of bioactive compounds from microalgae. Dr. Syed is passionate about green technologies and interdisciplinary innovation in chemical engineering.